FROM NUMBERS
TO ANALYSIS

FROM NUMBERS TO ANALYSIS

Inder K. Rana
Department of Mathematics
Indian Institute of Technology, Bombay
India

World Scientific
Singapore • New Jersey • London • Hong Kong

Published by

World Scientific Publishing Co. Pte. Ltd.
P O Box 128, Farrer Road, Singapore 912805
USA office: Suite 1B, 1060 Main Street, River Edge, NJ 07661
UK office: 57 Shelton Street, Covent Garden, London WC2H 9HE

Library of Congress Cataloging-in-Publication Data
Rana, Inder K.
    From numbers to analysis / Inder K. Rana.
      p.   cm.
    Includes bibliographical references and index.
    ISBN 9810233043
    1. Numbers, Real.   2. Set theory.   I. Title.
  QA255.R36   1998
  512'.7--dc21   97-52724
                                                                             CIP

British Library Cataloguing-in-Publication Data
A catalogue record for this book is available from the British Library.

Copyright © 1998 by World Scientific Publishing Co. Pte. Ltd.

All rights reserved. This book, or parts thereof, may not be reproduced in any form or by any means, electronic or mechanical, including photocopying, recording or any information storage and retrieval system now known or to be invented, without written permission from the Publisher.

For photocopying of material in this volume, please pay a copying fee through the Copyright Clearance Center, Inc., 222 Rosewood Drive, Danvers, MA 01923, USA. In this case permission to photocopy is not required from the publisher.

This book is printed on acid-free paper.

Printed in Singapore by Uto-Print

ॐ पूर्णमदः पूर्णमिदं
     पूर्णात् पूर्णमुदच्यते ।
पूर्णस्य पूर्णमादाय
     पूर्णमेवावशिष्यते ॥

- ईशावास्योपनिषद

The Whole is there. The Whole is here.
From the Whole emanates the Whole.
Taking away the Whole from the Whole,
What remains is still the Whole.

- Ishavasyopanishad

Preface

This text is an enlarged version of the notes entitled "Real Number Systems: Construction and Properties", which were written for the participants of the first Workshop in Mathematics. The workshop was organized in November 1994 by the Department of Mathematics, Indian Institute of Technology Bombay, Mumbai (India) and the participants were final year Bachelor of Science (Mathematics) students. The topic 'Real Number System' was chosen for the workshop for the following reasons:

(i) A rigorous construction of real and complex number systems is of historical importance.

(ii) Construction of the number systems lays the logical foundation of analysis and provides basis for algebraic structures.

(iii) The material, though basic and of fundamental importance, does not form part of Bachelor/Master level course curriculum. Consequently the transition from calculus to analysis presupposes this material and it leaves a gap in students' knowledge.

(iv) Most of the modern books on real/mathematical analysis start with the axiomatic definition of the real number system and only occasionally refer the reader to some literature for its construction.

Keeping in mind the above reasons, it was felt that there is need for a self-contained text which gives a detailed account of the construction of real and complex numbers based on the axiomatic theory of sets and derives their basic properties needed in higher analysis and various other branches of mathematics. This text will hopefully fulfill this need.

The logical order needed for the foundation of analysis is essentially the reverse of the historical order in which this was achieved. William R. Hamilton grounded complex numbers on real numbers in 1837. Starting with rationals, Richard Dedekind and Georg Cantor constructed real numbers in 1872. Karl Weierstrass and Guiseppe Peano constructed rationals from integers in 1889. Integers were constructed from natural numbers by Dedekind in 1888 and by Peano in 1889. Peano also gave the construction of integers and the axiomatic definition of natural numbers in 1889. Finally, natural numbers were constructed in the axiomatic set theory after the work of Ernst Zermelo in 1908. In this text we follow the logical order for the construction of real and complex numbers.

Each chapter (except the first and the last) has mainly three parts: in the first part we give some historical comments about the number system which is developed in the second part of the chapter and the third part, called abstractions, includes discussion about the abstract concepts evolving out of the construction of the number system. The aim of the historical comments is to make the reader aware of the fact that polished mathematical concepts are not discovered overnight. These are the culmination of the efforts of great minds often spanning many years. Each chapter also includes brief biographical sketches of some of the contributors of the concepts discussed in that chapter.

Overview of the text

The aim of **Chapter 1** is to give an introduction to the Zermelo-Fraenkel axiomatic set theory. After a brief discussion of some paradoxes of the naive set theory, axioms of set theory are presented. Basic concepts of set theory, needed for further discussions, are developed based on these axioms. The final section of

Preface ix

the chapter points out the limitations of the axiomatic approach to set theory. Set theorists and logicians may find this introduction to axiomatic set theory too short and not very satisfactory, my apologies to them. Natural numbers are constructed in the axiomatic setup in **Chapter 2,** and their basic properties are developed. The main ideas discussed are: the principles of mathematical induction, Dedekind's iteration theorem and a characterization of finite sets. The section on abstractions includes concepts of partially ordered sets, binary operations, semigroups, and groups. In **Chapter 3,** the integers are constructed and their uniqueness is established, assuming the natural number system. After discussing their arithmetical properties, the division algorithm and the prime factorizaton theorem are proved. Abstractions include concepts of rings, integral domains, ordered integral domains, and ordered fields. While the preceding chapters introduce and elaborate the algebraic properties of numbers, analysis makes its appearance in chapters 4, 5 and 6 in the form of convergence of sequences and the least upper bound property. These chapters form the core of the text. **Chapter 4** starts with the construction of rationals, based on the existence of integers. After developing the algebraic and the order properties of rationals, we discuss shortcomings of the rationals: order incompleteness and Cauchy incompleteness. Abstractions of this chapter include a discussion on the Archimedean property and the completeness of ordered fields. In **Chapter 5** real numbers are constructed by both Dedekind's method and by Cantor's method. These methods assume the existence of rational numbers. That both these methods yield the same object, namely a complete ordered field which includes rationals as a dense subfield, is established. The abstractions of this chapter include a discussion on the equivalence of order completeness and Cauchy completeness in ordered fields. Finally it is shown that any two complete ordered fields are isomorphic. This completes the foundation part of analysis. In **Chapter 6,** we start with the axiomatic definition of real numbers and derive various consequences of the completeness property. We develop the basic concepts of real analysis e.g., intervals, open sets, closed sets, compact sets and connected subsets of $\mathbb{R}$. Also the extreme value properties of continuous functions on compact subsets of $\mathbb{R}$ are discussed. The section on abstractions develops the concept of metric space. In **Chapter 7**, the final chapter of the text, we develop complex numbers. The aim is to give a self-contained proof of the fundamental theorem of algebra, based on the concepts developed in chapter 6.

The text is sprinkled with a sufficient number of exercises to familiarize the reader with the concepts discussed.

Audience and suggestions for using the text

Though no prerequisites are demanded on the part of the reader, it is expected that the reader has had encounters of some kind with set theory (naive) and numbers. Some degree of mathematical maturity is also expected on the part of the reader. A senior undergraduate or a fresh graduate student can use this book for a self-study program. It will be most suitable as a textbook for a one-semester course entitled "Foundations of Analysis" (which unfortunately does not exist in most of the colleges/universities). The text can also be adopted for a seminar course.

Logically, each chapter depends upon the ideas developed in the previous chapter, and hence it seems a natural corollary that chapter 1 must be read before the rest of the text. This is true in a sense, for chapter 1 contains the minimum amount of axiomatic set theory that every student of mathematics should go through once. However, some readers, not familiar with the language and methodology of logic, may find the contents of chapter 1 too technical. In that case, one can just read through chapter 1 to convince oneself that set theory can be developed without paradoxes. Similarly, those interested in knowing only the construction of real numbers can assume rational numbers and their properties (i.e., chapter 4) and read only chapter 5.

The three digit system is used to number the definitions, lemmas, theorems, etc. For example, theorem 3.2.4 means it is the 4th numbered statement in section 2 of chapter 3. Equations needed for later references are numbered by two digits, e.g., (1.2) means it is the 2nd numbered equation of chapter 1. The symbol ■ is used to indicate the end of a proof. The symbol $A := B$ or $B =: A$ means that this equality is the definition of A by B.

Acknowledgements

As stated earlier, the text is based on the material I had written for the participants of the first workshop in mathematics held in 1994 at I.I.T. Bombay. I thank the Mathematics Department of I.I.T. Bombay for giving me an opportunity to conduct the workshop. I am grateful for the financial support received under the Curriculum Development Programme of I.I.T. Bombay to prepare the material for the workshop and the present text.

The historical comments and the biographical sketches of the mathematicians in the text are based on facts from the books: [Bel], [Boy], [Die], [Ebb], [Eve], [Kli], [Mao] and [Tre]. Some of the texts referred to for the subject matter are: [Bur], [Cou], [Coh-1], [Hen], [Lan] and [Men]. The English translation of the Sanskrit verse at the beginning of the text is from [Kri]. The sketches of the mathematicians were prepared by Kaladhar Bapu (I.I.T. Bombay) and M. Ravishankar (I.I.T. Bombay) prepared 'bmp files' of this sketches. (The sketches of Brahamagupta and Mahaviracharya are as imagined by the artist.) The hard part of drawing the figures in the text was done by P. Devraj. Though, I have tried to give due credit to the pioneer who discovered an idea/result, no effort has been made to be exhaustive. Any omissions are purely unintentional. Also I am under no illusion as to originality, for the subject matter is of historical importance and has been written by many experts. My contribution is in choosing the topics and arrangement of the subject matter so as to make it interesting and easily accessible. I hope I have been able to do so.

I am grateful to S. Kumaresan (University of Mumbai), S. Purkayastha (I.S.I. Calcutta), A. Ranjan (I.I.T. Bombay) and S.M.

Srivastava (I.S.I. Calcutta) for going through parts of the manuscript, weeding out many typos, and for making suggestions which helped to improve the text. I thank Nandita Bose for editing the manuscript.

I thank Charles Swartz (New Mexico State University) for his comments on the manuscript and for putting me in contact with World Scientific Publishing Company. I am happy to thank Joy Marie, Editor WSPC, for her suggestions for preparing the camera-ready copy and for being very cooperative and understanding. I appreciate and thank E.H. Chionh, Editor WSPC, for going through the manuscript. Her comments and suggestions helped in improving the final version of the manuscript.

I thank C.L. Anthony for processing the whole manuscript in $\LaTeX$. I appreciate the help and cooperation I received from Jones Doss (I.I.T. Bombay) and P. Devraj in preparing the camera-ready copy of the manuscript. I thank the Department of Mathematics, I.I.T. Bombay, for the use of computer laboratory.

Finally, many thanks are due to my wife Lalita for her help in more ways than one. She helped in going through the final copy of the manuscript and spotted many typos, inconsistencies and ambiguities. I am solely responsible for any shortcomings still left in the text. I would be grateful for comments and suggestions from the readers.

Mumbai, May 1998   **Inder K. Rana**

Contents

PREFACE	**vii**
1 SET THEORY	**1**
1.1 Historical comments	1
1.2 Prologue: Paradoxes of naive set theory	6
1.3 Axiomatic set theory	8
1.4 Epilogue	27
2 NATURAL NUMBERS	**31**
2.1 Historical comments	31
2.2 Existence and uniqueness of natural numbers	34
2.3 Arithmetic of natural numbers	41
2.4 Order on natural numbers	49
2.5 An extension of the iteration theorem	68
2.6 Abstractions	72
3 INTEGERS	**81**
3.1 Historical comments	81
3.2 Construction and uniqueness of integers	84
3.3 Order on integers	93
3.4 Prime factorization of integers	96
3.5 Representation of integers: Numeration	102
3.6 Abstractions	106

4 RATIONAL NUMBERS 123
- 4.1 Historical comments 123
- 4.2 Construction and uniqueness of rational numbers . 132
- 4.3 Order on rationals 137
- 4.4 Order incompleteness of $\mathbb{Q}$ 141
- 4.5 Geometric representation of rational numbers ... 146
- 4.6 Decimal representation of rationals 149
- 4.7 Absolute value of rationals 152
- 4.8 Sequences of rationals 153
- 4.9 Abstractions 173

5 REAL NUMBERS: CONSTRUCTION AND UNIQUENESS 179
- 5.1 Historical comments 179
- 5.2 Construction of real numbers by Dedekind's method 185
- 5.3 Construction of real numbers by Cantor's method . 205
- 5.4 Uniqueness of the real number system 224
- 5.5 Abstractions 229

6 PROPERTIES OF REAL NUMBERS 233
- 6.1 Historical Comments 233
- 6.2 Axiomatic definition of reals and its consequences . 237
- 6.3 Sequences of real numbers and their applications . . 246
- 6.4 Decimal representation of real numbers 255
- 6.5 Special subsets of $\mathbb{R}$. 260
- 6.6 Heine-Borel property and compact subsets of $\mathbb{R}$ .. 269
- 6.7 Nested interval property 273
- 6.8 Uncountability of $\mathbb{R}$ and the continuum hypothesis 276
- 6.9 Bolzano-Weierstrass property 287
- 6.10 Connected subsets of $\mathbb{R}$ 290
- 6.11 Extreme values of functions on $\mathbb{R}$ 293
- 6.12 Abstractions: metric spaces 307

7	**COMPLEX NUMBERS**	**319**
	7.1 Historical comments	319
	7.2 Construction of complex numbers	322
	7.3 Impossibility of ordering complex numbers	327
	7.4 Geometric representation of complex numbers	327
	7.5 Cauchy completeness of $\mathbb{C}$	332
	7.6 Algebraic completeness of $\mathbb{C}$	337
	7.7 Beyond complex numbers	346
SUGGESTIONS FOR FURTHER READING		**353**
SYMBOL INDEX		**355**
REFERENCES		**359**
INDEX		**363**

Chapter 1

SET THEORY

1.1 Historical comments

Set theory lies at the foundation of all modern mathematics. It gives a very general framework in which almost every branch of mathematics can be discussed.

Mathematicians have been using sets since the very beginning of the subject. For example, Greek mathematicians defined a circle as the set of points at a fixed distance r from a fixed point P. However, the concepts of 'infinite' and 'infinite set' eluded mathematicians and philosophers over the centuries. For example, Hindu minds conceived of infinite in their scriptural text Ishavasyopanishad as follows: "The Whole is there. The Whole is here. From the Whole emanates the Whole. Taking away the Whole from the Whole, what remains is still the Whole". Pythagoras ($\sim$ 585–500 B.C.), a Greek mathematician, associated good and evil with the limited and the unlimited, respectively. Aristotle (384–322 B.C.) said, "The infinite is imperfect, unfinished and therefore, unthinkable; it is formless and confused." The Roman Emperor and philosopher Marcus Aqarchus (121–180 A.D.) said, "Infinity is a fathomless gulf, into which all things vanish". English philosopher Thomas Hobbes (1588–1679) said, "When we say anything is infinite, we signify only that we are not able to conceive the ends and bounds of the thing named". The German mathematician Carl Friedrich Gauss (1777–1855) said, "Infinity is only a figure of speech, meaning a limit to which certain ratios may approach as closely as desired, when others are permitted to increase indefinitely".

In the 18th century when efforts were being made to institute rigor in analysis, it was realized that one can no longer avoid the concept of infinite sets. Though a study of infinite sets was initiated by Bernard Bolzano (1781–1848), his work was only published three years after his death. Study of set theory started with the work of Georg Cantor (1845–1918), who was led to it in the course of his work on trigonometric series. He soon realized that sets and their properties need to be studied for their own sake. Cantor defined a set as 'any collection into a whole of definite distinguishable objects of our intuition or thought'. An infinite set was defined as one which can be put into one-to-one correspondence with a part of itself. Using these ideas, Cantor was able to prove that the sets of natural numbers, integers and rationals are infinite sets, and have same 'size'. He also gave a construction of real numbers from rationals. However, Cantor's results were not immediately accepted by his contemporaries. Also, it was discovered that his definition of a set leads to contradictions and logical paradoxes. The most well known among these was given in 1918 by Bertrand Russell (1872–1970), now known as Russell's paradox (see 1.2.1). Also, despite his best efforts, Cantor was not able to resolve problems in his analysis of 'cardinal numbers' (now known as Continuum Hypothesis) and the Well-Ordering Principle. While Russell's paradox raised doubts about Cantor's concept of a set on one hand, it also made mathematicians realize the importance of logic in mathematical arguments and hence in set theory.

In an effort to resolve these paradoxes, the first reaction of mathematicians was to 'axiomatize' Cantor's intuitive set theory (motivation coming from the fact that axiomatization of geometry by Euclid had resolved its logical problems). Axiomatization means the following: starting with a set of unambiguous statements

1.1 Historical comments

called axioms, whose truth is assumed, one is able to deduce all the remaining propositions of the theory from these axioms using the axioms of logical inference. Russell and Alfred North Whitehead (1861–1947) in 1903 proposed an axiomatic theory of sets in their three-volume work called *Principia Mathematica*. Being heavily dependent upon formal logic, mathematicians found it awkward to use. An axiomatic set theory which is workable and is fully logistic was given in 1908 by Ernst Zermelo (1871–1953). This was improved in 1921 by Abraham A. Fraenkel (1891–1965) and T. Skolem (1887–1963) and is now known as 'Zermelo–Frankel (ZF)-axiomatic theory' of sets. Axiomatic set theory can be viewed as the foundation of mathematics. From the point of view of logic, the 'consistency' of the axioms of the ZF-theory cannot be proved. Also, there exist statements which can neither be proved nor be disproved from these axioms. (These are consequences of the Kurt Gödel's (1906–78) work in 1931 on mathematical logic.) For ZF-axiomatic set theory this means that the Axiom of Choice (see 1.3.41) and the Continuum Hypothesis (see section 6.8) are both not disprovable from the axioms of ZF-theory. In 1963, Paul Cohen (1934–) proved that the Axiom of Choice and the Continuum Hypothesis cannot be deduced from the ZF-axioms. In fact, both are independent of the ZF-axioms and if either or both are included in the ZF-axioms, then the enlarged systems will remain consistent if one assumes the consistency of the ZF-axioms. The set theory based on ZF-axioms, including the axiom of choice, is called ZFC-set theory. In this chapter we describe briefly these axioms and develop the basic concepts of set theory needed for further discussions.

Georg Cantor (1845–1918)

Cantor was born in Russia to German parents. He showed early signs of a great talent and interest in mathematics. His father tried to persuade him to make a career in engineering but consented finally to

pursuing his love for mathematics. He joined the University of Berlin in 1863 and studied mathematics, physics and philosophy. There his teachers were Ernst Eduard Kummer, Leopold Kronecker and Karl Weierstrass. Cantor's first love was the Gaussian theory of numbers, the topic of his Ph.D. dissertation. Then, under the influence of Weierstrass, he turned to foundations of mathematics. From 1869 to 1905 he taught at the University of Halle. He was the founder president of the Association of German Mathematicians. In 1897 he organized the first International Congress of Mathematicians at Zürich. He contributed to the theory of Fourier series (1870–71); construction of irrational numbers (1883); theory of sets and transfinite numbers (1874–97). His path breaking work on infinite sets and transfinite numbers lead to crises in set theory and was one of the greatest stimuli towards giving mathematics a firm logical foundation. However, his work aroused the hostility of his own teacher Kronecker, a powerful mathematician. As a result of non-acceptance of his ideas by fellow mathematicians and their criticism, Cantor suffered from bouts of deep depression in the later years of his life. David Hilbert described Cantor's contribution as 'the most admirable fruit of the mathematical mind, and one of the highest achievements of man's intellectual process'. He died at the Psychiatric Clinic of University of Halle in 1918.

Ernst Zermelo (1871–1953)

Zermelo was born in Berlin. He graduated from Berlin in 1894 and was a faculty member at Göttingen (1899–1910), Zürich (1910–16) and Freiburg (1916–35). In protest against Hitler's régime, he left the University of Freiburg in 1935 and rejoined it in 1946. His most important works are the creation of axiomatic set theory (1908) and the proof of the fact that every set can be well-ordered (1908, 1909). He also made contributions in mechanics and calculus of variation.

Abraham Fraenkel (1891–1965)

Fraenkel was born in Münich and studied at the University of Münich, Marburg, Berlin and Breslau. He taught at the University of Marburg (1916–25), University of Kiel (1928–29) and at the University of Jerusalem (1925–59). He was known for his work in logic and improvements in the Zermelo's axiomatic set theory.

1.2 Prologue: Paradoxes of naive set theory

The naive set theory is based on the following concept of a set: "A set is a collection of definite distinguishable objects of our perception which can be conceived as a whole". The members of a collection x are called elements of the set. If y is a member of x, we write this as $y \in x$. Two sets are said to be equal if they have same members. Another interpretation could be that if we can describe a property of 'objects', then we can also conceive a set of all the objects having that property. This came to be known as the 'comprehension principle'. Let us see how this leads to paradoxes.

1.2.1 Russell's paradox. Let $\phi(x)$ denote the property that x is not a member of itself. Then by the comprehension principle, we can construct a set y whose elements are those sets x which are not

1.2 Paradoxes of naive set theory

member of themselves. Symbolically we can write this as

$$y := \{x|\phi(x)\} = \{x|x \notin x\}.$$

Now we ask the question whether $y \in y$? Then by the equality of sets

$$y \in y \text{ if and only if } y \in \{x|x \notin x\}.$$

Thus, $y \in y$ iff $y \notin y$. This is the paradox.

1.2.2 Barry's paradox. Assume we can form sentences in English by picking words from a given dictionary. Consider those natural numbers n whose description require maximum of 50 words from the dictionary. All such numbers will form a set, say S, by the comprehension principle. Since the dictionary has only finite number of words, the set S has only finite number of elements. Let k be the greatest element of S. Then $(k+1) \notin S$, i.e.,

"$(k+1)$ is a natural number which requires more than 50 words to describe it."

But, we have just described $(k+1)$ in less than 50 words. Hence $(k+1) \in S$. Thus $(k+1) \in S$ and also $(k+1) \notin S$, a paradox.

Russell's paradox and others of that type are called logical paradoxes, while Barry's paradox and others of that type are called semantic paradoxes. Logical paradoxes arise because the set under consideration is too large, the semantic paradoxes arise because of the defining property of the objects. Thus, where on one hand one has to restrict the sizes of the sets that can be considered, one also has to look carefully at the properties that can be used to define sets. Occurrences of paradoxes raised the following questions:

(i) What properties defined sets?

(ii) What kind of properties can be used to define sets?

One looked for the answers to these questions so that one could avoid the paradoxes of set theory and yet could prove most of the results of naive set theory. This was achieved by the axiomatic theory of sets proposed by Zermelo and later improved upon by Fraenkel and Skolem. Axiomatic set theory is discussed briefly in the next section. For a detailed exposition, refer to [Coh], [Dra], [van], [Sup] and [Sto].

1.3 Axiomatic set theory

Axiomatic set theory begins with two **primitive** notions: concept of a **set** and its **member**. In terms of these primitive concepts, other concepts are defined using a list of axioms. An **axiom** is a statement involving primitive notions whose truths are assumed. To avoid the paradoxes in naive set theory, symbolic notations of mathematical logic are used in place of sentences of the English language. These symbols not only give precision but also provide brevity. However, once the concepts are understood and there is no chance of confusion, formal logic is replaced by intuitive logic. We first describe the language of our theory.

The objects of our theory are sets. The concept of a set remains undefined. We use mostly letters of English alphabet to denote sets. A set is conceived to have elements or members which are again sets. This may look strange at this stage but as we develop various concepts, we shall see that function, integers, rationals, reals, complex numbers, polynomials, sequences, etc., are all sets. Almost all mathematical concepts can be reduced to construction of a set. If y is a member of x, we write $y \in x$. If the elements $a, b, c, \ldots$ of a set x can be explicitly listed, we write it as $\{a, b, c, \ldots\}$. Also, an element is listed only once.

In our language **variables** run over objects, namely sets. When a variable stands for a specific object, it is called a **constant**. The role of a variable is similar to that of a pronoun in English and that of a constant is similar to that of a proper noun.

1.3 Axiomatic set theory

The logical symbols used in our language, with their interpretation in common language given below each, are the following:

/	⇒	⇔	∧	∨	∀
not	implies	if and only if, iff	and	or	for all

∃	∃!	∈	=	(,)
there exists	there is exactly one	member of	equal	parenthesis

We shall add more symbols to our theory which will be coined using these symbols.

The symbol '/' is also denoted by the symbol '¬'. The symbols $/, \Rightarrow, \Leftrightarrow, \vee,$ and $\wedge$ are known as **propositional connectives**; the symbols $\forall$ and $\exists$ are called **quantifiers**. The symbols $\in$ and $=$ are primitive symbols. Statement $x \in y$ means x is a member of y and $x = y$ means x is same as y or x is identical with y or x equals y.

Parentheses are used as punctuation symbols in common language. All these symbols help us to express statements in mathematics. We give some examples. Consider the statement:

"Every integer is a prime number."

We can write this as:

"For every x, if x is an integer, then x is a prime".

If we write $Z(x)$ for "x is an integer" and $P(x)$ for "x is a prime", then the above statement can be written symbolically as:

$$(\forall x)(Z(x) \Rightarrow P(x)). \tag{1.1}$$

Similarly the statement

"For every prime number there exists an even number which divides it" can be written symbolically as:

$$\forall x(P(x)) \Rightarrow (\exists y)(E(y) \wedge D(y,x)), \qquad (1.2)$$

where $E(y)$ stands for "y is an even integer" and $D(x,y)$ stands for "x divides y".

Expressions of the type (1.1) and (1.2) are called 'formulae'. Formally, any finite string of symbols is called a **formula**. An inductive way of describing all the formulae is the following:

1. $x = y$ and $x \in y$ are formulae, where x, y are variables. These are called **atomic formulae**.

2. If Φ and Ψ are formulae, then so are $(\Phi \wedge \Psi), (\Phi \vee \Psi), (\neg \phi)$, $(\Phi \Rightarrow \Psi)$ and $(\Phi \Leftrightarrow \Psi)$.

3. If Φ is a formula and x is any variable then $\forall x(\Phi)$ and $\exists x(\Phi)$ are also formulae.

4. Every formula is obtained from atomic formulae by using steps 2 and 3 above.

In the beginning of our discussions, we shall use logical symbols extensively. As we progress and there is less chance of confusion, formal logic will be replaced by intuitive logic.

Sometimes we shall denote a formula Φ by $\Phi(x)$ or $\Phi(x,y)$ indicating that x or x, y are among the variables involved in the formula and these are the variables under discussion. We assume that the atomic formulae are either true or false and hence each formula is either true or false for the variables involved. If Φ holds for x, we write $\Phi(x)$. However, it is not demanded that the truth value of a formula be effectively determined. For example the formula $\Phi(m,n)$: "n is a natural number and there exist a number $m > n$ such that m and $m+2$ are both prime numbers" cannot be verified effectively for every natural number n. Still $\Phi(n)$ will be a formula in our theory. Also note that the formula Φ above is not written as a string of symbols. The reason being that we are yet to

1.3 Axiomatic set theory

define natural numbers, prime numbers, and so on. All meaningful statements in mathematics can be interpreted in terms of the formulae described above. We describe next the axioms of set theory and develop various concepts needed for our future discussions.

1.3.1 A(1) Axiom of extensionality. *Two sets are equal iff they have the same elements.*

Symbolically: $\forall\ z(z \in x \Leftrightarrow z \in y) \Rightarrow x = y$.

This axiom means that a set is determined by its elements. An equivalent way of saying this is that $\forall\ z(z \notin x \Leftrightarrow z \notin y) \Rightarrow x = y$.

1.3.2 A(2) Axiom of empty set. *There exists a set which has no elements.*

Symbolically : $\exists\ x\ \forall\ y\ (y \notin x)$.

1.3.3 Theorem. *There exists a unique set with no elements. We denote it by $\emptyset$ and call it the **empty set**.*

Proof. Let x and z be two sets having no elements. Then for $y \notin x$ obviously $y \notin z$. Similarly $y \notin z$ implies that $y \notin x$. Hence $\forall\ y, y \notin x \Leftrightarrow y \notin z$ and hence by axiom **A(1)**, $x = z$. ∎

1.3.4 Definition. Let x and y be two sets. We say x is a **subset** of y, denoted by $x \subseteq y$, if $\forall\ u(u \in x \Rightarrow u \in y)$, i.e., every u which is an element of x is also an element of y. This can also be written as: $\forall\ u(u \notin y \Rightarrow u \notin x)$.

1.3.5 Theorem. *The empty set is a subset of every set.*

Proof. If x is any set and $u \notin x$, we have $u \notin \emptyset$, hence $\emptyset \subseteq x$. ∎

1.3.6 Theorem. *Let x and y be two sets. Then $x = y$ iff $x \subseteq y$ and $y \subseteq x$.*

Proof. If $x = y$, then $\forall\ z \in x \Leftrightarrow z \in y$ by axiom **A(1)** and hence $z \in x \Rightarrow z \in y$, i.e., $x \subseteq y$ and $z \in y \Rightarrow z \in x$, i.e., $y \subseteq x$. Conversely, suppose $x \subseteq y$ and $y \subseteq x$. Then $\forall\ z \in x$, we have

$z \in y$ and $\forall z \in y$ we have $z \in x$. Hence $\forall z, z \in x \Leftrightarrow z \in y$, i.e., $x = y$ by axiom **A(1)**. ∎

1.3.7 A(3) Axiom of pairing. *For every pair of sets x and y there exists a set z whose only elements are x and y.*

Symbolically: $\exists z \, \forall u (u \in z \Leftrightarrow u = x \vee u = y)$.

It follows as an application of axiom **A(1)** that for given sets x and y, the set z as given by axiom **A(3)** is unique. We denote it by $\{x, y\}$. In case $x = y$, we abbreviate $\{x, y\}$ to $\{x\}$ and call it a **singleton** set.

1.3.8 A(4) Axiom of union. *For every set x there exists a set y whose elements are exactly those which occur in at least one element of x.*

Symbolically: $\exists y \, \forall z (z \in y \Leftrightarrow \exists u (u \in x \wedge z \in u))$.

For a given set x, it follows from axiom **A(1)** that the set y as given by axiom **A(4)** is unique. We write this as $y = \bigcup_{u \in x} u$ or $\cup x$. In case $x = \{v, w\}$, we denote $\bigcup_{u \in x} u$ by $v \cup w$ and call it the **union** of the sets v and w.

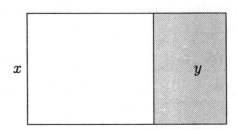

Figure 1.1 : Subset, $y \subseteq x$.

Sometimes for better understanding, the relationships between sets can also be described pictorially as follows. A set is

1.3 Axiomatic set theory

normally expressed as a region in plane. For example if x and y are two sets and y is a subset of x, we express this pictorially as in figure 1.1, the biggest rectangle representing the set x and the shaded smaller rectangle inside it representing the set y.

If x and y are two sets, then $x \cup y$ can be represented, as in figure 1.2, by the region enclosed by both the rectangles.

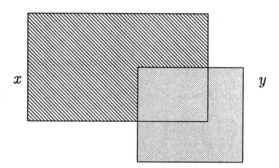

Figure 1.2 : Set union, $x \cup y$.

1.3.9 Theorem. *Let x and y be sets. Then the following hold:*

(i) $x \cup \emptyset = x$.

(ii) $x \cup x = x$.

(iii) $\cup \{x\} = x$.

(iv) $x \cup y = y \cup x$.

Proof. (i) By definition of $x \cup \emptyset$, $\forall\, z, z \in x \cup \emptyset$ iff $z \in x$ or $z \in \emptyset$. Since $\forall\, z, z \notin \emptyset$, we have $z \in x \cup \emptyset$ iff $z \in x$ and hence by axiom **A(1)**, $x \cup \emptyset = x$.

(ii) $\forall\, z, z \in x \cup x \Leftrightarrow z \in x$. Thus $x \cup x = x$.

(iii) $\forall\, z, z \in \cup\{x\}$ iff $z \in x$ and hence $\cup\{x\} = x$.

(iv) $\forall\, z, z \in x \cup y \;\Leftrightarrow\; z \in x$ or $z \in y$
$\Leftrightarrow\; z \in y$ or $z \in x$
$\Leftrightarrow\; z \in y \cup x$.

Hence $x \cup y = y \cup x$. ∎

1.3.10 Exercise. Prove the following for sets X, Y and Z:

(i) $X \cup (Y \cup Z) = (X \cup Y) \cup Z$.

(ii) If $A \in X$, then $A \subseteq \cup X$.

(iii) If $A \subseteq B$, then $A \cup B = A$.

(iv) If $A \subseteq B \,\forall\, A \in X$, then $\cup X \subseteq B$.

1.3.11 A(5) Axiom of power set. *For every set x there exists a set y the elements of which are the subsets of x.*

Symbolically: $\exists\, y\, \forall\, z (z \in y \Leftrightarrow \forall\, u(u \in z \Rightarrow u \in x))$.

If we use the subset symbol $\subseteq$, already defined, we can write the axiom as:

$$\exists\, y\, \forall\, z (z \in y \Leftrightarrow z \subseteq x).$$

From now onwards we shall follow this convention. A symbol, term or notation which has already been defined will be included in the list of symbols of our language and we will use them freely with an understanding that each such symbol, term or notation can be expressed in terms of the symbols of logic as stated in the beginning. By axiom **A(1)**, the set given by axiom **A(5)** is unique. We call it the **power set** of x and denoted by $\mathcal{P}(x)$.

1.3.12 Definition. Let a and b be sets. Consider the sets $\{a\}$ and $\{a, b\}$ which are defined by axiom **A(3)**. Again by axiom **A(3)**, the set $(a, b) := \{\{a\}, \{a, b\}\}$ is well-defined and is called the **ordered pair** of a and b.

1.3.13 Theorem.

(i) For sets a and b, the ordered pair $(a, b) \in \mathcal{P}(\mathcal{P}(a \cup b))$.

(ii) Let a_1, a_2, b_1 and b_2 be sets. Then $(a_1, b_1) = (a_2, b_2)$ iff $a_1 = a_2$ and $b_1 = b_2$.

1.3 Axiomatic set theory

Proof. Proof of (i) is obvious. To prove (ii) let a_1, a_2, b_1, b_2 be sets such that $(a_1, b_1) = (a_2, b_2)$, i.e.,

$$\{\{a_1\}, \{a_1, b_1\}\} = \{\{a_2\}, \{a_2, b_2\}\}.$$

We first show that $a_1 = b_1$ iff $a_2 = b_2$. Let $a_1 = b_1$ but $a_2 \neq b_2$. Then $\{\{a_1\}\} = (a_1, b_1) = (a_1, b_2) = \{\{a_2\}, \{a_2, b_2\}\}$ is not possible by axiom **A(1)**. Similarly $a_2 = b_2$ but $a_1 \neq a_2$ is also not possible. Hence only two cases can arise: either $a_1 = b_1$ and $a_2 = b_2$ or $a_1 \neq b_1$ and $a_2 \neq b_2$. In the first case $\{\{a_1\}\} = (a_1, b_1) = (a_2, b_2) = \{\{a_2\}\}$ implies $a_1 = a_2 = b_1 = b_2$ and the proof is complete. In the second case, by axiom **A(1)**,

$$\{a_1\} \in \{\{a_2\}, \{a_2, b_2\}\}.$$

Thus, either $\{a_1\} = \{a_2\}$ or $\{a_1\} = \{a_2, b_2\}$. But $\{a_1\} = \{a_2, b_2\}$ is not possible as $a_2 \neq b_2$. Thus $\{a_1\} = \{a_2\}$ and hence by axiom **A(1)**, $a_1 = a_2$. Similarly, $\{a_1, b_1\} = \{a_2, b_1\} \in \{\{a_2\}, \{a_2, b_2\}\}$. Thus, either $\{a_2, b_1\} = \{a_2\}$ or $\{a_2, b_1\} = \{a_2, b_2\}$. But $\{a_2, b_1\} = \{a_2\}$ is not possible since $a_2 = a_1 \neq b_1$. Thus $\{a_2, b_1\} = \{a_2, b_2\}$. But then by axiom **A(1)**, $b_1 = b_2$.

Conversely, if $a_1 = a_2$ and $b_1 = b_2$, then $\{a_1\} = \{a_2\}$ and $\{a_1, b_1\} = \{a_2, b_2\}$ hence $\{\{a_1\}, \{a_1, b_1\}\} = \{\{a_2\}, \{a_2, b_2\}\}$. ∎

1.3.14 Exercise. Let a and b be sets. Let $\langle a, b \rangle_1 := \{a, b\}, \langle a, b \rangle_2 := \{\{a\}, \{b\}\}, \langle a, b \rangle_3 := \{a, \{b\}\}$ and $\langle a, b \rangle_4 := \{a, b, \{a, b\}\}$. Does any one of these have the property that the ordered pairs have in theorem 1.3.13?

1.3.15 A(6) Axiom of comprehension or separation. *For every set x and every formula Φ there exists a set whose elements are exactly those of x for which Φ holds.*

Symbolically: $\forall y \, \forall z [z \in y \Leftrightarrow z \in x \wedge \Phi(x)]$, where y does not occur in the formula Φ.

Firstly, the above is not one axiom but a collection of infinite axioms: one corresponding to each formula Φ and variables x, y and z. Also, by axiom **A(1)**, the set whose existence is given by axiom **A(6)** for a given x and Φ is unique. We write this as $\{z \in x | \Phi(x)\}$ and read as: set of those elements z of x such that $\Phi(x)$. We give some simple applications of the axiom of comprehension.

1.3.16 Example. A set X is said to be **nonempty** if $X \neq \emptyset$. Let A and B be two nonempty sets. For every $a \in A$ and $b \in B$, consider the ordered pair (a, b), as defined in 1.3.12. We claim that by axiom **A(6)**, we can consider the set of all ordered pairs (a, b), for $a \in A$ and $b \in B$. To see this, recall $(a, b) \in \mathcal{P}(\mathcal{P}(A \cup B))$ and the set we want to consider is

$$\{w \in \mathcal{P}(\mathcal{P}(A \cup B)) | \Phi(w)\}$$

where the formula $\Phi(w)$ is that $w = (a, b)$ for some $a \in A$ and $b \in B$. We can express $\Phi(w)$ in terms of our logical symbols as

$$(\exists a \exists b (a \neq b \wedge a \in A \wedge b \in B \wedge \forall z (z \in w \Leftrightarrow z = \{a\} \vee z = \{a, b\}))) \vee (\exists a (a \in A \wedge a \in B \wedge \forall z (z \in w \Leftrightarrow z = \{a\}))).$$

Once again, this emphasizes the fact that though formulae can be expressed in terms of logical symbols, they are best understood when expressed informally. Thus the set of all ordered

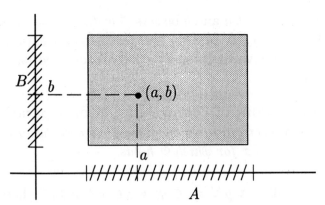

Figure 1.3 : Cartesian product, $A \times B$.

pairs (a,b) for $a \in A$ and $b \in B$ can be written as $\{(a,b) | a \in A, b \in B\}$ and denoted by $A \times B$. In case $A = \emptyset$ or $B = \emptyset$, we define $A \times B$ to be the empty set. The set $A \times B$ is called the **Cartesian product** of A with B. The set $A \times B$ can be visualized pictorially as the set of all points in the shaded rectangle in figure 1.3.

1.3.17 Exercise. For sets A, B and C, prove the following:

(i) $(A \cup B) \times C = (A \times C) \cup (B \times C)$.

(ii) $A \times (B \cup C) = (A \times B) \cup (A \times C)$.

(iii) If $A \neq \emptyset$ then $B \subseteq C$ iff $B \times A \subseteq C \times A$.

(iv) In general, $A \times B \neq B \times A$ and $A \times (B \times C) \neq (A \times B) \times C$.

1.3.18 Exercise. Let A and B be nonempty sets. Find conditions on A and B which imply and are implied by the statement $A \times B = B \times A$.

1.3.19 Example. Let x be a nonempty set. By axiom **A(6)**, there exists a set y that contains exactly those elements that belong to every element of x. We can write this set as follows:

$$y = \{z \in \cup x \mid \forall u(u \in x \Rightarrow z \in u)\}.$$

The set y is uniquely defined by axiom **A(1)** and is called the **intersection** of elements of x, denoted by $y = \cap x$. If $x = \{A, B\}$, then $\cap x$ is denoted by $A \cap B$, called the **intersection** of A and B. We can write this as $A \cap B = \{x \in A | x \in B\}$. Pictorially, $A \cap B$ can be visualized as the region common to both A and B – the region shaded in figure 1.4.

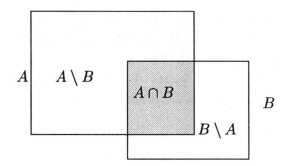

Figure 1.4 : Set intersection, $A \cap B$.

1.3.20 Exercise. For sets X, Y, A, B and C, prove the following:

(i) $\cap X \subseteq A \, \forall \, A \in X$.

(ii) $A \cap B = B \cap A$.

(iii) $A \cap (B \cap C) = (A \cap B) \cap C$.

(iv) $A \subseteq B$ iff $A \cap B = A$.

(v) If $A \subseteq B \, \forall \, B \in X$ then $A \subseteq \cap X$.

(vi) $A \cap \emptyset = \emptyset$.

(vii) $(A \cap B) \times X = (A \times X) \cap (B \times X)$.

(viii) $A \times (B \cap C) = (A \times B) \cap (A \times C)$.

1.3.21 Exercise. Let A and B be sets. Show that there exists a unique set whose elements are those of A which are not elements of B, i.e., $\{x \in A | x \notin B\}$ exists and is unique. Let this set be denoted by $A \setminus B$, called the **relative complement** of B in A. For sets A, B and C, show that

$$A \setminus (B \cup C) = (A \setminus B) \cap (A \setminus C).$$

1.3.22 Definition.

(i) Let A and B be nonempty sets. A subset F of $A \times B$ is called a **function** from A to B if it has the following property: whenever for some $a \in A$ $\exists\, b, c \in B$ such that $(a, b) \in F$ and $(a, c) \in F$ then $b = c$.

When $F = \emptyset$, the function is called the **empty function**. Let $F \subseteq A \times B$ be a function. The set A is called the **codomain** of F and $\mathcal{D}(F) := \{a \in A | (a, b) \in F \text{ for some } b \in B\}$, which is a well-defined subset of A by axiom **A(3)**, called the **domain** of F. The function $F \subseteq A \times B$ is also written as $F : \mathcal{D}(F) \to B$ and if for $a \in \mathcal{D}(F), (a, b) \in F$, we write $F(a) = b$ or $a \longmapsto b$. Again, by axiom **A(3)**, the set $\{b \in B | (a, b) \in F \text{ for some } a \in A\}$ is a well-defined set, called the **range** of F, and is denoted by $\mathcal{R}(F)$.

(ii) A function $F \subseteq A \times B$ is said to be **one-one** or **injective** if $F(a) = F(a')$ for $a, a' \in A$ implies $a = a'$.

(iii) A function $F \subseteq A \times B$ is said to be **onto** or **surjective** if $B = \mathcal{R}(F)$.

(iv) A function $F \subseteq A \times B$ is said to be **bijective** if it is both one-one and onto.

(v) The function $F \subseteq A \times A$ defined by $F(a) = a, a \in A$ is called the **identity function** on A and is denoted by Id_A. Note that $Id_A = \{(a, a) \,|\, a \in A\}$.

When we consider a function $F \subseteq A \times B$, such that $A = \mathcal{D}(F)$, we write it as $F : A \to B$. It can be visualized as a correspondence which assigns to every element $a \in A$ some element $b \in B$ if $(a, b) \in F$, taking care that no element at A gets assigned to two different elements of B. Whenever $(a, b) \in F$, we write it as $F(b) = a$ or $a \longmapsto F(a) = b$. The function is one-one if each element of $b \in \mathcal{R}(F)$ gets assigned to at most one element of A, i.e., for $a_1, a_2 \in A$, if $F(a_1) = F(a_2)$ implies that $a_1 = a_2$. Finally, F is onto if every element $b \in B$ gets assigned to some element of A, i.e., $\forall\, b \in B$ there exists some $a \in A$ such that $F(a) = b$. Pictorially, a function $F : A \to B$ can be represented as in figure 1.5 below:

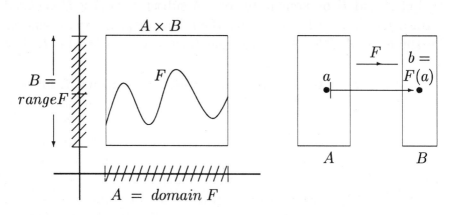

Function $F \subseteq A \times B$ $\qquad$ F as a correspondence

Figure 1.5

1.3.23 Theorem. *Let $F : A \to B$ be a one-one and onto function. Let $F^{-1} := \{(b,a) \in B \times A | (a,b) \in F\}$. Then F^{-1} is a function from B to A. Further F^{-1} is also one-one and onto.*

Proof. Since F is onto, for every $b \in B$ there exists some $a \in A$ such that $(a,b) \in F$. Also this a is unique since F is one-one. Thus, $(b,a) \in F^{-1}$ and if $(b,a') = (b,a)$, then $a' = a$. Thus, F^{-1} is a function. Further $\forall\, a \in A$, since $\exists\, b \in B$ such that $(a,b) \in F$, we have $(b,a) \in F^{-1}$, i.e., F^{-1} is onto. Finally, if $(b,a) = (b',a) \in F^{-1}$, then again $(a,b) = (a,b') \in F$ and hence $b = b'$ since F is a function. Thus F^{-1} is also one-one. ∎

1.3.24 Theorem. *Let $F : A \to B$ and $G : B \to C$ be functions. Let $G \circ F := \{(a,c) \in A \times C |\, \exists\, b \in B \text{ such that } (a,b) \in F, (b,c) \in G\}$. Then $G \circ F$ is a function, called the* **composite** *of F with G.*

Proof. Let $a \in A$. Then $\exists\, b \in B$ such that $(a,b) \in F$ and also $\exists\, c \in C$ such that $(b,c) \in G$. But then $(a,c) \in G \circ F$.

1.3 Axiomatic set theory

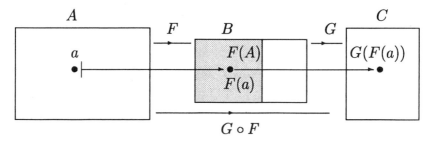

Figure 1.6 : Composite function, $G \circ F$.

Also, if $(a,c) = (a,c') \in G \circ F$, then $\exists\ b$ and $b' \in B$ such that $(a,b'), (a,b) \in F$ and $(b,c), (b',c') \in G$. But then $b = b'$ and hence $c = c'$. This proves that $G \circ F$ is a function. ∎

1.3.25 Exercise. Let $A \neq \emptyset, B \neq \emptyset$ and F be a one-one function from A onto B. Show that $F^{-1} \circ F = \{(a,a) | a \in A\} = Id_A$ and $F \circ F^{-1} = \{(b,b) | b \in B\} = Id_B$.

1.3.26 Exercise. Let $F : A \to B, G : B \to C$ and $H : C \to D$ be functions. Show that $H \circ (G \circ F) = (H \circ G) \circ F$ and $(H \circ G)^{-1} = G^{-1} \circ H^{-1}$.

1.3.27 Definition. Let A and B be nonempty sets.

(i) Any subset R of $A \times B$ is called a **relation** from A to B. If $(a,b) \in R \subseteq A \times B$, we write this as aRb. The set $\{a \in A | (a,b) \in R$ for some $b \in A\}$ is called the **domain** of the relation R, and the set $\{b \in B | (a,b) \in R$ for some $a \in A\}$ is called the **range** of the relation R. Whenever $R \subseteq A \times A$ and the domain of R is A, we say that R is a relation on A.

(ii) Let R be a relation on A. We say that R is **reflexive** if $aRa\ \forall\ a \in A$. We say R is **irreflexive** if aRa is not true for any $a \in A$. The relation R is said to be **symmetric** if $aRb \Rightarrow bRa\ \forall\ a,b \in A$. The relation R is said to be **transitive** if $(aRb \wedge bRc) \Rightarrow aRc\ \forall\ a,b,c \in A$. A relation which is reflexive, symmetric and transitive is called an **equivalence relation** on A.

1.3.28 Exercise. Show that a relation $C \subseteq A \times B$ is a function iff aRb and aRc imply $b = c$.

1.3.29 Theorem. *Let R be an equivalence relation on A. For every $a \in A$, let $[a] := \{b \in A | aRb\}$. Then the following hold:*

(i) $[a]$ *is a set* $\forall\, a \in A$ *and* $a \in [a]$, *called an* **equivalence class**.

(ii) $[a] = [b]$ *iff* aRb.

(iii) $[a] \cap [b] = \emptyset$ *iff* $(a, b) \notin R$.

(iv) $\bigcup_{a \in A} [a] = A$.

(v) $\{[a] | a \in A\}$ *is a set and is denoted by* A/R.

(In other words, given an equivalence relation on a set, every element of the set belongs to one and only one equivalence class and two elements belong to the same equivalence class iff they are related. Then the set gets 'partitioned' into equivalence classes.)

Proof. (i) It follows from axiom **A(3)** that $[a] := \{b \in A | aRb\}$ is a set. Also $a \in [a]$ follows from the fact that R is reflexive. Next, let $[a] = [b]$. Then $a \in [a]$ and hence by axiom **A(3)**, $a \in [b]$, i.e., bRa and by symmetry of R we get aRb. Conversely, if aRb then $b \in [a]$. Let $c \in [b]$. Then bRc and by transitivity, aRc, i.e., $c \in [a]$. Hence $\forall\, c \in [b], c \in [a]$. Similarly $\forall\, c \in [a]$ we will have $c \in [b]$. Hence by Axiom **A(3)**, $[a] = [b]$. This proves (ii). Suppose next $[a] \cap [b] \neq \emptyset$. Then $\exists\, c \in [a] \cap [b]$ and hence $c \in [a]$ and $c \in [b]$, i.e., aRc and bRc. But then aRc and cRb and we have aRb by transitivity of R. Thus $(a, b) \notin R$ implies $[a] \cap [b] = \emptyset$. Conversely, if $[a] \cap [b] = \emptyset$, then $(a, b) \notin R$ for if $(a, b) \in R$ then by (ii) $[a] = [b]$. This proves (iii). Consider the set $\bigcup_{a \in A} [a]$ given by the axiom **A(5)**. Then $\forall\, c \in \bigcup_{a \in A} [a]$, clearly $c \in A$. Conversely if $c \in A$, then $c \in [c]$, thus $c \in \bigcup_{a \in A} [a]$. This proves (iv). Finally, note that $[a] \in \mathcal{P}(A)$ $\forall\, a \in A$. Thus $\{[a] | a \in A\} = \{X \subset \mathcal{P}(A) | X = [a]$ for some $a \in A\}$ is a set by axiom **A(3)**. ∎

1.3.30 Definition.

(i) A relation R on a nonempty set A is called a **partial order**, or simply an **order**, on A if R is reflexive, transitive and anti-symmetric, i.e., if aRb and bRa then $a = b$. If R is a partial order and aRb, we normally write it as $a \leq b$.

(ii) A partial order R on A is said to be **total** (or **linear**) if $\forall\, a,b \in A$ one and only one of the following holds: $a = b$, aRb, or bRa.

(iii) A set X, with a partial order $\leq$ on it, is called a **partially ordered** set and is denoted as the pair $(X, \leq)$. A partially ordered set $(X, \leq)$ is said to be **totally ordered** if $\leq$ is a total order.

1.3.31 Definition. A subset B of a set A is said to be a **proper subset** of A if $B \neq A$. We normally write this as $B \subsetneq A$.

1.3.32 Definition. A nonempty set is said to be an **infinite** set if $\exists$ a proper subset B of A and a one-one, onto map $f : A \to B$. If a set is empty or is not infinite, it is called a **finite set**.

1.3.33 Examples. The set $\emptyset$ whose existence is given by axiom **A(1)** is a finite set by definition. Consider the set $\{\emptyset\}$ as given by axiom of pairing. Clearly it is a finite set. Using axiom of extensionality, pairing and union, we can construct sets $\{\emptyset, \{\emptyset\}\}, \{\emptyset, \{\emptyset\}, \{\emptyset, \{\emptyset\}\}\}$, and so on. All these are finite sets. The idea is that given a finite set S, if we consider the set $S \cup \{S\}$, as given by our axioms, then $S \cup \{S\}$ is also a finite set. Suppose not, then there exists a one-one map $\phi : S \cup \{S\} \to S \cup \{S\}$ which is not onto. In case $\phi(s') = \phi(S)$ for some $s' \in S$, define $\psi : S \to S$ by $\psi(s) := \phi(s)\ \forall\, s \neq s'$ and $\psi(s') := \phi(\{S\})$. Then ψ is a one-one map which is not onto, not possible as S is a finite set. Similarly, if $\phi(\{S\}) = \{S\}$, then $\phi(S) \subsetneq S$ and hence the restrictions of ϕ to S, i.e., ϕ considered as a map only on S, is a map from $S \to S$ which is one-one but not onto, not possible again. Hence $S \cup \{S\}$ is a finite set. Thus enough finite sets exist and given a finite set S we can construct another finite set $S \cup \{S\} \neq S$. However, our axioms do not imply the existence of an infinite set. This motivates the next axiom of set theory.

1.3.34 A(7) Axiom of infinity. *There exists a set S such that $\emptyset \in S$ and $x^+ := x \cup \{x\} \in S$ whenever $x \in S$. Such a set is called a* **successor** *set.*

Symbolically : $\exists S(\emptyset \in S \land \forall x(x \in S \Rightarrow x \cup \{x\} \in S))$.

1.3.35 Theorem. *Every successor set is an infinite set.*

Proof. Consider a successor set S. Define $\Phi : S \to S$ by

$$\Phi(x) := x^+ := x \cup \{x\}, \quad x \in S.$$

Then Φ is a one-one map, for if $x, y \in S$ and $x \neq y$, then clearly $\{x\} \neq \{y\}$ by axiom **A(1)**. Thus $x^+ \neq y^+$. Since $\emptyset \neq x \cup \{x\}$ for any $x \in S$, Φ is not onto. Hence S is infinite. ∎

1.3.36 Exercise. Let S be a set. Prove the following:

(i) If every $x \in S$ is a successor set, $\cap S$ is also a successor set.

(ii) If S is a successor set, is $S^+ := S \cup \{S\}$ also a successor set? Is it an infinite set?

1.3.37 Theorem. *Let A and B be sets such that $A \subseteq B$. Then the following hold:*

(i) *If A is an infinite set, so is B.*

(ii) *If B is a finite set, so is A.*

1.3 Axiomatic set theory

Proof. (i) Let B be an infinite set and $A \subseteq B$. By definition, there exists a proper subset C of A and a one-one, onto map $\phi : A \to C$. Define $\psi : B \to (B \setminus A) \cup C$ as follows:

$$\psi(b) := \begin{cases} b & \text{if } b \in B \setminus A, \\ \phi(b) & \text{if } b \in A. \end{cases}$$

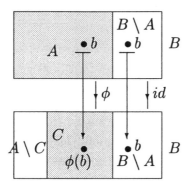

Figure 1.7 : Definition of ψ.

It is easy to see that ψ is a one-one, onto map. Since $(B \setminus A) \cup C \subsetneq B$, B is an infinite set.

(ii) Let B be a finite set and $A \subseteq B$. If A is an infinite set, so also is B, by (i), not true. Hence A is a finite set. ∎

We close this section by stating the remaining axioms of set theory which we shall use infrequently in our discussions.

1.3.38 A(8) Axiom of replacement. *If $\Phi(x, y)$ is a formula such that for every x in a set A there is exactly one y for which $\Phi(x, y)$, then there exists set B such that $y \in B$ iff there exists $x \in A$ such that $\Phi(x, y)$.*

Symbolically : $\forall \, x \, \exists! \, y(\Phi(x, y)) \Rightarrow$
$$\exists \, B \, \forall \, y(y \in B \Leftrightarrow \exists \, x(x \in A \land \Phi(x, y))).$$

For a given Φ and a set A, the set given by axiom **A(8)** is denoted by $\{y | \Phi(x, y), x \in A\}$.

1.3.39 Examples.

(i) Consider the formula $\Phi(x,y)$: $\forall u(u \subset x \Rightarrow u \subset y)$. Then Φ does not determine a function. This is because $\Phi(x,y)$ and $\Phi(x,z)$ mean $x \subseteq y$ and $x \subseteq z$, but it need not imply $y = z$. Similarly the formula $\Phi(x,y)$: $x \in \cup y$ does not have the property required by the replacement axiom.

(ii) The formula $\Phi(x,y)$: $x = y$ obviously has the required property. Also the formulae $\Phi(x,y)$: $y = \cup x$; $\Phi(x,y)$: $y = x \times A$, where A is a fixed set; and $\Phi(x,y)$: $y = \mathcal{P}(x)$ have the property required of the replacement axiom.

We shall give an application of the axiom of replacement in section 2.5. Axiom of replacement provides the existence of sets needed for Cantor's transfinite arithmetic. For more details refer to [Hal].

1.3.40 Remark. Axiom **A(6)**, the axiom of comprehension, can be deduced from the axiom of replacement. Indeed, given a set A and a formula $\Phi(x)$, consider the formula $\Phi(x,y)$: "$x = y$ and $\Phi(x)$". Then $\Phi(x,y)$ satisfies the required property of the axiom of replacement and we infer the existence of a set $B = \{y | \exists x \in A$ with $x = y$ and $\Phi(x)\} = \{x \in A | \Phi(x)\}$, which is axiom **A(6)**. Similarly, the axiom of pairing, axiom **A(3)**, can be deduced from the axiom of replacement and the axiom of power set, refer to [Sto].

1.3.41 A(9) Axiom of choice. *For every set A there exists a function f whose domain is the collection of nonempty subsets of A and for every $B \subseteq A, B \neq \emptyset, f(B) \in B$.*

We shall not bother to give the symbolic representation of this axiom. Intuitively, this says that if A is a nonempty set whose elements are nonempty sets, then we can choose one element of each member B of A and form a set of these chosen elements. The function f is normally called a **choice function**. Note that if $A = \emptyset$, then the choice function is the empty function. We shall see applications of this axiom in sections 2.4 and 6.7. Here we give a simple illustration from naive set theory for the need of such an axiom. Consider a set A of pairs of shoes. Then we can select a member of each pair, say the left shoe from the pair and form a set. Now suppose B is a set of pairs of socks and we are asked to pick one from each pair and form the set of socks so selected. Since each pair of socks comprises of two exactly identical objects, difficulty comes in choosing one of them and forming the set when the collection B is infinite. The axiom of choice guarantees such a choice.

1.3.42 A(10) Axiom of regularity. *For every nonempty set A, there exists $b \in A$ such that $b \cap A = \emptyset$.*

1.3.43 Remark. A simple consequence of the axiom of regularity is that for every set A, $A \notin A$. For if $A \in A$ then $A \in \{A\} \cap A$ and also by axiom **A(10)**, there exists $b \in \{A\}$ such that $b \cap \{A\} = \emptyset$. Since the only possibility for b is A, we have $A \cap \{A\} = \emptyset$. Thus $A \in \{A\} \cap A$ as well as $A \cap \{A\} = \emptyset$, not possible. A similar argument will show that for no two sets A and B is it true that $A \in B$ and $B \in A$. In fact, axiom **A(10)** rules out the possibility of a collection of sets $a_1, a_2, \ldots, a_n$ such that $a_1 \in a_2, a_2 \in a_3, \ldots, a_{n-1} \in a_n$, $a_n \in a_1$. Thus this axiom helps us to exclude the existence of sets which give rise to paradoxes in naive set theory.

1.4 Epilogue

In section 1.3 we outlined axioms **A(1)** to **A(10)** of set theory and developed some of the familiar concepts needed for further discussions. Axioms **A(2)** and **A(7)** are existential axioms. They postulate the existence of certain sets independent of any assumptions concerning the existence of other sets. Axioms **A(3)**, **A(4)**, **A(5)**, **A(6)**, **A(8)**, and **A(9)** are axioms of conditional existence, i.e., they permit us to construct certain sets from some given sets. The sets constructed using **A(3)**, **A(4)**, **A(5)**, **A(6)**, and **A(8)** are unique. Axiom **A(9)**, the axiom of choice, asserts the existence of a set not necessarily unique. The axiom of choice is one of the most debatable axioms, accepted by some mathematicians and rejected by others. One needs this axiom to establish some classical results in analysis, algebra and topology. The axiomatic set theory developed on the basis of these axioms, excluding the axiom of choice, is called Zermelo–Fraenkel set theory, in short ZF-set theory. Whenever the axiom of choice is also included in the list of axioms, it is denoted by ZFC-set theory. We shall work with ZFC-set theory and mention the use of axiom of choice explicitly.

1.4.1 Absence of paradoxes. In section 1.2, we had stated some of the paradoxes of naive set theory and said that to avoid these paradoxes the axiomatic theory of sets was developed. We show how this happens. First of all in naive set theory, the comprehension principle states that every property of objects gives rise to a set of all the objects having that property. This leads to the construction of very large sets like that in the Russell's paradox. In axiomatic set theory the comprehension principle is modified to state that given a set and a property (formula) Φ, there exists a set of these elements of the given set which have the property Φ. The set which leads to Russell's paradox in naive set theory is the set $R = \{x | x \notin x\}$. Russell's paradox does not appear in axiomatic set theory for R is not a set in this theory. For if R is a set, then $R \in R$ iff $R \notin R$, an impossibility. In fact in axiomatic set theory, one can at best consider the set $\{x \in A | x \notin A\}$ for a given set A, which is the empty set. It follows from the axiom of regularity that $A \notin A$ for any set A, as shown in remark 1.3.43. Also semantic paradoxes (as that of Barry's, see 1.2.2) are avoided for in axiomatic set theory a formula can only be written using formal language in terms of the logical symbols. Thus a statement like "a natural number describable in less than 50 words of a dictionary" cannot appear in the list of formulae. Thus in some sense the ZF-theory is very restrictive. For example given a set x, we cannot construct a set $\{y : x \subseteq y\}$, or one cannot talk of the set of "all men in India".

1.4 Epilogue

(Consideration of such objects is possible in the theory proposed by von Neumann and Paul Bernays, called the axiomatic class theory. But that also has its limitations. For details refer to [Coh] and [End].)

1.4.2 Consistency and completeness. The axioms of ZF/ZFC-set theory allow us to develop the concepts of naive set theory without encountering paradoxes. One can ask the question: what is the surety that new paradox will not appear? Well, till now no one has been able to discover any paradox in ZF-theory. But that does not mean that the 'consistency' of the axiom of ZF-theory is proved. Further, one might expect that all statements in an axiomatic set theory should either be true or false. If we assume that the axioms of ZF-theory are consistent, then there are statements P such that neither P nor the negation of P are provable from the ZF-axioms. It was shown by Kurt Gödel in 1936 that it is impossible to disprove the axiom of choice using the ZF-axioms. In 1963 Paul Cohen showed that it is impossible to prove it too. Thus axiom of choice is independent of the ZF-axioms. Similarly, Continuum Hypothesis (see section 6.8) is neither provable nor refutable in the ZFC-set theory. In other words the continuum hypothesis is independent of the ZFC-axioms. As far as consistency is concerned, it can be proved that ZFC-axioms are consistent if ZF-axioms are consistent. It follows from the work of Gödel that one of the statements that remain undecided in ZF-theory is the statement that the ZF-axioms are consistent. Thus the consistency of ZF-axioms remains undecided. One alternative to resolve this difficulty would be to enlarge the set of axioms by including an underivable statement or its negation as an additional axiom. But the work of Gödel applies not only to ZF-set theory, but to any consistent extension of ZF-theory that satisfies the following two very mild conditions: (i) it is as powerful as a small fragment of number theory, and (ii) whose set of axioms is such that there is an algorithm to decide if a formula is an axiom or not. Thus the new system will still have some true but unprovable statements. The outcome of the story is that an axiomatic approach in set theory will never be able to answer all questions of the theory. In the words of Herman Weyl, "The question of ultimate foundations and ultimate mathematics remains open; we do not know in what direction it will find its final solution or even if a final objective answer can be expected at all." However, ZFC-theory is strong enough for most of mathematics.

Chapter 2

NATURAL NUMBERS

2.1 Historical comments

It is very difficult to pinpoint the origin of numbers and the process of counting. One can only conjecture that the need to distinguish between one sheep and a herd of sheep; or between one tree and a forest; or between one man and a group of men; and the need to identify the similarity between one tree, one sheep, one man or the similarity between a pair of eyes, a pair of ears, a pair of hands, etc., must have given birth to the primitive concept of numbers. Of course, this must have been a process of gradual awareness developing over centuries, rather than a discovery. Further, the need to express these similarities or differences in symbols − first orally and then written − must have given rise to the process of counting and to number symbols. Anthropologists have discoveries to their credit to claim that this process must have begun at least about 30,000 years back. As the civilizations progressed, symbols were designed to represent these numbers. For example, discoveries about the stone-age indicate that man used to put notches in bones or marks on the wall of his cave to represent numbers: the notches ||| probably meant the number of kills he had made in a day. As man started living in groups the need to build homes, agriculture, trade, etc., must have given birth to arithmetic. The systematic study of numbers and arithmetic was developed by the Babylonians around 2000 B.C. The symbols they used for the numbers were as follows:

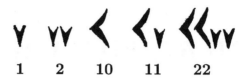

Figure 2.1 : Babylonian number symbols.

Around 1700 B.C. Egyptians had also developed symbols for numbers and had the knowledge of arithmetic. Some of the symbols used by them were:

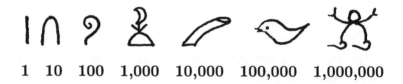

Figure 2.2 : Egyptian number symbols.

The origin of our story (the number systems) lies in the Mathematics developed by the Greeks (600 B.C.–200 A.D.). The Greeks had developed a system of number symbols which could be used to denote very large numbers. In fact, around 450 B.C. they used the 24 letters of the Greek alphabet for number symbols:

$$\alpha \quad \beta \quad \gamma \quad \delta \quad \epsilon \quad \chi \quad \zeta \quad \eta \quad \theta \quad \iota$$

1 2 3 4 5 6 7 8 9 10

Figure 2.3 : Greek number symbols.

They also had a good knowledge of arithmetic. The belief of the Greek mathematician Pythagoras and his followers (600 B.C.–300

2.1 Historical comments

B.C.) that 'Everything is Number', led to many discoveries, e.g., pairs of **amicable** numbers (two numbers are called amicable if the sum of all proper divisors of one is the other and vice versa, e.g., 284 and 220); **perfect numbers** (a number is called perfect if the sum of all its proper divisors is the number itself, e.g., 6); and **Pythagorean triplets** (numbers a, b, c are called Pythagorean triplets if $a^2 + b^2 = c^2$, e.g., $a = (2m)^2, b = (m^2 - 1)^2, c = (m^2 + 1)^2, m = 1, 2, \ldots$).

The number symbols $1, 2, \ldots, 9$ (to which 0 was added later) used in almost all parts of the globe can be directly traced to the nine characters of the Sanskrit language used by the Hindus. These reached Europe (around 1400 A.D.) via the Arabs.

Even though symbols were used to represent numbers - they were only regarded as units for numbers. No effort was made to define a number. This state of affairs continued until the 19th century. In an effort to give a secure foundation to analysis, questions were raised about the concept of number itself. It was only after the works of Dedekind, Frege, Cantor and finally Peano that the natural numbers were constructed based on the axioms of set theory (as described in chapter 1). We first define the natural numbers and then show their existence and uniqueness.

Giuseppe Peano (1858–1932)

Peano was born in Italy where he studied and followed his academic career. In 1890 he became a professor of calculus at the University of Turin. From 1896 to 1901 he was also teaching at the Military Academy. Though he believed 'logic is the servant of mathematics', he used a lot of symbolism in his work (notably in *Formulario mathematico*, 5 Vols., 1895–1908) and teaching. Students protested against this and he was obliged to resign from the Military Academy.

His works include, *Arithmetices Principia Nove Methodo Exposita* (1889) giving the axiomatization of natural numbers and the construction of integers; *Applicazioni geometriche del calcolo infinitesimale* (1887)

on the notion of curve and the area enclosed by it; *I Principii di geometria* (1889) is his work on projective geometry; discovery of square filling curves (1890); and *Sui fondamenti della geometria* (1899) on the axiomatization of Euclidean geometry.

2.2 Existence and uniqueness of natural numbers

2.2.1 Definition. A set $\mathbb{N}$ is called a set of **natural numbers** if it has the following properties:

N(i) *There exists an element* $1 \in \mathbb{N}$.
This distinguished element is called **one**. This ensures that the set $\mathbb{N}$ is nonempty.

N(ii) *For every $n \in \mathbb{N}$, there exists an element $S(n) \in \mathbb{N}$ such that $\{(n, S(n)) | n \in \mathbb{N}\}$ is a function.*
The function $n \longmapsto S(n)$ is called the **successor function** and the element $S(n)$ is called the **successor** of n.

2.2 Existence and uniqueness

We denote this function by S itself.

N(iii) $1 \notin S(\mathbb{N})$, *i.e., 1 is not the successor of any element.*

N(iv) *S is one-one, i.e., $S(n) = S(m)$ for $n, m \in \mathbb{N} \Rightarrow n = m$.*

N(v) *If A is any subset of $\mathbb{N}$ such that $1 \in A$ and $S(n) \in A$ $\forall\, n \in A$, then $A = \mathbb{N}$.*

Another way of denoting natural numbers is by the triple $(\mathbb{N}, 1, S)$ such that $\mathbb{N}$ is a nonempty set, $1 \in \mathbb{N}$ and $S : \mathbb{N} \to \mathbb{N}$ is a function with properties (iii), (iv) and (v). The triple $(\mathbb{N}, 1, S)$ is also called a **natural number system** or a **Peano system**.

We can visualize the set $\mathbb{N}$ as the symbols and S as the counting process. Then **N(i)** says that we can start counting, **N(ii)** gives the method of counting, **N(iv)** says that in the process of counting, a number is encountered only once and **N(v)** is known as the **"Principle of Mathematical Induction"**, i.e., if you can start counting and have counted upto n and can count upto $S(n)$ then you can count everything. We shall see applications of all these properties. (Many authors denote the special element of $\mathbb{N}$ by 0, which plays the role of conventional zero in the number system.)

2.2.2 Remark. It is easy to show that the properties **N(iii)**, **N(iv)** and **N(v)** of a triple $(\mathbb{N}, 1, S)$ are independent of each other. For example, consider any natural number system $(\mathbb{N}, 1, S)$ and define $S'(n) := S(S(n))\ \forall\, n \in \mathbb{N}$. Then the triple $(\mathbb{N}, 1, S')$ will have properties **N(iii)** and **N(iv)** but not **N(v)**. To see this, note that the set $A := \mathbb{N} \setminus \{1\}$ has the property that $1' := S'(1) \in A$ and whenever $n \in A, S'(n) \in A$, but $A \neq \mathbb{N}$. Similarly, let $\mathbb{N}' := \{1, S(1), S(S(1))\}$ and define $S'' : \mathbb{N}' \to \mathbb{N}'$ by $S''(1) := S(1), S''(S(1)) := S(S(1))$ and $S''(S(S(1))) := 1$. Then the triple $(\mathbb{N}', 1, S'')$ has properties **N(iv)** and **N(v)** but not **N(iii)**. Let $S''' : \mathbb{N}' \to \mathbb{N}'$ be defined by $S'''(1) := S(1), S'''(S(1)) := S(S(1))$ and $S'''(S(S(1))) := S(1)$. Then the system $(\mathbb{N}', 1, S''')$ will have properties **N(iii)** and **N(v)** but not **N(iv)**.

Before we proceed further, we show that in the axiomatic theory of sets a natural number system exists and is unique.

2.2.3 Theorem (Existence of $\mathbb{N}$). *There exists a natural number system.*

Proof. Consider a successor set A whose existence is given by the axiom of infinity. By the axiom of power set and the axiom of separation, there exists a set $\mathcal{F} := \{B \in \mathcal{P}(A) \mid B \text{ is a successor set}\}$. The set $\mathcal{F}$ is nonempty as $A \in \mathcal{F}$. By example 1.3.19, $\mathbb{N} := \cap \mathcal{F}$ is a well-defined set and, by exercise 1.3.36, $\mathbb{N}$ is also a successor set. Using the power set axiom, we can define a function $S : \mathbb{N} \to \mathbb{N}$ by $S(x) := x \cup \{x\}$ $\forall x \in \mathbb{N}$. Clearly $S(x) \neq \emptyset$ for every $x \in \mathbb{N}$. If $S(n) = S(m)$ we claim that $n = m$. Suppose not, i.e., $n \neq m$ but $n \cup \{n\} = m \cup \{m\}$. Let $x \in n \cup \{n\}$. If $x = n$, then $x \in m \cup \{m\}$ and hence $x \in m$, for $n \neq m$. Thus $n \in m$. Similarly $m \in n$. But that is not possible in view of the axiom of regularity. Thus S is one-one. Finally, let B be any subset of $\mathbb{N}$ such that $\emptyset \in B$ and $S(x) \in B$ whenever $x \in B$. Then B is a successor set and hence $B \in \mathcal{F}$. Thus $(\mathbb{N}, \emptyset, S)$ is a natural number system. ∎

2.2.4 Remarks.

(i) In fact $\mathbb{N}$, as defined in theorem 2.2.3, is the smallest successor set, i.e., if V is any successor set then $\mathbb{N} \subseteq V$ and $\mathbb{N}$ includes no successor set. To see this note that $\mathbb{N}$ being the intersection of all successor subsets of A is itself a successor set by 1.3.36. Also, if V is any successor set, then $\mathbb{N} \cap V$ is a successor set. Since $\mathbb{N} \cap V \subseteq \mathbb{N} \subseteq A$, hence $\mathbb{N} \cap V \in \mathcal{F}$, where $\mathcal{F}$ is as defined in theorem 2.2.3. Thus $\mathbb{N} \subseteq \mathbb{N} \cap V$, i.e., $\mathbb{N} \subseteq V$. Further, let D be a successor set and $D \subseteq \mathbb{N}$. Then $D \subseteq \mathbb{N} \subseteq A$, i.e., D is a successor subset of A and hence $\mathbb{N} \subseteq D$. Thus $D = \mathbb{N}$.

(ii) In the proof of theorem 2.2.3, to prove that $S(x) := x \cup \{x\}$ is a one-one map on $\mathbb{N}$ we used the axiom of regularity. This can be avoided. We first note that $\forall x \in \mathbb{N}$, $x = \cup S(x)$. For

2.2 Existence and uniqueness

this consider the set $\mathcal{F} := \{x \in \mathbb{N} \mid x = \cup S(x)\}$ which exists by the axiom of comprehension. This set $\mathcal{F}$ is nonempty for $S(\emptyset) = \emptyset \cup \{\emptyset\}$ implies that $\cup S(\emptyset) = \emptyset$, i.e., $\emptyset \in \mathcal{F}$. Also if $x \in \mathcal{F}$, i.e., $x = \cup S(x)$, then

$$\begin{aligned} \cup S(S(x)) &= \cup (S(x) \cup \{S(x)\}) \\ &= (\cup S(x)) \cup \{S(x)\} \\ &= x \cup \{S(x)\} \\ &= x \cup \{x \cup \{x\}\} \\ &= x \cup \{x\} \\ &= S(x). \end{aligned}$$

Hence $\mathcal{F}$ is a successor subset of $\mathbb{N}$. This implies, by remark (i) above, $\mathcal{F} = \mathbb{N}$. Thus $x = \cup S(x)$ $\forall\, x \in \mathbb{N}$. Hence if $S(x) = S(y)$ then $x = \cup S(x) = \cup S(y) = y$, i.e., S is one-one.

(iii) In our model of natural number system, $\emptyset$ corresponds to number 1. In case 0 is to be included in the set of natural numbers, as many authors do, it is represented by $\emptyset$ and $1 := S(\emptyset)$.

2.2.5 Theorem. *Let $(\mathbb{N}, 1, S)$ be a natural number system. Then $\mathbb{N}$ is an infinite set.*

Proof. If $\mathbb{N}$ were finite, then every one-one map $\phi : \mathbb{N} \to \mathbb{N}$ must also be onto. Since $S : \mathbb{N} \to \mathbb{N}$ is one-one but not onto, $\mathbb{N}$ is an infinite set. ∎

2.2.6 Theorem. *Let A be any infinite set. Then there exists a natural number system $(\mathbb{N}', a, \phi)$ such that $\mathbb{N}' \subseteq A$.*

Proof. Since A is infinite, there exists a one-one map $\phi : A \to A$ which is not onto, i.e., $\phi(A)$ is a proper subset of A. Let $a \in A \setminus \phi(A)$. Let $\mathcal{G} := \{N \subseteq A \mid a \in N \text{ and } \phi(N) \subseteq N\}$. Then $\mathcal{G}$ is a nonempty set as $A \in \mathcal{G}$. Let $\mathbb{N}' := \bigcap_{N \in \mathcal{G}} N$. Then $a \in \mathbb{N}'$ and $\phi(\mathbb{N}') \subseteq \mathbb{N}'$. Thus ϕ can be restricted to the set $\mathbb{N}'$. Clearly, this restriction is one-one and $a \notin \phi(\mathbb{N}')$, for $a \notin \phi(A)$. Finally, let B be a subset of $\mathbb{N}'$ such that $a \in B$ and $\phi(n) \in B$ for every $n \in B$. Then $B \in \mathcal{G}$ and hence $\mathbb{N}' \subseteq B$, i.e., $\mathbb{N}' = B$. Thus $(\mathbb{N}', a, \phi)$ is a natural number system. ∎

As pointed earlier, a set $\mathbb{N}$ with properties **N(i)** to **N(v)** essentially says that there exists a method of counting. It is natural to ask the question: does there exist some other method of counting? Or, are any two methods of counting essentially the same? This motivates our next definition.

2.2.7 Definition. Let $(\mathbb{N}, 1, S)$ and $(\mathbb{N}', 1', S')$ be two natural number systems satisfying properties **N(i)** to **N(iv)**. We say $\mathbb{N}$ and $\mathbb{N}'$ are **isomorphic** if there exists a bijective function $\phi : \mathbb{N} \to \mathbb{N}'$ such that $\phi(1) = 1'$ and $\phi \circ S = S' \circ \phi$.

To prove that any two natural number systems $(\mathbb{N}, 1, S)$ and $(\mathbb{N}', 1', S')$ are isomorphic, we have to show the existence of a function $\phi : \mathbb{N} \to \mathbb{N}'$ with the following properties:

(i) $\phi(1) = 1'$.

(ii) $\phi(S(n)) = S'(\phi(n)) \; \forall \, n \in \mathbb{N}$.

One can take the two properties stated above as the definition of ϕ and claim that ϕ is well-defined. That is, consider the set $A := \{n \in \mathbb{N} \mid \phi(n) \text{ is defined}\}$. Since $\phi(1) = 1', 1 \in A$ and if $n \in A$, i.e., $\phi(n)$ is defined then, by (ii), $\phi(S(n)) = S'(\phi(n))$ implies that $S(n) \in A$. Hence by property **N(v)**, $A = \mathbb{N}$, i.e., ϕ is defined everywhere. The flaw with this argument is that it assumes the existence of ϕ which is yet to be defined. The right justification for this comes from the next theorem.

2.2.8 Iteration Theorem (Dedekind). *Let $(\mathbb{N}, 1, S)$ be any natural system and A be any nonempty set. Given $a \in A$ and a function $\psi : A \to A$, there exists a unique function $\phi : \mathbb{N} \to A$ such that $\phi(1) = a$ and $\phi(S(n)) = \psi(\phi(n)) \; \forall \, n \in \mathbb{N}$.*

2.2 Existence and uniqueness

Proof. We first establish the uniqueness of ϕ. Suppose there exist two mappings $\phi_i : \mathbb{N} \to A, i = 1, 2$, such that $\phi_i(1) = a$ and $\phi_i(S(n)) = \psi(\phi_i(n))\ \forall\ n \in \mathbb{N}$. Let $C := \{n \in \mathbb{N} | \phi_1(n) = \phi_2(n)\}$. By the given hypothesis, $1 \in C$ and if $n \in C$, then $\phi_1(S(n)) = \psi(\phi_1(n)) = \psi(\phi_2(n)) = \phi_2(S(n))$ implies that $S(n) \in C$. Hence by axiom $\mathbf{N(v)}$, $C = \mathbb{N}$. This proves the uniqueness of the required map.

To prove the existence of a function ϕ with the required properties is equivalent to prove the existence of a set $F \subseteq \mathbb{N} \times A$ having the following properties:

(0) If $(n, m), (n, m') \in F$ then $m = m'$, i.e., F is a function.

(i) $(1, a) \in F$.

(ii) If $(n, m) \in F$, then $(S(n), \psi(m)) \in F$.

Clearly, the set $F = \mathbb{N} \times A$ has the properties (i) and (ii). Let $\mathcal{F}$ denote the collection of all those subsets F of $\mathbb{N} \times A$ which have properties (i) and (ii). Then $\mathcal{F} \neq \emptyset$ as the set $\mathbb{N} \times A \in \mathcal{F}$. We claim that the 'smallest' element of $\mathcal{F}$ is the required set having properties (0), (i) and (ii). Let

$$E := \bigcap_{F \in \mathcal{F}} F.$$

Then E is the smallest subset of $\mathbb{N} \times A$ having the properties (i) and (ii), i.e., E has properties (i) and (ii) and $E \subseteq F\ \forall\ F \in \mathcal{F}$. To prove that E also has the property (0), let $P := \{n \in \mathbb{N} | (n, m) \in E$ for some $m \in A$ and if $(n, m') \in E$ for some $m' \in A$ also, then $m = m'\}$. We first show that $1 \in P$. For this, note that $(1, a) \in E$. Suppose $(1, b) \in E$ for some $b \in E, b \neq a$. Then the set $E \setminus \{(1, b)\}$ will still have properties (i) and (ii), i.e., $E \setminus \{(1, b)\} \in \mathcal{F}$, contradicting the fact that E is the smallest set with these properties. Hence, if $(1, b) \in E$ then $a = b$, i.e., $1 \in P$. Next, let $n \in P$ and $m \in A$ be the unique element such that $(n, m) \in E$. By property (ii), $(S(n), \psi(m)) \in E$. If possible, let $m' \in A$ be such that $(S(n), m') \in E, m' \neq \psi(m)$. Then again $E \setminus \{(S(n), m')\} \in \mathcal{F}$ and this contradicts the fact E is the smallest set in $\mathcal{F}$. Hence $n \in P$ implies that $S(n) \in P$. Thus by axiom $\mathbf{N(v)}, P = \mathbb{N}$. This proves that E is the required function. ∎

The above theorem will be used time and again to define functions $\psi : \mathbb{N} \to \mathbb{N}$ with special properties. We shall specify $\psi(1)$ and $\psi(S(n))$ in terms of $\phi(n)$ and say, by Iteration theorem, that ϕ is well-defined everywhere or just say that ϕ is defined **inductively** or **recursively**. As an application of this, we prove the uniqueness of the natural number system.

2.2.9 Theorem (Uniqueness of $\mathbb{N}$). *Let $(\mathbb{N}, 1, S)$ and $(\mathbb{N}', 1', S')$ be two natural number systems. Then $\mathbb{N}$ is isomorphic to $\mathbb{N}'$.*

Proof. By the iteration theorem applied to the function $S' : \mathbb{N}' \to \mathbb{N}'$ with $a = 1$, we get a function $\phi : \mathbb{N} \to \mathbb{N}$, such that $\phi(1) = 1'$ and if $\phi(n) = n'$ then $\phi(S(n)) = S'(n')$, i.e., $(\phi \circ S)(n) = (S' \circ \phi)(n) \ \forall \ n \in \mathbb{N}$. By a similar reasoning we will get a map $\psi : \mathbb{N}' \to \mathbb{N}$ such that $\psi(1') = 1$ and $\psi \circ S' = S \circ \psi$. But then $(\psi \circ \phi) \circ S = \psi \circ (\phi \circ S) = \psi \circ (S' \circ \phi) = (\psi \circ S') \circ \phi = (S \circ \psi) \circ \phi = S \circ (\psi \circ \phi)$. We shall show that this implies $\psi \circ \phi : \mathbb{N} \to \mathbb{N}$ is the identity mapping. For this let $B := \{n \in \mathbb{N} | (\psi \circ \phi)(n) = n\}$. Since $(\psi \circ \phi)(1) = \psi(1') = 1$, we have $1 \in B$. Suppose $n \in B$, i.e., $(\psi \circ \phi)(n) = n$. Then $(\psi \circ \phi)(S(n)) = S((\psi \circ \phi)(n)) = S(n)$. Hence $S(n) \in B$. Once again by property $\mathbf{N(v)}, B = \mathbb{N}$. Similarly, we can show that $\phi \circ \psi : \mathbb{N}' \to \mathbb{N}'$ is the identity mapping. Hence ψ is one-one and onto. This completes the proof. ∎

2.2.10 Remarks.

(i) Theorem 2.2.9 says that any two methods of counting are compatible - knowledge from one can be translated to the other and vice versa.

(ii) In theorem 2.2.8, property $\mathbf{N(v)}$ of $\mathbb{N}$ is crucially used. We

2.3 Arithmetic of natural numbers

will see how this property and theorem 2.2.8 are used in defining various concepts on $\mathbb{N}$.

(iii) By construction $\mathbb{N}$ is an infinite set and in view of the proof of theorems 2.2.3 and 2.2.6, every infinite set includes a copy of $\mathbb{N}$. Thus the existence of natural number system is equivalent to the existence of an infinite set.

In view of theorem 2.2.9, every set $\mathbb{N}$ with properties $N(i)$ to $N(v)$ is essentially the same and is called the set of **Natural Numbers** denoted by $(\mathbb{N}, 1, S)$. We close this section with some elementary properties of the natural number system.

2.2.11 Theorem. *Let $(\mathbb{N}, 1, S)$ be a natural number system. Then the following hold:*

(i) $S(n) \neq n \ \forall \, n \in \mathbb{N}$.

(ii) *For every $n \in \mathbb{N}, n \neq 1$, there exists some $m \in \mathbb{N}$ such that $n = S(m)$.*

(iii) *If $n \in \mathbb{N}$ is such that $n \notin S(\mathbb{N})$, then $n = 1$.*

Proof. (i) Let $A := \{n \in \mathbb{N} \mid S(n) \neq n\}$. Since $1 \notin S(\mathbb{N}), 1 \neq S(1)$ and hence $1 \in A$. Next, let $n \in A$, i.e., $S(n) \neq n$. Then $S(S(n)) \neq S(n)$ for S is one-one. Hence by induction $A = \mathbb{N}$.

(ii) Let $A := \{n \in \mathbb{N} \mid n = S(m) \text{ for some } m \in \mathbb{N}\}$. Then for $n \in A, S(n) = S(S(m))$ and hence $S(n) \in A$ also. Thus by induction $A \cup \{1\} = \mathbb{N}$, i.e., $\forall \, n \in \mathbb{N}$ if $n \neq 1$, then $n = S(m)$ for some m.

(iii) Let $n \in \mathbb{N}$ be such that $n \notin S(\mathbb{N})$. Then it follows from (ii) that $n = 1$. ∎

2.3 Arithmetic of natural numbers

As yet another application of theorem 2.2.8, we define binary operations of addition and multiplication on natural numbers.

2.3.1 Theorem. *There exists a unique function* $P : \mathbb{N} \times \mathbb{N} \to \mathbb{N}$ *with the following properties:*

P(i): $P(n, 1) = S(n) \; \forall \, n \in \mathbb{N}$.
P(ii): $P(n, S(m)) = S(P(n, m)) \; \forall \, n, m \in \mathbb{N}$.

Proof. We first show that if a map P with properties **P(i)** and **P(ii)** exists, then it is unique. Suppose there exists another map $P' : \mathbb{N} \times \mathbb{N} \to \mathbb{N}$ such that **P(i)** and **P(ii)** hold for P'. Then for every fixed $n \in \mathbb{N}$, consider

$$A_n := \{m \in \mathbb{N} | P(n, m) = P'(n, m)\}.$$

Since $P(n, 1) = S(n) = P'(n, 1), 1 \in A_n$. Also suppose $m \in A_n$, for some $m \in \mathbb{N}$. Then

$$P(n, S(m)) = S(P(n, m)) = S(P'(n, m)) = P'(n, S(m)).$$

Hence $S(m) \in A_n$. Thus by property **N(v)** of $\mathbb{N}$, $A_n = \mathbb{N}$, implying that $P = P'$. This proves the uniqueness. We next show the existence of a map P with properties **P(i)** and **P(ii)**. Let $n \in \mathbb{N}$ be fixed arbitrarily. For every $n \in \mathbb{N}$, let $P_n : \mathbb{N} \to \mathbb{N}$ be the unique map given by theorem 2.2.8 such that $P_n(1) = S(n)$ and $P_n(S(m)) = S(P_n(m)) \; \forall \, m \in \mathbb{N}$. For $n, m \in \mathbb{N}$, define the map $P : \mathbb{N} \times \mathbb{N} \to \mathbb{N}$ by $P(n, m) := P_n(m)$. Clearly the map P has the required properties **P(i)** and **P(ii)**. ∎

2.3.2 Remarks.

(i) In modern terminology a map $P: \mathbb{N} \times \mathbb{N} \to \mathbb{N}$ is called a **binary operation** on $\mathbb{N}$ (see 2.6.15 for more details).

(ii) The map P is nothing but a formal way of saying that we can add natural numbers. We denote $P(n, m)$ by $n + m$, and call it the **sum** or the **addition** of the natural numbers n and m. In this notation **P(i)** and **P(ii)** take the following form: $\forall \, n, m \in \mathbb{N}$

2.3 Arithmetic of natural numbers

P*(i): $S(n) = n + 1$,

P*(ii): $n + (m + 1) = (n + m) + 1$.

2.3.3 Exercise. Show that $\forall\ m, n \in \mathbb{N}$,

(i) $1 + n = n + 1$.

(ii) $(m + 1) + n = (m + n) + 1$.

2.3.4 Theorem (Properties of addition). *For every* $m, n, k \in \mathbb{N}$, *the following hold:*

(i) $m + n = n + m$, *called the* **commutative law** *of addition.*

(ii) $(m + n) + k = m + (n + k)$, *called the* **associative law** *of addition.*

Proof. (i) Let $m \in \mathbb{N}$ be fixed arbitrarily. Let $C_m := \{n \in \mathbb{N} \mid m + n = n + m\}$. Then $1 \in C_m$ by exercise 2.3.3. Also for $n \in C_m$,

$$\begin{aligned} m + S(n) &= m + (n + 1) \\ &= (m + n) + 1 \quad \text{(by } \mathbf{P*(ii)}\text{)} \\ &= (n + m) + 1 \quad \text{(since } n \in C_m\text{)} \\ &= (n + 1) + m \quad \text{(by exercise 2.3.3)} \\ &= S(n) + m. \end{aligned}$$

Hence $S(n) \in C_m$ whenever $n \in C_m$. Thus by property $\mathbf{N(v)}$, $C_m = \mathbb{N}$.

(ii) Let $m, n \in \mathbb{N}$ be fixed and let

$$A := \{k \in \mathbb{N} \mid (m + n) + k = m + (n + k)\}.$$

Clearly $1 \in A$ by **P*(ii)**. Let $k \in A$. Then

$$\begin{aligned}
(m+n) + S(k) &= (m+n) + (k+1) \\
&= ((m+n) + k) + 1 \quad \text{(by \textbf{P*(ii)})} \\
&= (m + (n+k)) + 1 \quad \text{(as } k \in A\text{)} \\
&= m + ((n+k) + 1) \quad \text{(by \textbf{P*(ii)})} \\
&= m + (n + (k+1)) \quad \text{(by \textbf{P*(ii)})} \\
&= m + (n + S(k)).
\end{aligned}$$

Thus $S(k) \in A$ if $k \in A$. Hence by property **N(v)**, $A = \mathbb{N}$. ∎

2.3.5 Exercise. Show that $\forall\, m, n, k \in \mathbb{N}$,

(i) $m \neq m + n$.

(ii) $m + k = m + n$ implies $k = n$ (this is called the **cancellation law**).

2.3.6 Theorem (Multiplication of natural numbers). *There exists a unique well-defined map*

$$M : \mathbb{N} \times \mathbb{N} \to \mathbb{N}$$

having the following properties: $\forall\, n, m \in \mathbb{N}$,

M(i) : $M(n, 1) = n$.
M(ii) : $M(n, S(m)) = M(n, m) + n$.

The element $M(n, m)$ is called the **product** or **multiplication** of the natural numbers n and m, denoted by nm.

Proof. We first prove the uniqueness of the map M. Suppose there exists another map $M' : \mathbb{N} \times \mathbb{N} \to \mathbb{N}$ having the properties **M(i)** and **M(ii)**. For $n \in \mathbb{N}$ fixed, define

$$A_n := \{m \in \mathbb{N} \,|\, M(n, m) = M'(n, m)\}.$$

2.3 Arithmetic of natural numbers

Clearly $1 \in A_n$. Suppose $m \in A_n$. Then

$$\begin{aligned} M(n, S(m)) &= M(n, m) + n \\ &= M'(n, m) + n \\ &= M'(n, S(m)). \end{aligned}$$

Hence $S(m) \in A_n$. Thus by property **N(v)**, $A_n = \mathbb{N}$. This proves the uniqueness of M. To show the existence of M, let $n \in \mathbb{N}$ be fixed. Let $M_n : \mathbb{N} \to \mathbb{N}$ be the unique map given by the iteration theorem such that $M_n(1) = n$ and $\forall n \in \mathbb{N}, M_n(S(m)) = M_n(m) + n$. Define $\forall n \in \mathbb{N}, M(n, m) = M_n(m)$. Then $M(n, 1) = M_n(1) = n$ and we have

$$M(n, S(m)) = M_n(S(m)) = M_n(m) + n = M(n, m) + n. \blacksquare$$

2.3.7 Definition. In view of the property **M(i)**, i.e., that $M(n, 1) = n \ \forall \ n \in \mathbb{N}$, the element $1 \in \mathbb{N}$ is called the **multiplicative identity** of $\mathbb{N}$. The property $M(n, S(m)) = M(n, m) + n$ is called the **distributive property** of multiplication over addition. A more general form is given in the next theorem.

2.3.8 Theorem. *For every* $n, m, k \in \mathbb{N}$,

(i) $n1 = 1n$.

(ii) $(m + 1)n = mn + n$.

(iii) $nm = mn$, *called the* **commutative** *law of multiplication.*

(iv) $m(n + k) = mn + mk$, *called the* **distributive** *law of multiplication.*

(v) $m(nk) = (mn)k$ *called the* **associative** *law of multiplication.*

Proof. (i) The proof of (i) is easy and is left as an exercise.

(ii) Let $m \in \mathbb{N}$ be fixed. Consider $A_m := \{n \in \mathbb{N} | (m+1)n = mn + n\}$. Clearly $1 \in A_m$. Suppose $n \in A_m$. Then

$$\begin{aligned}
(m+1)S(n) &= (m+1)(n+1) \\
&= (m+1)n + (m+1) &&\text{(by property } \mathbf{M(ii)}) \\
&= (mn+n) + (m+1) &&\text{(as } n \in A_m) \\
&= mn + (n + (m+1)) &&\text{(by theorem 2.3.4(ii))} \\
&= mn + ((n+m) + 1) &&(\quad " \quad) \\
&= mn + ((m+n) + 1) &&(\quad " \quad) \\
&= (mn+m) + (n+1) &&(\quad " \quad) \\
&= m(n+1) + (n+1) &&\text{(by property } \mathbf{M(ii)}) \\
&= mS(n) + S(n).
\end{aligned}$$

Hence $S(n) \in A_m$. Thus by property $\mathbf{N(v)}$, $A_m = \mathbb{N}$, proving (ii).

(iii) Let $n \in \mathbb{N}$ be fixed and consider $C_n := \{m \in \mathbb{N} | mn = nm\}$. Clearly $1 \in C_n$ by (i). Also for $m \in C_n$,

$$\begin{aligned}
S(m)n &= (m+1)n \\
&= mn + n &&\text{(by (ii))} \\
&= nm + n &&\text{(as } m \in C_n) \\
&= nS(m).
\end{aligned}$$

Thus $S(m) \in C_n$ if $m \in C_n$. Hence by property $\mathbf{N(v)}, C_n = \mathbb{N}$.

(iv) We fix $m, n \in \mathbb{N}$ and define

$$D := \{k \in \mathbb{N} | m(n+k) = mn + mk\}.$$

Clearly $1 \in D$. If $k \in D$, then

$$\begin{aligned}
m(n + S(k)) &= m(n + (k+1)) \\
&= m((n+k) + 1) &&\text{(by theorem 2.3.4(ii))} \\
&= m(n+k) + m &&\text{(by (ii))} \\
&= (mn + mk) + m &&\text{(as } m \in D) \\
&= mn + (mk + m) &&\text{(by theorem 2.3.4(ii))} \\
&= mn + m(k+1) &&\text{(by (ii))} \\
&= mn + m(S(k)).
\end{aligned}$$

2.3 Arithmetic of natural numbers

Hence $S(k) \in D$ and by property **N(v)**, $D = \mathbb{N}$.
(v) Let $n, k \in \mathbb{N}$ be fixed. Define
$$A := \{m \in \mathbb{N} | m(nk) = (mn)k\}.$$

Clearly $1 \in A$. If $m \in A$, then

$$\begin{aligned}(S(m)(nk)) &= (m+1)(nk) \\ &= m(nk) + nk \quad \text{(by (ii))} \\ &= (mn)k + nk \quad \text{(as } m \in A) \\ &= (mn+n)k \quad \text{(by (iv))} \\ &= ((m+1)n)k \quad \text{(by (i))} \\ &= (S(m)n)k.\end{aligned}$$

Hence $S(m) \in A$ and by property **N(v)**, $A = \mathbb{N}$. ∎

2.3.9 Exercise. For $n, m, k, r \in \mathbb{N}$ show that

(i) $(mn)(kr) = m(n(kr)) = (mk)(rn)$.

(ii) $nm = nk$ iff $m = k$ (called cancellation law of multiplication).

Using multiplication and the iteration theorem, we can define powers of natural numbers as follows:

2.3.10 Theorem (Exponentiation). *There exists a unique function* $E : \mathbb{N} \times \mathbb{N} \to \mathbb{N}$ *with the following properties:*

E(i): $E(n, 1) = n \quad \forall\, n \in \mathbb{N}$
E(ii): $E(n, S(m)) = M(E(n, m), n) \quad \forall\, n, m \in \mathbb{N}$,

where the function M is as defined in theorem 2.3.6.

Proof. For every fixed $n \in \mathbb{N}$, the iteration theorem gives a unique map ϕ_n such that $\phi_n(1) = n$ and $\phi_n(S(m)) = M(\phi_n(m), n)$ $\forall\, m \in \mathbb{N}$. Define $\forall\, n, m \in \mathbb{N}$,

$$E(n, m) := \phi_n(m).$$

Clearly $E(n,1) = \phi_n(1) = n$ and $E(n, S(m)) = \phi_n(S(m)) = M(\phi_n(m), n) = M(E(n,m), n) \ \forall \ n, m \in \mathbb{N}$. This proves the existence of E.

We now prove the uniqueness of E. Let there exist two maps E_1 and E_2 with properties **E(i)** and **E(ii)**. For every $n \in \mathbb{N}$ fixed, consider the set

$$A := \{m \in \mathbb{N} | E_1(n,m) = E_2(n,m)\}.$$

Clearly $1 \in A$ and if $m \in \mathbb{N}$, then by the given properties of E_1 and E_2 it follows easily that $S(m) \in A$. Hence by property **N(v)**, $A = \mathbb{N}$. ∎

2.3.11 Notation. For $n, m \in \mathbb{N}$, we shall denote $E(n,m)$ by n^m. Then property **E(ii)**, i.e., $E(n, S(m)) = M(E(n,m), n) \ \forall \ n, m \in \mathbb{N}$, can be written as

$$n^{S(m)} = (n^m)n, \ \text{i.e.,} \ n^{m+1} = (n^m)n.$$

2.3.12 Theorem (Laws of exponentiation). *For every $n, m, k \in \mathbb{N}$, the following hold:*

(i) $(1)^n = 1$.

(ii) $(n^m)(n^k) = n^{m+k}$.

(iii) $(n^m)^k = n^{mk}$.

(iv) $(nm)^k = (n^k)(m^k)$.

Proof. (i) Let $A := \{n \in \mathbb{N} | (1)^n = 1\}$. By definition $(1)^1 = E(1,1) = 1$, i.e., $1 \in A$. Suppose $n \in A$. Then $(1)^{n+1} = (1)^n 1 = (1)^n = 1$. Hence $(n+1) \in A$. Thus by property **N(v)**, $A = \mathbb{N}$. This proves (i).

(ii) Let $n, m \in \mathbb{N}$ be fixed arbitrarily. Consider the set

$$B := \{k \in \mathbb{N} | (n^m)(n^k) = n^{m+k}\}.$$

Since $n^{m+1} = (n^m)n, 1 \in B$. Also if $k \in B$, then using properties of multiplications we have

$$\begin{aligned}(n^m)(n^{k+1}) &= (n^m)(n^k n) \\ &= (n^m n^k)n \\ &= (n^{m+k})n \\ &= n^{(m+k)+1} \\ &= n^{m+(k+1)}.\end{aligned}$$

Hence $k+1 \in B$. Thus by property $\mathbf{N(v)}$, $B = \mathbb{N}$. This proves (ii). Proofs of (iii) and (iv) are similar and are left as exercises. ■

2.4 Order on natural numbers

2.4.1 Definition. Let $m, n \in \mathbb{N}$. We say that m is **greater than** n or n is **less than** m if $\exists\, k \in \mathbb{N}$ such that $m = n + k$. We write this as $m > n$ or as $n < m$.

2.4.2 Examples.

(i) Since $S(n) = n + 1$, clearly $S(n) > n$. By the same reason $S(n) > 1$.

(ii) For every $1 \neq n \in \mathbb{N}, n > 1$. To see this let $A := \{n \in \mathbb{N} | n > 1\}$. Clearly $A \neq \emptyset$, for example $S(1) \in A$. Also if $n \in A$, by theorem 2.2.11(ii), $n = 1 + m$ for some m. Hence $S(n) = n + 1 = 1 + m + 1 > 1$. Thus $S(n) \in A$. Hence by principle of induction $A = \mathbb{N} \setminus \{1\}$.

2.4.3 Theorem. *Let $m, n \in \mathbb{N}$, then one and only one of the following statements hold:*

(i) $m = n$.

(ii) $m > n$.

(iii) $m < n$.

Proof. We first show that at most one of (i), (ii) and (iii) can hold. Suppose both (i) and (ii) hold. Then $m = n$ and $m = n + k$ for some $k \in \mathbb{N}$, i.e., $n + k = n$, which is not possible (exercise 2.3.5). Similarly both (i) and (iii) cannot hold at the same time. Suppose (ii) and (iii) hold. Then $m = n + k$ for some $k \in \mathbb{N}$ and $m + r = n$ for some $r \in \mathbb{N}$, i.e., $m = n + k = (m + r) + k = m + (r + k)$, which is again not possible (exercise 2.3.5). Hence only one of (i), (ii) and (iii) can hold. We show next that at least one of the (i), (ii) and (iii) does hold. Let us choose and fix $m \in \mathbb{N}$ arbitrarily. Let $O_m := \{n \in \mathbb{N} |$ for the pair of numbers m and n one and exactly one of (i), (ii) and (iii) hold$\}$. If $m = 1$, then $1 \in O_m$ with $m = 1 = n$, i.e., (i) holds for $n = 1$. If $m \neq 1$, then by theorem 2.2.11 (ii), $\exists\, n$ such that $S(n) = m$, i.e., $m = n + 1$. Thus $m > 1$ and hence $1 \in O_m$. Thus $1 \in O_m$ always. Also if $n \in O_m$, then either $m = n$, in which case $S(m) = S(n)$ and hence $S(n) = m + 1$, i.e., $S(n) > m$, implying that $S(n) \in O_m$. In case $m > n$, i.e., $m = n + k$ for some $k \in \mathbb{N}$, then $m = n + 1 = S(n)$ if $k = 1$. If $k \neq 1$, then $k = r + 1$ for some $r \in \mathbb{N}$ by theorem 2.2.11 (ii), i.e., $m = (n + r) + 1 = (n + 1) + r$. Hence $m > S(n)$, i.e., $S(n) \in O_m$. Finally, if $m < n$ then $m + k = n$ for some $k \in \mathbb{N}$. Thus $n + 1 = (m + k) + 1 = m + (k + 1)$, i.e., $S(n) > m$. Hence $S(n) \in O_m$ in this case also. Thus $S(n) \in O_m$ if $n \in O_m$ and by property $\mathbf{N}(\mathbf{v}), O_m = \mathbb{N}$. This proves the theorem completely. ∎

2.4.4 Definition. For m and $n \in \mathbb{N}$, we say that m is **greater than or equal** to n, written as $m \geq n$, if either $m = n$ or $m > n$. In case $m \geq n$, we write it also as $n \leq m$ and say that n is **less than or equal** to m.

2.4.5 Theorem (Trichotomy law). *The relation $\leq$ on $\mathbb{N}$ is a partial order which is also total.*

Proof. Obviously $m = m$, i.e., $\leq$ is reflexive. Next let $m \leq n$ and $n \leq m$. If $m \neq n$, then $m < n$ and $n < m$, not possible by theorem 2.4.3. Hence $m = n$, i.e., $\leq$ is also anti-symmetric.

2.4 Order on natural numbers

Finally, let $m \leq n$ and $n \leq k$. If $m = n = k$ then there is nothing to prove. Suppose $m < n$ and $n = k$. Again $m \leq k$ is obvious. The case $m = n, n < k$ is similar. Finally if $m < n$ and $n < k$, then by definition $m + r = n$ and $n + r = k$ for some $r, s \in \mathbb{N}$. Thus $(m + r) + s = k$, i.e., $m < k$. Hence $\leq$ is also transitive. Thus $\leq$ is a partial order on $\mathbb{N}$. That the partial order $\leq$ is also total, follows from theorem 2.4.3. ∎

We next describe the behaviour of the order $\leq$ with respect to the operations of addition and multiplication.

2.4.6 Theorem. *For all $m, n, k, r \in \mathbb{N}$, the following hold:*

(i) *If $n \neq m$, then one and only one of the equations $x + m = n$ and $x + n = m$ has a solution in $\mathbb{N}$ and the solution is unique.*

(ii) $m + n > m$ *and* $m + n > n$.

(iii) $m > n$ *iff* $m + k > n + k$.

(iv) $m > n$ *and* $k > r$ *imply* $m + k > n + r$.

(v) $m < n + 1$ *iff* $m \leq n$.

(vi) *There does not exist $m \in \mathbb{N}$ such that $n < m < n + 1$.*

(vii) $n < m$ *iff* $nr < mr$.

(viii) *If $nm = nk$, then $m = k$ (**cancellation law**).*

(ix) *The statement $n < n$ is not true for any $n \in \mathbb{N}$.*

Proof. (i) Let $n \neq m$. Then by theorem 2.4.3, either $n > m$ or $m > n$, i.e., either $n = m + k$ or $m = n + k$ for some $k \in \mathbb{N}$. Thus either $n = x + m$ or $m = x + n$ has a solution in $\mathbb{N}$. Suppose $n = x_1 + m = x_2 + m$ for $x_1, x_2 \in \mathbb{N}$. If $x_1 \neq x_2$, then either $x_1 > x_2$ or $x_2 > x_1$. Thus by theorem 2.4.3, either $x_1 + m > x_2 + m$ or $x_2 + m > x_1 + m$. But neither is possible again by theorem 2.4.3. Hence $x_1 = x_2$. This proves (i).

(ii) By definition $m + n > n$ and $m + n > m$.

(iii) Let $m > n$ and $r \in \mathbb{N}$ be such that $m = n + r$. Then $m+k = n+r+k = n+k+r$ and hence $m+k > n+k$. Conversely, let $m + k > n + k$ and $m = n$. Then $m + k = n + k$, contradicting theorem 2.4.3. Similarly $m < n$ is not possible. Hence $m > n$.

(iv) If $m > n$ and $k > r$ then $m = n + l$ and $k = r + s$ for some $l, s \in \mathbb{N}$. Thus $m + k = n + l + r + s = (n + r) + (l + s)$ and hence $m + k > l + s$.

(v) Suppose $m < n + 1$ and $m \neq n$. Then either $m < n$ or $m > n$. If $m > n$, then $m = n + k$ for some $k \in \mathbb{N}$. But then $m = n + k > n + 1$ by (iv). Thus $m < n + 1$ and $m > n + 1$, which contradicts theorem 2.4.3. Hence $m < n$. Conversely, let $m \leq n$. Then clearly $m \leq n < n + 1$. Hence $m < n + 1$.

(vi) Suppose for some $n \in \mathbb{N}$ there exists $m \in \mathbb{N}$ such that $n < m < n + 1$. Then by (v), $m < n + 1$ implies that $m \leq n$. Then either $m < n$ and $n < m$, or $m = n$ and $n < m$. Either one of them is not possible by theorem 2.4.3. Hence $\forall n$ there does not exist $m \in \mathbb{N}$ such that $n < m < n + 1$.

(vii) Suppose $n < m$ and $l \in \mathbb{N}$ is such that $n + l = m$. Then for every $r \in \mathbb{N}$, $mr = (n + l)r = nr + lr$. Hence $nr < mr$. Conversely, let $r \in \mathbb{N}$ be such that $nr < mr$. Suppose that $n < m$ is not true. Then by theorem 2.4.3, either $n = m$, in which case $nr = mr$ or $n > m$ in which case, by earlier part, $nr > mr$. However, either of them is contradictory as $nr < mr$. Hence $n < m$.

(viii) Suppose $nm = nk$ but $m \neq k$. Then either $m < k$ or $m > k$ and by (vii), either $nm < nk$ or $nm > nk$, not true for either will be a contradiction to theorem 2.4.3. Hence $m = k$.

(ix) If $n < n$ for some $n \in \mathbb{N}$ then $n = n + k$ for some $k \in \mathbb{N}$. But also $n < n + k$, by definition. Thus $n = n + k$ and $n < n + k$, contradicting theorem 2.4.5. Thus $n < n$ is not true for any n. ■

2.4.7 Exercise. Prove the following statements:

(i) $n \geq 1 \quad \forall\, n \in \mathbb{N}$.

(ii) If $n < m$, then $S(n) \leq m$.

2.4 Order on natural numbers

(iii) If $\phi : \mathbb{N} \to \mathbb{N}$ is any function such that $\phi(n) < \phi(m)$ whenever $n < m$, then $\phi(n) \geq n \ \forall \ n$.

2.4.8 Exercise. For every $n, m, k \in \mathbb{N}$, prove the following:

(i) $n < m$ iff $n^k < m^k$.

(ii) $n > 1$ iff $n^m > 1$.

(iii) $(n+1)^m \geq 1 + nm$.

2.4.9 Theorem (Archimedean property). *Let $m, n \in \mathbb{N}$. Then there exists $k \in \mathbb{N}$ such that $mk > n$.*

Proof. Let $m \in \mathbb{N}$ be fixed and let
$$A_m := \{n \in \mathbb{N} \mid mk > n \ \text{for some} \ k \in \mathbb{N}\}.$$
Since $m \geq 1, n = 1 \in A_m$ with $k = 1 + 1$. Suppose $n \in A_m$. Then $mk > n$ for some $k \in \mathbb{N}$. Then $mk + 1 > n + 1$ and hence $m(k+1) = mk + m \geq mk + 1 > n + 1$. Thus $(n+1) \in A_m$ and by property **N(v)**, $A_m = \mathbb{N}$. ∎

2.4.10 Exercise. For $m \in \mathbb{N}, m > 1$, prove the following:

(i) $m^k \geq k + 1 \ \forall \ k \in \mathbb{N}$.

(ii) Given $n \in \mathbb{N}, \exists k \in \mathbb{N}$ such that $m^k > n$.

As noted in remark 2.2.10 (iii), $\mathbb{N}$ is an infinite set. We describe next some finite subsets of $\mathbb{N}$.

2.4.11 Theorem (Pigeonhole principle). *For every $n \in \mathbb{N}$, the set $[1, n] := \{m \in \mathbb{N} \mid 1 \leq m \leq n\}$ is a finite set, i.e., every one-one map $f : [1, n] \to [1, n]$ is also onto.*

Proof. Let $A := \{n \in \mathbb{N} \mid [1, n] \ \text{is finite}\}$. We shall show that $A = \mathbb{N}$. Clearly $1 \in A$, for $[1, 1] = \{1\}$ and any one-one map

$f : \{1\} \to \{1\}$ will be onto. Suppose $n \in A$, i.e., every one-one map $f : [1, n] \to [1, n]$ is also onto. We shall show $S(n) \in A$. Let $f : [1, S(n)] \to [1, S(n)]$ be a one-one map. Suppose $f[1, n] \subseteq [1, n]$. Consider the restriction of f to $[1, n]$ and call it $\tilde{f}$, i.e., $\tilde{f}$ is defined by $\tilde{f}(m) = f(m) \ \forall \ m \in [1, n]$. Then $\tilde{f} : [1, n] \to [1, n]$ is a one-one map and as $n \in A$, $\tilde{f}$ must be onto, i.e., $\tilde{f}[1, n] = [1, n]$. But then $f(S(n)) = S(n)$, as f is one-one, and thus f is also onto. In case $f[1, n] \not\subseteq [1, n]$, i.e., $f(k) = S(n)$ for some $1 \leq k \leq n$, we must have $f(S(n)) \neq S(n)$, for f is one-one. We define $\tilde{f} : [1, n] \to [1, n]$ as follows: $\tilde{f}(m) := f(m)$ if $1 \leq m \leq n, m \neq k$ and $\tilde{f}(k) := f(S(n))$, see figure 2.4.

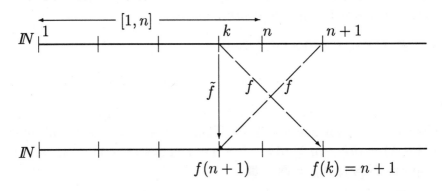

Figure 2.4 : Definition of $\tilde{f}$.

Then $\tilde{f} : [1, n] \to [1, n]$ is one-one and thus onto. We claim that f itself is onto, because $\forall \ 1 \leq m \leq n$, $\exists \ r \in [1, n]$ such that $\tilde{f}(r) = m$. But $\tilde{f}(r) = f(r)$ if $r \neq k$ and $\tilde{f}(k) = f(S(n))$. Thus there exists $r \in [1, S(n)]$ such that $f(r) = m$. Also $f(k) = S(n)$. Hence f is onto. This proves that $S(n) \in A$ whenever $n \in A$. Hence by property $\mathbf{N(v)}, A = \mathbb{N}$. ∎

2.4.12 Definition.

(i) A subset $\emptyset \neq S \subseteq \mathbb{N}$ is said to be **bounded above** if there exists some $n \in \mathbb{N}$ such that $s \leq n \ \forall \ s \in S$. In that case n is called an **upper bound** of S. Similarly a subset $\emptyset \neq S \subseteq \mathbb{N}$ is said to

2.4 Order on natural numbers

be **bounded below** if $\exists\, n \in \mathbb{N}$ such that $n \leq s$ for every $s \in S$. In that case, n is called a **lower bound** for S.

(ii) Let $\emptyset \neq S \subseteq \mathbb{N}$ be such that S is bounded above. Let $U := \{n \in \mathbb{N} \mid n \text{ is an upper bound for } S\}$. Suppose $\exists\, \alpha \in U$ such that $\alpha \leq n\ \forall\ n \in U$. Then we call α a **least upper bound** for S. Similarly, if $\emptyset \neq S \subseteq \mathbb{N}$ is a set which is bounded below, we call $\beta \in \mathbb{N}$ the **greatest lower bound** of S, if β is a lower bound for S and $\beta \geq n\ \forall$ lower bounds n for S.

2.4.13 Theorem. *Let S be a nonempty subset of $\mathbb{N}$. Then the following hold:*

(i) If a greatest lower bound of S exists then it is unique.

(ii) If a least upper bound of S exists then it is unique.

We denote the greatest lower bound of S by $glb(S)$ and the least upper bound by $lub(S)$. The numbers $glb(S)$ and $lub(S)$ are also denoted by $\inf(S)$ and $\sup(S)$, respectively.

Proof. (i) Suppose if possible there exists $n, m \in \mathbb{N}$ such that both are greatest lower bounds of S. Then n is also a lower bound of S and hence $n \leq m$. Similarly $m \leq n$. Hence $m = n$.
  (ii) Proof is similar to that of (i). ∎

2.4.14 Example. The set $\mathbb{N}$ is bounded below by 1 because $1 \leq n\ \forall\ n \in \mathbb{N}$. The set $\mathbb{N}$ is not bounded above because $\forall\ n \in \mathbb{N}$, $(n+1) \in \mathbb{N}$ and $(n+1) > n$. Consider the set $[1, n] := \{m \in \mathbb{N} \mid 1 \leq m \leq n\}$. Clearly $[1, n]$ is both bounded below and bounded above. Further $lub[1, n] = n$ and $glb[1, n] = 1$.

2.4.15 Theorem. *Let $\emptyset \neq S \subseteq \mathbb{N}$ be arbitrary. Then the following hold:*

(i)   *S is always bounded below and $glb(S) \in S$. (This is called the **well-ordering** property of the natural numbers.) Further, $1 \in S$ iff $glb(S) = 1$.*

(ii) If S is bounded above then $lub(S)$ exists and $lub(S) \in S$.

(iii) S is infinite iff S is not bounded above.

Proof. (i) Let $L := \{n \in \mathbb{N} | n \leq m \; \forall \; m \in S\}$. Clearly $1 \in L$. Note that if $k \in S$ then $k+1 \notin L$, for $(k+1) > k$. Thus $L \neq \mathbb{N}$. Hence $\exists \; m_0 \in L$ such that $(m_0+1) \notin L$, for otherwise L would be $\mathbb{N}$. Since $m_0 \in L, m_0 \leq m \; \forall \; m \in S$ and if $m_0 \notin S$, then $m_0 < m \; \forall \; m \in S$. However, $m_0 + 1 \leq m \; \forall \; m \in S$, i.e., $(m_0 + 1) \in L$, which is not true. Thus $m_0 \in S$. If $l \in \mathbb{N}$ and $l \leq m \; \forall \; m \in S$ then clearly $l \leq m_0$ also. Thus $m_0 = glb(S)$. It is easy to show that $1 \in S$ iff $glb(S) = 1$.

(ii) Let $\emptyset \neq S \subseteq \mathbb{N}$ be such that S is bounded above. Then $\exists \; n_0 \in \mathbb{N}$ such that $n_0 \geq m \; \forall \; m \in S$. Thus $n_0 + k \notin S \; \forall \; k \in \mathbb{N}$. Let $U := \{n \in \mathbb{N} | n + k \notin S \; \forall \; k \in \mathbb{N}\}$. Then U is a nonempty set as $n_0 \in U$. By (i), $\alpha := glb(U)$ exists and $\alpha \in U$. Clearly α is an upper bound for S. We claim that $\alpha \in S$. Suppose not. Then $\alpha > m \; \forall \; m \in S$. As $S \neq \emptyset$, we get $\alpha > 1$. Thus by theorem 2.2.11(ii), $\exists \; \beta$ such that $\beta + 1 = \alpha$. Then $\beta \geq m \; \forall \; m \in S$ by theorem 2.4.6(v). Thus $\beta \in U$ and $\beta < \beta + 1 = \alpha$ contradicting the fact that $\alpha = glb(U)$. Hence $\alpha \in S$. Finally, if $n \in \mathbb{N}$ is such that $n \geq m \; \forall \; m \in S$ then in particular we will have $n \geq \alpha$. This shows that $\alpha = lub(S)$.

(iii) Suppose S is not bounded above. For $n \in S$, let $C_n := \{m \in S | m > n\}$. Then $C_n \neq \emptyset$, because $\forall \; n \in \mathbb{N}$, $\exists \; m \in S$ such that $m > n$, as S is not bounded above. Let $\alpha_n := glb(C_n)$. Since $\alpha_n \in C_n, \alpha_n \geq n + 1$ and clearly $\alpha_n \in S$. Define $f : S \to S$ by $f(n) := \alpha_n \; \forall \; n \in S$. Let $\alpha_n = \alpha_{n'}$. If $n < n'$, then $\alpha_n \leq n' < n' + 1 \leq \alpha'_n$. Thus $n < n'$ is not possible. Similarly $n' < n$ is also not possible. Hence $n = n'$ by theorem 2.4.3. Thus f is a one-one map. Let $n_0 := glb(S)$. Since $\alpha_n \geq n + 1$, $\forall \; n \in S$, we have $\forall \; n \in S, n_0 \leq n < n+1 \leq \alpha_n$. Hence $n_0 \neq \alpha_n$ for any n. Thus $f(S) \subsetneq S$, showing that S in an infinite set. Conversely, suppose S is an infinite set and is bounded above. Let $\alpha := lub(S)$. Then $S \subseteq [1, \alpha]$, which contradicts theorem 1.3.37(ii) as $[1, \alpha]$ is a finite

2.4 Order on natural numbers

set by theorem 2.4.11. On the other hand since S is an infinite set, $\exists$ a one-one map $f : S \to S$ such that $f(S) \subsetneq S$. Define $\tilde{f} : [1, \alpha] \to [1, \alpha]$ by $\tilde{f}(n) = n$ if $n \in [1, \alpha] \setminus S$ and $\tilde{f}(n) = f(n)$ if $n \in S$. Then $\tilde{f}$ is also a one-one map and $\tilde{f}([1, \alpha]) \subsetneq [1, \alpha]$, showing that $[1, \alpha]$ is an infinite set, which is a contradiction. Hence S cannot be bounded above. ∎

2.4.16 Exercise. Show that for $n, m \in \mathbb{N}$ there exists a bijective map $\Phi : [1, n] \to [1, m]$ iff $n = m$.

We show next how $\mathbb{N}$ and $[1, n]$ can be used to compare 'sizes' of sets.

2.4.17 Definition. A set A is said to be **countably finite** if either $A = \emptyset$ or there exists an $n \in \mathbb{N}$ and a bijective map $\phi : A \to [1, n]$. In that case we say A has n elements. A nonempty set B is said to be **countably infinite** if there exists a bijective map $\psi : B \to \mathbb{N}$. A set C is said to be **countable** if either it is countably finite or countably infinite. A set which is not countable is called an **uncountable set**.

2.4.18 Examples.

(i) For every $n \in \mathbb{N}$, the set $[1, n] := \{k \in \mathbb{N} | 1 \leq k \leq n\}$ is obviously countably finite.

(ii) The set $\mathbb{N}$ itself is countably infinite. Let $n \in \mathbb{N}$ be called an **even number** if $n = k + k$ for some $k \in \mathbb{N}$. If we define $2 := 1 + 1$, then n is even if $n = 2k$ for some k. A number $n \in \mathbb{N}$ is called **odd** if it is not even. Clearly 1 is an odd number for $1 < 2k$ $\forall k \in \mathbb{N}$. Also, $\forall k \in \mathbb{N}, 2k + 1$ is an odd number. For if possible, let $2k + 1 = 2m$ for some $m \in \mathbb{N}$. Then $k \neq m$ as $2k + 1 > 2k$. If $m > k$, let $m = k + r$ for some $r \in \mathbb{N}$. Then $2k + 1 = 2(k + r) = 2k + 2r$ and hence by the cancellation law (exercise 2.3.5), $1 = 2r$, which is not true. Thus $m < k$. But again $m + r = k$ for some $r \in \mathbb{N}$ and hence $2(m + r) + 1 = 2m$, i.e.,

$2m + 2r + 1 = 2m$, which is not true. Hence $2k + 1 \neq 2m$ for any m. Thus $2k + 1$ is an odd number for every k. Finally let

$$A := \{n \in \mathbb{N} | \text{ either } n = 2k \text{ or } n = 2k + 1 \text{ for some } k \in \mathbb{N}\}.$$

If $n \in A$ and $n = 2k$, then $S(n) = n + 1 = 2k + 1$ and if $n = 2k + 1$ then $S(n) = n + 1 = 2k + 1 + 1 = 2(k + 1)$. Hence if $n \in A$ then $S(n) \in A$. Thus by induction $A \cup \{1\} = \mathbb{N}$. Thus $n > 1$ is odd iff $n = 2k + 1$ for some $k \in \mathbb{N}$. Let $\mathbb{N}_e$ denote the set of all even numbers in $\mathbb{N}$ and $\mathbb{N}_o$ denote the set of all odd numbers in $\mathbb{N}$. Then $\mathbb{N}_o \cap \mathbb{N}_e = \emptyset$ and

$$\mathbb{N}_e = \{2k \mid k \in \mathbb{N}\}, \quad \mathbb{N}_o = \{1\} \cup \{2k + 1 \mid k \in \mathbb{N}\}.$$

Let $\phi : \mathbb{N}_e \to \mathbb{N}$ and $\psi : \mathbb{N}_o \to \mathbb{N}$ be defined by $\phi(2k) := k$ and $\psi(2k - 1) := k \ \forall \ k \in \mathbb{N}$. It is easy to check that ϕ and ψ are bijective maps and hence $\mathbb{N}_e$ and $\mathbb{N}_o$ are both countably infinite sets.

(iii) Consider $\mathcal{P}(\mathbb{N})$ the power set of $\mathbb{N}$ as given by the power set axiom. The set $\mathcal{P}(\mathbb{N})$ consists of all subsets of $\mathbb{N}$. We first show that there does not exist any bijective map between $\mathbb{N}$ and $\mathcal{P}(\mathbb{N})$. Suppose if possible, there exists a bijective map $\phi : \mathbb{N} \to \mathcal{P}(\mathbb{N})$. Then $\forall \ n \in \mathbb{N}, \phi(n) \in \mathcal{P}(\mathbb{N})$, i.e., $\phi(n)$ is a subset of $\mathbb{N}$. Let $A := \{n \in \mathbb{N} \mid n \notin \phi(n)\}$. Then A is a subset of $\mathbb{N}$ and, ϕ being onto, there exists $n \in \mathbb{N}$ such that $A = \phi(n)$. Now either $n \in A$ or $n \notin A$. If $n \in A = \phi(n)$, then by definition of $A, n \notin \phi(n)$ which is a contradiction. Also if $n \notin A = \phi(n)$ then again by definition of A, $n \in \phi(n)$, a contradiction. Hence there does not exist a bijective map between $\mathbb{N}$ and $\mathcal{P}(\mathbb{N})$. (Note that this proof works for any set S and its power set $\mathcal{P}(S)$. This proof was first given by G. Cantor.) Since $\mathcal{P}(\mathbb{N})$ includes the subset $\{\{n\}|n \in \mathbb{N}\}$ it cannot be countably finite, by theorem 1.3.3(ii). Thus $\mathcal{P}(\mathbb{N})$ is an uncountable set.

2.4.19 Theorem. *Let A and B be arbitrary sets. Then the following hold:*

2.4 Order on natural numbers

(i) If A is countably finite, then it is finite.

(ii) If A is countably infinite, then A is infinite.

(iii) If $A \subseteq B$ and B is countably finite, so also is A.

(iv) If $A \subseteq B$ and A is countably infinite, then B is not countably finite.

(v) If B is countably infinite, then every subset of it is either countably finite or countably infinite.

Proof. (i) Let A be countably finite. Suppose A is not finite. Then there exists a one-one map $\phi : A \to A$ which is not onto. Since A is countably finite, there exists a bijective map $\psi : A \to [1, n]$ for some $n \in \mathbb{N}$. Then the map $\psi \circ \phi \circ \psi^{-1} : [1, n] \to [1, n]$ is one-one but not onto. Hence $[1, n]$ is not finite, contradicting theorem 2.4.11. Thus A is finite.

(ii) Since A is countably infinite there exists a bijective map $\phi : A \to \mathbb{N}$. Let $S : \mathbb{N} \to \mathbb{N}$ be the successor function, $S(n) := n + 1$. Then $\phi^{-1} \phi \circ S \circ \phi : A \to A$ is a one-one map which is not onto, for S is not onto. Hence A is infinite.

(iii) Since $A \subseteq B$ and B is countably finite, B is a finite set by (i). Let $a_1 := glb(A)$. Then $a_1 \in A$. If $A \setminus \{a_1\} = \emptyset$, then $A = \{a_1\}$. If not, let $a_2 := glb(A \setminus \{a_1\})$. Then $a_2 \in A, a_2 > a_1$. We claim that by proceeding in this manner we shall find $r \in \mathbb{N}$ such that $A \setminus \{a_1, a_2, \ldots, a_r\} = \emptyset$. Suppose not. Then we will have $a_i \in A, i \geq 1$, such that $a_1 < a_2 < \ldots$. Thus $\{a_i | i \in \mathbb{N}\}$ is a countably infinite subset of A and hence by (ii) A is an infinite set. But then B is infinite by theorem 1.3.37, a contradiction. Hence $A = \{a_1, \ldots, a_r\}$ for some $r \in \mathbb{N}$. Clearly, A is countably finite.

(iv) Follows from (iii).

(v) Let B be countably infinite and $A \subseteq B$ be such that A is not finite. We show that A is countably infinite. Since B is countably infinite, there exists a bijective map $f : \mathbb{N} \to B$. Let $f(n) := b_n, n \in \mathbb{N}$. Then $B = \{b_n | n \in \mathbb{N}\}$. Let $n_1 := glb\{n \in \mathbb{N} | b_n \in A\}$. Suppose $n_1 < n_2 < \ldots < n_{k-1}$ have been defined.

Since A is not finite, it is not countably finite by (i). Thus $\exists\, n \in \mathbb{N}, n > n_{k-1}$ such that $b_n \in A$. Let $n_k := glb\{n \in \mathbb{N} | n > n_{k-1}$ and $b_n \in A\}$. Thus by induction we get natural numbers $1 \le n_1 < n_2 < \ldots < n_k < \ldots$ such that $b_{n_k} \in A$. In fact, if $a \in A$, then $a \in B$ and hence $a = b_m$ for some $m \in \mathbb{N}$. Let $n_k = lub\{n_i | n_i < m\}$. Then $m = n_{k+1}$, i.e., $a = b_{n_{k+1}}$. Hence $A = \{b_{n_i} | i \in \mathbb{N}\}$. Clearly $\phi : \mathbb{N} \to A$, defined by $\phi(i) := a_{n_i}$, is a one-one, onto map. Hence A is countably infinite. ∎

2.4.20 Corollary. *If A, B are countably infinite sets, then $A \times B$ is also countably infinite.*

Proof. Clearly $A \times B$ is not countably finite, for example $\{(a,b) | b \in B\}$ is a countably infinite subset of $A \times B$ for every $a \in A$ fixed. Thus in view of theorem 2.4.19 (v), to show that $A \times B$ is countably infinite, it is enough to give a one-one map $\phi : A \times B \to \mathbb{N}$. We define ϕ as follows. Since A, B are countably infinite, there exist bijective maps $f : A \to \mathbb{N}$ and $g : B \to \mathbb{N}$. Let $\phi : A \times B \to \mathbb{N}$ be defined by

$$\phi(a,b) := 2^{f(a)}(2g(b)+1).$$

We claim that ϕ is a one-one map. Let $\phi(a,b) = \phi(c,d)$. Then

$$2^{f(a)}(2g(b)+1) = 2^{f(c)}(2g(d)+1). \qquad (2.1)$$

If possible let $f(a) > f(c)$. Then $f(a) = f(c) + r$, for some $r \in \mathbb{N}$. Thus

$$2^{f(c)} 2^r (2g(b)+1) = 2^{f(c)}(2g(d)+1).$$

By the cancellation property, exercise 2.3.9, we get

$$2^r (2g(b)+1) = 2g(d)+1,$$

which is a contradiction as the left-hand side in the above equality is an even number and the right-hand side is an odd number. Thus $f(a) > f(c)$ is not true. Similarly $f(a) < f(c)$ is not possible.

2.4 Order on natural numbers

Thus $f(a) = f(c)$ and hence $a = c$. Further (2.1) gives $2g(b) + 1 = 2g(d) + 1$ and hence $g(b) = g(d)$. Thus $b = d$. ∎

2.4.21 Corollary. *The set $\mathbb{N} \times \mathbb{N}$ is countably infinite.*

Proof. Follows from corollary 2.4.20. ∎

2.4.22 Exercise. Use the function $g : \mathbb{N} \times \mathbb{N} \to \mathbb{N}$ defined below to show that $\mathbb{N} \times \mathbb{N}$ is countably infinite:

$$g(m,n) := 2^m(2n+1) - 1.$$

2.4.23 Theorem. *Let A and B be two sets. Then the following hold:*

(i) *If A is countably finite and B is countably finite, so also is $A \cup B$.*

(ii) *If A is countably finite and B is countably infinite, $A \cup B$ is countably infinite.*

(iii) *If A and B are countably infinite, so also is $A \cup B$.*

Proof. (i) First suppose A and B are disjoint sets, i.e., $A \cap B = \emptyset$. Let $\phi : A \to [1, n]$ and $\psi : B \to [1, m]$ be bijective maps. Let $g : [n+1, n+m] \to [1, m]$ be defined by $g(n+r) = r \ \forall \ 1 \leq r \leq m$. Let $f : [1, n+m] \to A \cup B$ be defined by

$$f(x) := \begin{cases} \phi^{-1}(x) & \text{if } 1 \leq x \leq n, \\ \psi^{-1}(g(x)) & \text{if } n+1 \leq x \leq n+m. \end{cases}$$

Clearly f is a bijective map. Hence $A \cup B$ is a finite set if $A \cap B = \emptyset$. In general, $A \cup B = A \cup (B \setminus A)$. Since $B \setminus A \subseteq B$, by theorem 1.3.37(ii), $B \setminus A$ is a finite set and hence by earlier case $A \cup B = A \cup (B \setminus A)$ is a finite set.

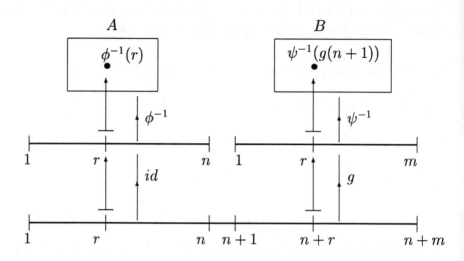

Figure 2.5 : Definition of f.

(ii) Let A be countably finite and B be countably infinite. Let $\phi : A \to \{1, 2, \ldots, n\}$ and $\psi : B \to \mathbb{N}$ be one-one, onto maps given by the definition. Let $\eta : \{k \in \mathbb{N} \mid k \geq n\} \to \mathbb{N}$

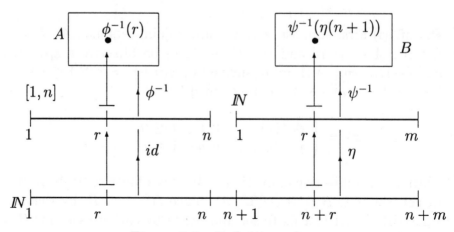

Figure 2.6 : Definition of g.

2.4 Order on natural numbers

be defined by

$$\eta(n+r) := r, \; r \in \mathbb{N}.$$

If A and B are disjoint, define $g : \mathbb{N} \to A \cup B$ by

$$g(k) := \begin{cases} \phi^{-1}(k) & 1 \le k \le n, \\ \psi^{-1}(\eta(k)) & \text{if } k > n. \end{cases}$$

It is easy to see that g is one-one and onto. Thus $A \cup B$ is countably infinite. In the general case write $A \cup B = B \cup (A \setminus B)$. Since A is countably finite and $(A \setminus B) \subseteq A$, $A \setminus B$ is countably finite by theorem 2.4.19(iii). Hence by earlier case $B \cup (A \setminus B) = A \cup B$ is countably infinite. This proves (ii).

(iii) It is enough to consider the case when both A and B are countably infinite with $A \cap B = \emptyset$, for in the general case we can consider $A \cup B = A \cup (B \setminus A)$. Then $(B \setminus A) \subseteq B$ and either it will be countably finite or it will be countably infinite by theorem 2.4.19(v). So, assume $A \cap B = \emptyset$. Let $\phi : A \to \mathbb{N}$ and $\psi : B \to \mathbb{N}$ be the one-one, onto maps given by the definition. Let $\eta_e : \mathbb{N}_e \to \mathbb{N}$ and $\eta_o : \mathbb{N}_o \to \mathbb{N}$ be defined by, $\forall \, m \in \mathbb{N}$,

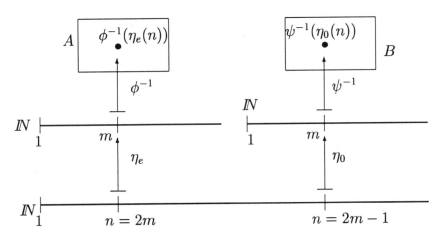

Figure 2.7 : Definition of f.

$$\eta_e(2m) := m,$$
$$\eta_o(2m-1) := m.$$

Define $f : \mathbb{N} \to A \cup B$ by

$$f(n) := \begin{cases} \phi^{-1}(\eta_e(n)) & \text{if } n \text{ is even,} \\ \psi^{-1}(\eta_o(n)) & \text{if } n \text{ is odd.} \end{cases}$$

It is easy to see that f is one-one and onto. ∎

2.4.24 Proposition. *For a set A, the following statements are equivalent:*

(i) A *is countable.*

(ii) *There exists a one-one map* $\phi : A \to \mathbb{N}$.

(iii) *There exists an onto map* $\psi : \mathbb{N} \to A$.

Proof. Suppose (i) holds. If A is countably finite, then by definition we have a one-one, onto map $\phi : A \to \{1, 2, \ldots, n\} \subseteq \mathbb{N}$. If A is countably infinite, then we have a one-one onto map $\phi : A \to \mathbb{N}$. Hence (i) $\Rightarrow$ (ii). Next suppose (ii) holds and $\phi : A \to \mathbb{N}$ is a one-one map. Define $\psi : \mathbb{N} \to A$ as follows:

$$\psi(n) := \begin{cases} a & \text{if } n = \phi(a), a \in A, \\ a_0 & \text{if } n \in \mathbb{N} \setminus \phi(A), \end{cases}$$

where a_0 is any fixed element of A. Then clearly ψ is onto. Thus (ii) $\Rightarrow$ (iii). Finally if $\psi : \mathbb{N} \to A$ is an onto map. Define $f : A \to \mathbb{N}$ by $f(a) := glb\{n \,|\, \psi(n) = a\}$. Then f is a one-one map. To see this, let $f(a) = f(b)$. Then $glb\{n \,|\, \psi(n) = a\} = glb\{n \,|\, \psi(n) = b\}$. Since glb of a set of natural numbers is attained in the set, $a = \psi(n) = b$

2.4 Order on natural numbers 65

for some $n \in \mathbb{N}$. Since f is one-one and $f(A)$ is a countable set, by theorem 2.4.19(v), A is countable. ∎

2.4.25 Corollary. *If A_n is a countable set for each $n \in \mathbb{N}$, then $A := \bigcup_{n=1}^{\infty} A$ is also a countable set.*

Proof. Since each A is countable, for every $n \in \mathbb{N}$ there exists a one-one map from $\mathbb{N}$ onto A_n. For each n, choose any such map ϕ_n (here we are using axiom of choice to choose ϕ_n's). Define $\phi : \mathbb{N} \times \mathbb{N} \to A$ by $\phi(n,m) = \phi_n(m)$. Clearly ϕ is onto. Since $\mathbb{N} \times \mathbb{N}$ is countable by corollary 2.4.21, it follows from proposition 2.4.24 that A is countable. ∎

2.4.26 Theorem. *A set X is infinite iff X has a subset which is countably infinite.*

Proof. First suppose X is infinite. Then there exists a proper subset Y of X and a bijective map $f : X \to Y$. Let $x \in X$ be such that $x \notin Y$. For any $n \in \mathbb{N}$, let $f^1(z) := f(z)$ and $f^{n+1}(z) := f(f^n(z))\ \forall\ z \in X$. Then $x \neq f^n(x)\ \forall\ n \in \mathbb{N}$, because $f^n(x) \in Y$. Also, for $n \neq m, f^n(x) \neq f^m(x)$. This is because, if $n > m$ then $f^n(x) = f^m(x)$ will imply $f^{n-m}(x) = x$, not true. Thus $\{f^n(x) | n \in \mathbb{N}\}$ is a countably infinite subset of X. Converse follows from theorem 2.4.19(ii) and theorem 1.3.37. ∎

2.4.27 Theorem. *Let X be a nonempty set. Then the following hold:*

(i) If X is countably finite, it is also finite.

(ii) If X is not countably finite, it includes a subset which is countably infinite.

Proof. (i) Already proved in theorem 2.4.19(i).

(ii) Since $X \neq \emptyset$, choose any $x_1 \in X$. If $X \setminus \{x_1\} = \emptyset$, i.e., $X = \{x_1\}$, then $f : [1,1] \to X$ defined by $f(1) := x_1$ is a one-one, onto map implying that X is countably finite, not true. Thus $X \setminus \{x_1\} \neq \emptyset$. Choose any element $x_2 \in X \setminus \{x_1\}$. If $X \setminus \{x_1, x_2\} = \emptyset$, then $f : [1,2] \to X$ defined by $f(1) := x_1, f(2) := x_2$ will be a one-one, onto map again implying that X is countably finite, a contradiction. Thus $X \setminus \{x_1, x_2\} \neq \emptyset$. In general, suppose for some $n \in \mathbb{N}$ we have chosen $x_1, \ldots, x_n \in X$. Consider $X_n := X \setminus \{x_1, x_2, \ldots, x_n\}$. If $X_n = \emptyset$, we can define $f : [1,n] \to X$ by $f(k) := x_k, 1 \leq k \leq n$. Then f will be a one-one, onto map and we will have a contradiction. Thus we can choose $x_{n+1} \in X_n$. Hence by induction, we get a subset $\{x_n | n \in \mathbb{N}\} := Y$ of X. Clearly Y is a countably infinite set. ∎

2.4.28 Remarks.

(i) The statement (i) of theorem 2.4.27, does not require axiom of choice. The statement that X is finite means if $f : X \to X$ is a one-one map then it is also onto. This is known as the **Pigeonhole principle** and is a basic result in the theory of counting.

(ii) In the proof of theorem 2.4.27 (ii) we implicitly assumed that $\forall \, n \in \mathbb{N}$ we can choose $x_{n+1} \in X \setminus \{x_1, \ldots, x_n\}$ and construct a set $\{x_n \, | \, x \in \mathbb{N}\}$. This is possible if we assume axiom of choice. Though the choice of x_n is dependent upon the choices $x_1, \ldots, x_{n-1}$, a more formal proof of (ii) can be given. See for example remark 2.5.6.

(iii) Let X be any set and $f : \mathbb{N} \to X$ be a map. Then f is called a **sequence** in X. If $f(n) := x_n, n \in \mathbb{N}$, we denote the sequence by $\{x_n\}_{n \geq 1}$. Note that x_n's will be distinct if f is one-one. Thus, theorem 2.4.27(ii) says that if X is not countably finite, then it includes the sequence of distinct elements.

2.4.29 Corollary. *Let X be a nonempty set. Then the following hold:*

(i) X is finite iff X is countably finite.

2.4 Order on natural numbers

(ii) X *is infinite iff* X *is not countably finite.*

Proof. (i) If $X \neq \emptyset$ and is finite, then it is not countably infinite, for if so then it will be infinite by theorem 2.4.26. Thus X is countably finite. Conversely, if X is countably finite, by theorem 2.4.27(i), it is finite.

(ii) If X is infinite, it is not countably finite because of 2.4.27 (i). Also if X is not countably finite, by theorem 2.4.27(ii), it has a countably infinite subset (assuming axiom of choice) and hence by Theorem 2.4.26, it is infinite. ∎

2.4.30 Note. The statement that X is finite iff X is countably finite does not require axiom of choice.

2.4.31 Remark (Variations of the 'Principle of Induction') In the previous discussions we have seen various applications of axiom $\mathbf{N(v)}$, the principle of induction. There are variations of this principle which find applications in many situations. One of them called the **'General Principle of Induction'**, is as follows: *Let A be a subset of $\mathbb{N}$ such that $1 \in A$ and whenever for some $n \in \mathbb{N}, \{k \in \mathbb{N} \mid k \leq n\} \subseteq A$ then $S(x) \in \mathbb{N}$. Then $A = \mathbb{N}$.* Clearly, general principle of induction implies principle of induction. Conversely, let $A \subseteq \mathbb{N}$ be such that $1 \in A$ and $S(n) \in \mathbb{N}$ whenever $\{k \in \mathbb{N} \mid k \leq n\} \subseteq A$ for some $n \in \mathbb{N}$. Consider the set $B := \{m \in \mathbb{N} \mid \{k \in \mathbb{N} \mid k \leq m\} \subseteq A\}$. Clearly $1 \in B$ and $S(n) \in B$ whenever $n \in B$ for some $n \in \mathbb{N}$. Thus $B = \mathbb{N}$, by the principle of induction. Hence $A = \mathbb{N}$. Thus the principle of induction also implies the general principle of induction, and hence the two are equivalent. Another version of the principle of induction, called the **'Modified Principle of Induction'**, is the following: *let A be a subset of $\mathbb{N}$ and there exists $n_0 \in A$ having the property that $S(n) \in A$ whenever $n \geq n_0$ and $n \in A$. Then $A = \mathbb{N}$.* The equivalence of the modified principle of induction and the principle of induction is easy to verify.

2.5 An extension of the iteration theorem

Recall that in theorem 2.2.8 we proved the iteration theorem which helped us to define concepts inductively. Let us consider the following problem. Let A be a nonempty set. We defined $A \times A$, the Cartesian product of A with itself in example 1.3.16. Let us denote this by $A^2 := A \times A$. Question arises: For every $n \in \mathbb{N}$ can we define by induction $A^n := A^{n-1} \times A$. To do that, let us try to apply theorem 2.2.8. For that we need a set B such that $A \in B$ and a map $\psi : B \to B$ with $\psi(X) := X \times A$ for $X \in B$. Then we will get a map ϕ such that

$$\phi(1) = A \text{ and } \phi(S(n)) = \psi(\phi(n)) = \phi(n) \times A.$$

But what should be the set B? In fact B can only be defined in terms of ϕ as follows: $B = \{\phi(n) \,|\, n \in \mathbb{N}\}$. To overcome this problem, we note that we can treat ψ as a formula. Consider the formula $H(x, y)$: $y = x \times A$. Then H has the property that it is functional in x, i.e., if $H(x, y)$ and $H(x, z)$ hold for y, z then $y = x \times A = z$. Thus for every x, there is a unique $h(x)$ such that $H(x, h(x))$ is true. Given H and a set A, we want a function ϕ on $\mathbb{N}$ such that the following hold:

(i) $\phi(1) = A$.

(ii) $\phi(S(n)) = h(\phi(n)) = \phi(n) \times A$.

This is guaranteed by the next theorem.

2.5.1 Theorem (Extended iteration theorem). *Let $H(x, y)$ be a formula which is functional in x and let $h(x)$ denote the (unique) y such that $H(x, h(x))$ holds. Then for every set A there exists a unique function Φ defined on $\mathbb{N}$ such that the following hold:*

(i) $\Phi(1) = A$.

2.5 An extension of the iteration theorem

(ii) $\Phi(S(n)) = h(\Phi(n))$ $\forall n \geq 1$.

Proof. We shall work with the representation of natural numbers as constructed in theorem 2.2.3, i.e., $1 := \emptyset$ and $\forall n \in \mathbb{N}, S(n) = n \cup \{n\}$. Then the order on $\mathbb{N}$ is given by

$$n \leq m \quad \text{iff} \quad n \subseteq m.$$

We first prove the uniqueness of the function Φ. Let Φ and Φ' be two functions satisfying the required properties. Let

$$B := \{n \in \mathbb{N} \mid \Phi(n) = \Phi'(n)\}.$$

Then $1 \in B$ and if $n \in B$, i.e., $\Phi(n) = \Phi'(n)$, then $\Phi(S(n)) = h(\Phi(n)) = h(\Phi'(n)) = \Phi'(S(n))$. Thus $S(n) \in B$ if $n \in B$. Hence by induction $S = \mathbb{N}$, i.e., $\Phi(n) = \Phi'(n)$ $\forall n \in \mathbb{N}$.

To prove the existence of Φ, we first show that $\forall n \in \mathbb{N}$ there exists a function Φ_n on $n \cup \{n\}$ such that

$$\Phi_n(1) = A \quad \text{and} \quad \Phi_n(S(m)) = h(\Phi_n(m)) \; \forall \, m \subseteq S(n). \quad (2.2)$$

For example, for $n = 1$, consider the function $\Phi_1(1) := A, \Phi_1(S(1)) := h(A)$. Also, if a function Φ_n exists with properties (2.2), it is unique. To see this, suppose there exists another function Φ'_n satisfying (2.2). Suppose $\Phi_n(m) \neq \Phi'_n(m)$ for some $m \subseteq S(n)$. Then $\mathbb{N}$ being well-ordered, we can choose the smallest $m, 1 < m \leq S(n)$, such that $\Phi_n(m) \neq \Phi'_n(m)$. Let $m = S(k)$. Then $\Phi_n(k) = \Phi'_n(k)$ and we get $\Phi_n(m) = h(\Phi s(k)) = h(\Phi'_n(s(k))) = \Phi'_n(m)$, a contradiction. Hence if there exists a function Φ_n satisfying (2.2), it is unique. Let

$$B := \{n \in \mathbb{N} \mid \exists \text{ a function } \Phi_n \text{ on } (n+1) \text{ satisfying } (2.2)\}.$$

Then $1 \in B$. If $n \in B$ and Φ_n is the corresponding function, define Φ_{n+1} on $S(n+1) = (n+1) \cup \{n+1\}$ as follows:

$$\Phi_{n+1}(x) := \begin{cases} \Phi_n(x) & \forall \, x \in S(n), \\ h(\Phi_n(n+1)) & \text{for } x = \{n+1\}. \end{cases}$$

Then Φ_{n+1} is a well-defined function on $S(n+1)$ satisfying property (2.2). Hence by induction $B = \mathbb{N}$. Thus $\forall\, n \in \mathbb{N}$, a function Φ_n on $S(n+1)$ satisfying (2.2) exists. Further, if $m \leq n$, it follows from the uniqueness of Φ_n that

$$\Phi_m(x) = \Phi_n(x) \ \forall\, x \leq m+1.$$

Now consider the function Φ on $\mathbb{N}$ defined by

$$\Phi(n) := \Phi_n(n) \ \forall\, n \in \mathbb{N}.$$

Then Φ is a well-defined function on $\mathbb{N}$ and has the required properties. ∎

2.5.2 Corollary. *Let A be a nonempty set. Then for every n there exists a function f on $\mathbb{N}$ such that*

$$f(1) = A \quad \text{and} \quad f(n+1) = f(n) \times A \ \forall\, n \in \mathbb{N}, n > 1.$$

For every $n > 1$, the set $f(n)$ is called the n-**fold Cartesian product** of A with itself and is denoted by A^n.

Proof. Consider the functional formula $H(x, y)$: $y = X \times A$. Then for y, z, if $H(x, y)$ and $H(x, z)$ hold, $y = x \times A = z$. An application of theorem 2.5.1 to A and H gives the required function. ∎

2.5.3 Corollary. *Let A be a nonempty set. Then there exists a set B whose elements are $A^n, n \in \mathbb{N}$.*

Proof. Let $B :=$ Image of the function f as in corollary 2.5.2. Then B is a set by the replacement axiom. ∎

2.5.4 Corollary. *There exists a function g on $\mathbb{N}$ such that*

$$g(1) = \mathcal{P}(\mathbb{N}) \quad \text{and} \quad g(n+1) = \mathcal{P}(g(n)) \ \forall\, n \in \mathbb{N}, n > 1.$$

(If we consider g as an operation of constructing power set of a given set, then $g(n)$ is the n-fold application of this operation and given the set $\mathcal{P}(\mathcal{P}(\ldots(\mathcal{P}(\mathbb{N}))\ldots))$, denoted by $\mathcal{P}^n(\mathbb{N})$.)

2.5 An extension of the iteration theorem

Proof. It is an application of theorem 2.5.1 to the set $\mathbb{N}$ and the formula $H(x,y)$: $y = \mathcal{P}(x)$. ∎

2.5.5 Corollary. *There exists a unique set whose elements are precisely the sets $\mathcal{P}^n(\mathbb{N}), n \in \mathbb{N}$.*

Proof. The required set is the image of the function g which exists by replacement axiom. ∎

2.5.6 Remarks.

(i) Note that for a set A, $A^2 = A \times A$ and for any $n \geq 3$, $A^n := A^{n-1} \times A$. We can describe the elements of A^n as follows: for example, elements of A^3 are ordered pairs of the type $((a_1, a_2), a_3)$ where $a_1, a_2, a_3 \in A$. We call such an element as **ordered triple** and denoted by (a_1, a_2, a_3). In general for any n, ordered $(n-1)$-tuples having been defined, we denote the element $((a_1, a_2, \ldots, a_{n-1}), a_n) \in A^{n-1} \times A := A^n$ by $(a_1, a_2, \ldots, a_n)$ and call it an **ordered n-tuple**. Thus $A^n := \{(a_1, a_2, \ldots, a_n) \mid a_i \in A \; \forall \; i\}$. Two ordered n-tuples $(a_1, \ldots, a_n)$ and $(b_1, \ldots, b_n)$ are equal iff $a_i = b_i \; \forall \; i$.

(ii) Now we can give a more formal proof of theorem 2.4.24(ii), as mentioned in remark 2.4.25(ii). Let X be a nonempty set which is not countably finite. For each $n \in \mathbb{N}$, consider the set X^n as constructed in corollary 2.5.2. The elements of X^n are represented as ordered n-tuples $(x_1, \ldots, x_n)$, $x_i \in X$. Let for every $n \geq 2$, $Y_n := \{(x_1, \ldots, x_n) \in X^n \mid x_1, \ldots, x_n \text{ are distinct}\}$. We first show that each Y_n is a nonempty set. Suppose for some n, $Y_n = \emptyset$ and $Y_{n-1} \neq \emptyset$. Then there does not exist any ordered n-tuple of distinct elements of X and there exists an ordered $(n-1)$-tuple of distinct elements of X, say $(x_1, \ldots, x_{n-1})$. Then $X := \{x_1, \ldots, x_{n-1}\}$ and clearly it will be a countably finite set. Thus $Y_n \neq \emptyset \; \forall \; n$. We choose $y_n \in Y_n$ using axiom of choice and construct the set $Y := \{y_n \mid n \in \mathbb{N}\}$. Let $S = \bigcup_n A_n$, where each $A_n = \{x \in X \mid y_n = (x_1, \ldots, x_n), x_i = x \text{ for some } i\}$. By corollary 2.4.29, S is a countably infinite subset of X.

2.6 Abstractions

We start by defining the concepts of upper bound, lower bound, greatest lower bound, and least upper bound for arbitrary partially ordered sets.

2.6.1 Definition. Let $(X, \leq)$ be a partially ordered set and let S be any nonempty subset of X. An element $x \in X$ is called an **upper bound** of S if $s \leq x$ for every $s \in S$. An element $y \in X$ is called a **lower bound** of S if $s \geq y$ for every $y \in S$. We say S is **bounded above (below)** if S has an upper (lower) bound. We say $\alpha \in X$ is a **least upper bound** of S, denoted by $lub(S)$, if α is an upper bound of S and $\alpha \leq x$ for every upper bound x of S. An element $\beta \in X$ is called a **greatest lower bound** of S, denoted by $glb(S)$, if β is a lower bound of S and $\beta \geq y$ for every lower bound y of S.

2.6.2 Exercise. Let $(X, \leq)$ be a partially ordered set and S be a nonempty subset of X. Prove the following:

(i) If $lub(S)$ exists, it is unique.

(ii) If $glb(S)$ exists, it is unique.

2.6.3 Example. Consider the set $\mathbb{N}$ of natural numbers with the order $\leq$ as given in definition 2.4.4. Then $\mathbb{N}$ itself is not bounded above. It is bounded below by 1 and $1 = glb(\mathbb{N})$. By theorem 2.4.15, every nonempty subset S of $\mathbb{N}$ is bounded below, $glb(S)$ exists and is an element of S. A subset S is bounded above iff S is finite and in that case $lub(S)$ exists and belongs to S.

2.6.4 Definition. Let $(X, \leq)$ be a partially ordered set and A be a nonempty subset of X. An element $x \in A$ is called a **maximum** of A if $x \geq a \ \forall \ a \in A$. Similarly, an element $y \in A$ is called a **minimum** of A if $y \leq a \ \forall \ a \in A$.

2.6 Abstractions

2.6.5 Example. By example 2.6.3, every nonempty subset A of $\mathbb{N}$ has a minimum and every nonempty set which is bounded above has a maximum.

2.6.6 Definition. Let $(X, \leq)$ be a partially ordered set. We say $\leq$ is a **total** (or **linear**) **order** or that $(X, \leq)$ is **totally** (or **linearly**) **ordered** if for every $x, y \in X$ either $x \leq y$ or $y \leq x$.

2.6.7 Examples.

(i) By theorem 2.4.5, the set of natural numbers $\mathbb{N}$ with the usual order (as in definition 2.4.4) is totally ordered.

(ii) Let $X = \{A \subseteq \mathbb{N} | A \text{ has at most 2 elements}\}$, where $2 := 1 + 1$. For $A, B \in X$, define $A \leq B$ if $A \subseteq B$, i.e., A is a subset of B. Clearly $(X, \leq)$ is a partially ordered set which is not totally ordered. For example $\{1\}$ and $\{2, 3\}$ are elements of X but neither $\{1\} \leq \{2, 3\}$ nor $\{2, 3\} \leq \{1\}$.

2.6.8 Definition. A partially ordered set $(X, \leq)$ is said to be **well-ordered** if for every nonempty subset S of X, $glb(S)$ exists and is an element of S.

2.6.9 Example. The set of natural number $\mathbb{N}$ with its usual order $\leq$ is well-ordered by theorem 2.4.15(i).

2.6.10 Proposition. *Let $(X, \leq)$ be a well-ordered set. Then $\leq$ is total.*

Proof. Suppose $\leq$ is not total. Then there exists $x, y \in X$ such that $x \neq y$ and neither $x \leq y$ nor $y \leq x$. Consider the set $S := \{x, y\} \subseteq X$. Then S is a nonempty subset of X and does not contain $glb(S)$. Thus S is not well-ordered, a contradiction. Hence $\leq$ is total. ∎

2.6.11 Example. Consider the set $\mathbb{N}$ with the order relation $\mathcal{R}$ defined as follows: for $n, m \in \mathbb{N}$,

$$n\mathcal{R}m \text{ iff } n \geq m.$$

Then it is easy to check that $\mathcal{R}$ is a partial order on $\mathbb{N}$ but it is not total, for example $\mathbb{N}$ itself does not have glb under this order.

2.6.12 Note. Example 2.6.11 tells us that not every partially ordered set is well-ordered. However, on $\mathbb{N}$ we have a partial order (the natural one) which makes it a well-ordered set. The question arises: can every set be well-ordered, i.e., can one define a partial order on a given set so that it is well-ordered? The claim that every set can be well-ordered is known as the **Well-Ordering Principle** and can be shown to be equivalent to the Axiom of Choice. For this equivalence and many other equivalent statements the reader may refer to [Hal].

2.6.13 Definition. Let $(X, \leq)$ be a partially ordered set. We say $(X, \leq)$ has the **least upper bound property** if for every nonempty subset S of $X, lub(S)$ exists whenever S is bounded above. We say $(X, \leq)$ has **greatest lower bound property** if for every nonempty subset S of $X, glb(S)$ exists whenever S is bounded below. If $(X, \leq)$ has both the least upper bound property and the greatest lower bound property, we say $(X, \leq)$ is **order complete**.

2.6.14 Example. The set $\mathbb{N}$ with its usual order is order complete.

The operations of addition and multiplication on $\mathbb{N}$ motivate our next definition.

2.6.15 Definition. Let X be a nonempty set. A function $\phi : X \times X \to X$ is called a **binary operation** on X.

2.6.16 Examples.

(i) Let $X = \mathbb{N}$ be the set of natural numbers. Then addition and multiplication, as defined in theorems 2.3.1 and 2.3.6 are binary operations on $\mathbb{N}$.

2.6 Abstractions

(ii) Let A be any nonempty set and let X denote the set of all functions $f : A \to A$. For $f, g \in X$ consider $(f, g) \mapsto f \circ g$, the composition of f and g. Then this is a binary operation on X. Consider $Y := \{f \in X | f \text{ is bijective}\}$. Then $(f, g) \mapsto f \circ g, (f, g) \mapsto f \circ g^{-1}, (f \circ g) \mapsto (f \circ g)^{-1}$ are binary operation on Y. For $x_0 \in X$ fixed, consider $F(f, g) := (f(x_0), g(x_0))$, $f, g \in X$. Then F is a function on $X \times X$ with values in $A \times A$. Thus F is not a binary operation on X.

2.6.17 Definition. Let X be a nonempty set and ϕ a binary operation on X.

(i) ϕ is said to be **associative** if $\forall\, x, y, z \in X$,

$$\phi(\phi(x, y), z) = \phi(x, \phi(y, z)).$$

(ii) ϕ is said to be **commutative** if $\forall\, x, y \in X$

$$\phi(x, y) = \phi(y, x).$$

(iii) An element $x_0 \in X$ is said to be an **identity element** for ϕ if

$$\phi(x, x_0) = x = \phi(x_0, x) \ \forall\, x \in X.$$

(iv) Let ϕ have an identity element x_0. An element $y \in X$ is said to be a ϕ-**inverse** of y with respect to the identity x_0 if

$$\phi(x, y) = \phi(y, x) = x_0.$$

2.6.18 Exercise. Let X be a set with a binary operation ϕ on it. Prove the following:

(i) An identity element for ϕ is unique, whenever it exists.

(ii) If ϕ has an identity, a ϕ-inverse of every element is unique, whenever it exists.

2.6.19 Examples.

(i) The binary operation of addition on $\mathbb{N}$ is associative, commutative (theorem 2.3.4) and has no identity element. The binary operation of multiplication on $\mathbb{N}$ is associative, commutative and has identity, namely $1 \in \mathbb{N}$ (theorem 2.3.6). For multiplication only $1 \in \mathbb{N}$ has multiplicative inverse, and the inverse is 1 itself.

(ii) Consider the set X of all functions on a nonempty set A with the binary operation ϕ of composition of functions (see example 2.6.16(ii)). Then ϕ is associative but not commutative in general. For example, let $A = \mathbb{N}, f(n) := n+1 \ \forall \ n \in \mathbb{N}$ and $g(n) := n^2 \ \forall \ n \in \mathbb{N}$. Then $(f \circ g)(n) = f(g(n)) = f(n^2) = n^2 + 1$ whereas $(g \circ f)(n) = g(f(n)) = g(n+1) = (n+1)^2$. Clearly $(f \circ g)(n) \neq (g \circ f)(n) \ \forall \ n \in \mathbb{N}$. The binary operation ϕ has the identity element, namely the identity map on $A, Id_A(a) := a \ \forall \ a \in A$. An element $f \in X$ has ϕ-inverse iff f is bijective.

Properties of addition and multiplication on $\mathbb{N}$ motivate our next definition.

2.6.20 Definition.

(i) A **semigroup** is a set X with a binary operation ϕ defined on it which is associative. We denote it by the pair (X, ϕ). We say a semigroup (X, ϕ) is **commutative** if the binary operation ϕ is commutative.

(ii) A semigroup (X, ϕ) is called a **group** if every $x \in X$ has ϕ-inverse with respect to its identity. We say a group (X, ϕ) is a **commutative** or **abelian group** if ϕ is commutative.

2.6.21 Examples.

(i) $\mathbb{N}$ under the binary operations of usual addition is a commutative semigroup and has no identity element. Also $\mathbb{N}$ is a commutative semigroup under the binary operation of usual multiplication having $1 \in \mathbb{N}$ as the identity element.

2.6 Abstractions

(ii) Clearly, every group is also a semigroup. $\mathbb{N}$ under usual multiplication is a semigroup which is also commutative, but it is not a group.

(iii) The set X, of example 2.6.19.(ii) is a semigroup under composition of functions but it need not be commutative in general. The set Y of example 2.6.19(ii) is a group which need not be commutative in general. In the particular case when A is a finite set having n elements (say $A = \{k \in \mathbb{N} | 1 \leq k \leq n\}$), every bijective map from A onto A is called a **permutation** on n symbols and the group Y under the composition of binary operations is called the **symmetric group of permutations** on n symbols, denoted by S_n.

2.6.22 Definition. Let ϕ, ψ be two binary operations on a set X. We say ψ is **right-distributive** over ϕ if $\forall\, x, y, z \in X$,

$$\psi(\phi(x,y), z) = \phi(\psi(x,z), \psi(y,z)).$$

Similarly ψ is said to be **left-distributive** over ϕ if $\forall\, x, y, z$,

$$\psi(z, \phi(x,y)) = \phi(\psi(z,x), \psi(z,y)).$$

2.6.23 Examples.

(i) The binary operation of multiplication on $\mathbb{N}$ is left as well as right distributive over the binary operation of addition on $\mathbb{N}$.

(ii) Let X be the set of all bijective maps $f : \mathbb{N} \to \mathbb{N}$. Let $\phi(f, g) = f \circ g$ and $\psi(f \circ g) = f \circ g^{-1}$ $\forall\, f, g \in X$. Then for $f, g, h \in X$,

$$\begin{aligned}
\psi(h, \phi(f,g)) &= h \circ (\phi(f,g))^{-1} \\
&= h \circ (f \circ g)^{-1} \\
&= h \circ (g^{-1} \circ f^{-1}) \\
&= (h \circ g^{-1}) f^{-1}.
\end{aligned}$$

And

$$\phi(\psi(h,f),\psi(h,g)) = \psi(h,f)\circ\psi(h,g)$$
$$= (h\circ f^{-1})\circ(h\circ g^{-1}).$$

In particular if $f = g = h \neq Id_{\mathbb{N}}$, we have

$$\psi(h,\phi(f,g)) = Id_{\mathbb{N}}\circ f^{-1} = f^{-1} \quad \text{and} \quad \phi(\psi(h,f),\psi(h,g)) = Id_{\mathbb{N}}.$$

Thus in general ψ is not distributive over ϕ.

2.6.24 Table of a binary operation. When X is a finite set and $\phi : X \times X \to X$ is a binary operation, it is convenient to describe $\phi(x,y)$ for $x, y \in X$ explicitly by making a table as follows: let $X = \{x_1, \ldots, x_n\}$. Then the table for the binary operation ϕ is written as follows:

ϕ	x_1	x_2	$\ldots$	x_j	$\ldots$	x_n
x_1	—	—		$\vdots$		—
x_2	—	—		$\vdots$		—
$\vdots$				$\vdots$		
x_i	—	—		$\phi(x_i, x_j)$		—
$\vdots$				$\vdots$		
x_n	—	—		—		—

Figure 2.8 : Table of a binary operation.

We assume that the elements of X are listed across the top of the table in the same order as they are listed on the left of the table, say as $x_1, \ldots, x_n$. The jth entry (from top) in the ith row of the table gives the value $\phi(x_i, x_j)$. For example, let $X = \{a, b, c\}$ and the table of ϕ be given by

2.6 Abstractions

ϕ	a	b	c
a	b	c	b
b	a	c	b
c	c	b	a

Figure 2.9 : Table of ϕ.

Then $\phi(c,a) = \phi(a,b) = c, \phi(a,c) = b$, and so on. Note that ϕ as defined above is not commutative for $\phi(a,b) = c \neq a = \phi(b,a)$. A binary operation ϕ is commutative iff the entries in its table are symmetric with respect to the diagonal.

2.6.25 Definition: We say a semigroup (group) $(X, \oplus)$ is **isomorphic** to a semigroup (group) $(Y, +)$ if there exists a bijective map $\phi : X \to Y$ such that $\phi(x \oplus y) = \phi(x) + \phi(y)$ for every $x, y \in X$. The map ϕ as above is called an **isomorphism** from the semigroup (group) X to the semigroup (group) Y. We write $(X, \oplus) \simeq (Y, +)$ if X is isomorphic to Y.

2.6.26 Example: Let $(X, \oplus)$ and $(Y, +)$ be groups and $\phi : X \to Y$ be an isomorphism. Let e_X, e_Y denote the identity elements of X and Y, respectively. For any $x \in X, y \in Y$, let $-x, -y$ denote their inverses, respectively. Let $y_0 = \phi(e_X)$. Then

$$y_0 + y_0 = \phi(e_X) + \phi(e_X) = \phi(e_X \oplus e_X) = \phi(e_X) = y_0.$$

Thus,
$$y_0 = y_0 + (y_0 - y_0) = (y_0 + y_0) - e_Y.$$

Hence $\phi(e_X) = e_Y$. Also,

$$e_Y = \phi(e_X) = \phi(x \oplus (-x)) = \phi(x) + \phi(-x).$$

Hence $-(\phi(x)) = \phi(-x)$ for every $x \in X$. Thus an isomorphism preserves the algebraic structures on the semigroups (groups). In other words, X and Y are identical except for a renaming. Or we can say they are copies of each other.

2.6.27 Exercises: For semigroups (groups) $(G_i, \oplus_i), i = 1, 2, 3$, the following hold:

(i) $G_1 \simeq G_1$

(ii) $G_1 \simeq G_2 \Leftrightarrow G_2 \simeq G_1$.

(iii) $G_1 \simeq G_2$ and $G_2 \simeq G_3 \Rightarrow G_1 \simeq G_3$.

Chapter 3

INTEGERS

3.1 Historical comments

We saw in chapter 1 that given any two numbers $n, m \in \mathbb{N}$, the equation $x + n = m$ need not have a solution in $\mathbb{N}$. For example for $n \geq m$ the above equation has no solution in $\mathbb{N}$ and for $n < m$ there exists a unique $x \in \mathbb{N}$ such that $x + n = m$. The need to find solutions of equation $x + n = m$ for m and n in $\mathbb{N}$ (which arise naturally in keeping records of business transactions like representing debts) must have motivated the Hindus to introduce negative numbers. The first known use of such numbers is by the Hindu mathematician Brahmagupta about 628 A.D. He is also credited for discovering arithmetical rules of such numbers. The concept of zero and arithmetic operations involving zero are attributed to the Hindu mathematician Mahavira in the 9th century. However, negative numbers were not accepted by mathematicians for centuries. Mathematicians of the fifteenth and sixteenth centuries called negative numbers as 'absurd' numbers. René Descartes (1596–1650) accepted them in parts and called such numbers as 'false' numbers. By the late nineteenth century these numbers (which came to be known as integers) gained acceptance among mathematicians. Leopold Kronecker (1823–91) accepted integers and his views were reflected in his famous quip "God made the integers and the rest is the work of man". However, efforts of mathematicians to 'Arithmetize Analysis' (see historical comments of chapter 5) finally gave rise to the construction of integers, assuming the existence of natural numbers and their properties. We give this construction (due to Richard Dedekind) in the next section.

Brahmagupta (598 A.D.–680 A.D.)

Brahmagupta was born in 598 A.D. in Western/Central-India. He wrote two important monographs on astrology (in those times in India, mathematics was treated as part of astrology) which were later translated into Arabic as 'Sind-Hind' and 'Al-Arkand'. Later ($\sim$ 665 A.D.) he also wrote a monograph called 'Khanda-Khadayah' which contained many facts about arithmetic calculations, geometry, algebra, mensuration and astrology. He came across negative numbers as roots of quadratic equations and gave a systematic treatment of arithmetic involving negative numbers. He gave a method of constructing 'Pythagorean-triads' of numbers. He was the first to give general solution of linear Diophantime equation $ax+by = c$ for a, b, c, to be integers. He also analyzed

3.1 Historical comments

Diophantine quadratics $x^2 = 1 + py^2$, usually attributed to John Pell (1611–1685). He died in 680 A.D.

Mahaviracharya ($\sim$ 850 A.D.)

Not much is known about the life of Mahaviracharya. From his writings it is guessed that he lived in the 9th century. He wrote monographs called 'Ganit-Sar', 'Jyotish-Patel' and 'Shatrishanka'. 'Ganit-Sar' was translated in south Indian languages and was used as a textbook in south India for many centuries. An English translation was introduced to the mathematical world in 1908 in the Fourth International Congress of Mathematics, held in Rome, and later was published in 1912. In 'Ganit-Sar' Mahavira gave arithmetic of numbers, introduced the concept of zero and arithmetical rules for operations with it, gave methods of computing square roots and cube roots. The concept of LCM is also attributed to him.

3.2 Construction and uniqueness of integers

We assume the existence of natural numbers and their properties, as developed in chapter 1. Given two natural numbers $n, m \in \mathbb{N}$, we would like to find the difference between them. If $n > m$, we know $n = m + k$ for some $k \in \mathbb{N}$, and we can write the difference as $n - m := k$. What to do when $n \leq m$? The way out is to associate with $n - m$ a new object and define 'arithmetic' of these objects. For example one can associate with $n - m$ the ordered pair $(n, m), n \in \mathbb{N}, m \in \mathbb{N}$. This motivates us to consider the set $\mathbb{N} \times \mathbb{N} := \{(n, m) | n, m \in \mathbb{N}\}$. Since $n - m$ could be the same as $r - s$, for certain $n, m, r, s \in \mathbb{N}$, we should not distinguish between such ordered pairs (n, m) and (r, s). To do this, we make the following definition.

3.2.1 Definition. For $(n, m), (r, s) \in \mathbb{N} \times \mathbb{N}$, we say (n, m) is **equivalent** to (r, s), and write it as $(n, m) \sim (r, s)$ if $n + s = m + r$, addition being that of natural numbers.

3.2.2 Proposition. *The relation $\sim$ as defined in 3.2.1, is an equivalence relation on the set $\mathbb{N} \times \mathbb{N}$.*

Proof. Obviously for $(m, n) \in \mathbb{N} \times \mathbb{N}, (m, n) \sim (m, n)$. For $(n, m), (r, s) \in \mathbb{N}$ if $(m, n) \sim (r, s)$ then it is easy to check that $(r, s) \sim (m, n)$. Finally, let $(m, n), (r, s)$ and $(k, \ell) \in \mathbb{N} \times \mathbb{N}$ be such that $(m, n) \sim (r, s)$ and $(r, s) \sim (k, \ell)$. Then $(m, n) \sim (k, \ell)$. To see this, note that $m + s = n + r$ and $r + l = k + s$. Thus $(m + s) + (r + l) = (n + r) + (k + s)$ and hence $(m + l) + (s + r) = (n + k) + (s + r)$. By the cancellation property for addition on $\mathbb{N}$, we get $m + l = n + k$ and hence $(m, n) \sim (k, \ell)$. ∎

The equivalence relation $\sim$ partitions the set $\mathbb{N} \times \mathbb{N}$ into equivalence classes. Let us denote the equivalence class containing the ordered pair (m, n) by $[m, n]$. Then $[m, n]$ is nothing but the

3.2 Construction and uniqueness of integers

set of all those elements of $\mathbb{N} \times \mathbb{N}$ which are equivalent to (m, n), i.e., $[m, n] = \{(r, s) \in \mathbb{N} \times \mathbb{N} | m + s = n + r\}$. We denote the set $\mathbb{N} \times \mathbb{N}/\sim$, of equivalence classes of $\mathbb{N} \times \mathbb{N}$ under this relation, by $\mathbb{Z}$.

3.2.3 Lemma. *For* $m, n, k \in \mathbb{N}$,

$$[m, n] = [m + 1, n + 1] = [m + k, n + k].$$

Proof. Follows from the equalities:

$$m + (n + 1) = n + (m + 1)$$

and

$$(m + 1) + (n + k) = (n + 1) + (m + k). \blacksquare$$

3.2.4 Theorem (Addition on $\mathbb{Z}$ and its properties). *For* $m, n, k, r, l, s \in \mathbb{N}$, *let*

$$[m, n] \oplus [k, r] := [m + k, n + r].$$

Then $[m, n] \oplus [k, r]$ *is a well-defined binary operation on $\mathbb{Z}$ and has the following properties:*

(i) $[m, n] \oplus [k, r] = [k, r] \oplus [m, n]$.

(ii) $([m, n] \oplus [k, r]) \oplus [l, s] = [m, n] \oplus ([k, r] \oplus [l, s])$.

(iii) $[m, n] \oplus [1, 1] = [m, n] = [1, 1] \oplus [m, n]$.

(iv) $[m, n] \oplus [n, m] = [1, 1] = [n, m] \oplus [m, n]$.

Proof: Let $[m, n] = [m', n']$ and $[k, r] = [k', r']$. Then $m + n' = n + m'$ and $k + r' = r + k'$. But then $(m + n') + (k + r') = (n + m') + (r + k')$, i.e., $(m + k) + (n' + r') = (n + r) + (m' + k')$. Hence $[m + k, n + r] = [m' + k', n' + r']$ proving that $[m, n] \oplus [k, r]$

is well-defined. Proofs of (i), (ii), (iii) are easy and are left as exercises. Proof of (iv) follows from lemma 3.2.3. ∎

3.2.5 Remark. Theorem 3.2.4 tells that $\oplus$, called **addition**, is a well-defined binary operation on $\mathbb{Z}$. Addition is commutative, associative, and has **identity** element [1,1] ([1,1] is playing the role of 'zero'). Further every element $[m, n] \in \mathbb{Z}$ has **additive inverse** $[n, m]$. One says $\mathbb{Z}$ is a commutative group under addition (see 2.6.20 for definition).

3.2.6 Exercise.

(i) Show that $[1, 1] \in \mathbb{Z}$ is the only element of $\mathbb{Z}$ such that $\forall\, n, m \in \mathbb{N}$
$$[m, n] \oplus [1, 1] = [m, n].$$

(ii) Show for $[m, n] \in \mathbb{Z}$, $[n, m]$ is the only element of $\mathbb{Z}$ such that
$$[m, n] \oplus [n, m] = [1, 1].$$

3.2.7 Theorem. *Let* $\mathbb{N}_{\mathbb{Z}} := \{[n + 1, 1] \in \mathbb{Z} \,|\, n \in \mathbb{N}\}$. *Then the following hold:*

(i) $(\mathbb{N}_{\mathbb{Z}}, \oplus)$ *is a sub-semigroup of* $(\mathbb{Z}, \oplus)$, *i.e.,* $\mathbb{N}_{\mathbb{Z}} \subseteq \mathbb{Z}$ *and* $(\mathbb{N}_{\mathbb{Z}}, \oplus)$ *is a semigroup.*

(ii) $(\mathbb{N}_{\mathbb{Z}}, \oplus)$ *is isomorphic to* $(\mathbb{N}, +)$ *as semigroups.*

(iii) *For every* $x \in \mathbb{Z}$, *there exist* $y, z \in \mathbb{N}_{\mathbb{Z}}$ *such that* $x = y \oplus (-z)$, *where* $-z$ *is the additive inverse of* $z \in \mathbb{Z}$.

Proof. (i) Since for $n, m \in \mathbb{N}$,
$$\begin{aligned}[n + 1, 1] \oplus [m + 1, 1] &= [n + m + 1 + 1, 1 + 1] \\ &= [n + m + 1, 1],\end{aligned}$$

3.2 Construction and uniqueness of integers

it follows that $\oplus$ is a binary operation on $\mathbb{N}_{\mathbb{Z}}$ and obviously it is associative (and commutative). Hence $(\mathbb{N}_{\mathbb{Z}}, \oplus)$ is a sub-semigroup of $(\mathbb{Z}, \oplus)$.

(ii) Define $\phi : \mathbb{N} \longrightarrow \mathbb{N}_{\mathbb{Z}}$ by $\phi(n) := [n+1, 1], n \in \mathbb{N}$. Clearly ϕ is onto. If $[n+1, 1] = [m+1, 1]$, then $n+1+1 = m+1+1$ and hence $n = m$. Thus ϕ is also one-one. Finally for $n, m \in \mathbb{N}$, since

$$[n+1, 1] \oplus [m+1, 1] = [n+m+1+1, 1+1]$$
$$= [n+m+1, 1],$$

it follows that $\phi(n+m) = \phi(n) \oplus \phi(m) \ \forall \ n, m \in \mathbb{N}$. Hence $(\mathbb{N}, +)$ is isomorphic to $(\mathbb{N}_Z, \oplus)$.

(iii) Let $x = [n, m] \in \mathbb{Z}$. Then

$$[n, m] = [n+1+1, m+1+1]$$
$$= [n+1, 1] \oplus [1, m+1]$$
$$= [n+1, 1] \oplus (-[m+1, 1]).$$

Let $y := [n+1, 1]$ and $z := [m+1, 1]$. Then $x = y \oplus (-z)$. ∎

3.2.8 Definition. A commutative group $(G, \oplus)$ is called an **integer system** if there exists a sub-semigroup $(H, \oplus)$ of $(G, \oplus)$ with the following properties:

(i) $(H, \oplus)$ is isomorphic to $(\mathbb{N}, +)$.

(ii) For every $g \in G$, there exist $h, f \in H$ such that $g = h \oplus (-f)$, where $-f$ is the inverse of f in G.

3.2.9 Theorem (Existence and uniqueness of integers). *Integer systems exist and any two integer systems are isomorphic as groups.*

Proof. That integer systems exist follows from theorem 3.2.7. Let $(G_1, \oplus_1)$ and $(G_2, \oplus_2)$ be two integer systems. Let $\mathbb{N}_{G_1}$ and

$\mathbb{N}_{G_2}$ be their respective sub-semigroups which are isomorphic to $(\mathbb{N}, +)$. Then $\mathbb{N}_{G_1}$ and $\mathbb{N}_{G_2}$ are themselves isomorphic, by exercise 2.6.27. Let ϕ be that isomorphism, $\phi : \mathbb{N}_{G_1} \longrightarrow \mathbb{N}_{G_2}$. For $g \in G_1$, let $g = h_1 \oplus_1 (-f_1) = h_2 \oplus_1 (-f_2)$ for $h_1, h_2, f_1, f_2 \in \mathbb{N}_{G_1}$. Then $h_1 \oplus_1 f_2 = h_2 \oplus_1 f_1 \in \mathbb{N}_{G_1}$ and we have $\phi(h_1 \oplus_1 f_2) = \phi(h_2 \oplus_1 f_1)$. Thus $\phi(h_1) \oplus_2 \phi(f_2) = \phi(h_2) \oplus_2 \phi(f_1)$ and hence $\phi(h_1) \oplus_2 (-\phi(f_1)) = \phi(h_2) \oplus_2 (-\phi(f_2))$. Thus for $g \in G_1$, if $g = h \oplus_1 (-f)$ and we define $\psi(g) = \phi(h) \oplus_2 (-\phi(f))$, then ψ is a well-defined map from G_1 to G_2. It is easy to check that ψ is a group isomorphism. ∎

3.2.10 Exercise. Let $(G, \oplus)$ be an integer system with $(H, \oplus)$ being the sub-semigroup isomorphic to $(\mathbb{N}, +)$ and let $\phi : \mathbb{N} \to H$ be this isomorphism. Let $1_H := \phi(1)$ and $S_H(n) = \phi(n) + 1$ $\forall n \in \mathbb{N}$. Show that $(H, 1_H, S_H)$ is a natural number system.

3.2.11 Note. In view of theorem 3.2.9, we can say that there is only one integer system $(\mathbb{Z}, \oplus)$ as constructed above. We shall call elements of $\mathbb{Z}$ as **integers** and identify $\mathbb{N}_{\mathbb{Z}}$ with $\mathbb{N}$. Note that $\forall k \in \mathbb{N}$, $[k+1, k] \in \mathbb{N}_{\mathbb{Z}}$ is identified with $1 \in \mathbb{N}$. Also under the identification of $[n+1, 1]$ with $n \in \mathbb{N}$, every element $x \in \mathbb{Z}$ has the representation $x = n \oplus (-m)$, for some $n, m \in \mathbb{N}$. Note that this representation need not be unique, e.g., $n \oplus (-m) = (n \oplus k) \oplus (-m \oplus k)$ $\forall k \in \mathbb{N}$. We shall write $x \oplus y$ as $x + y$ and $x + (-y)$ as $x - y$ from now onwards. Elements of $\mathbb{N}$ are called **positive integers** and elements of $-\mathbb{N} := \{-n | n \in \mathbb{N}\}$ are called **negative integers**. The additive identity in $\mathbb{Z}$ is denoted by 0, called **zero**. Note that 0 corresponds to the element $[1, 1] \in \mathbb{Z}$. In a sense we have extended the semigroup $(\mathbb{N}, +)$ to a group $(\mathbb{Z}, +)$.

3.2.12 Exercise.

(a)　　For $x, y, z \in \mathbb{Z}$, prove the following:

　　(i)　$-0 = 0$.

　　(ii)　$-(-x) = x$.

3.2 Construction and uniqueness of integers

(iii) $-(x+y) = (-x) + (-y)$.

(iv) $x + y = x + z$ iff $y = z$.

(v) $x - y = z$ iff $x = y + z$.

(vi) $-(x - y) = y - x$.

(vii) $x - y = 0 \Leftrightarrow x = y$.

(b) For $x, y \in \mathbb{N}$, $x + y = 0$ iff $x = y = 0$.

(c) The sets $\mathbb{N}, \{0\}$ and $-\mathbb{N}$ are pairwise disjoint.

3.2.13 Exercise. Let $(G, \oplus)$ be any commutative group such that $(G, \oplus)$ has a sub-semigroup isomorphic to $(\mathbb{N}, +)$. Show that $(G, \oplus)$ includes a subgroup $(H, \oplus)$ isomorphic to $(\mathbb{Z}, +)$. Thus $(\mathbb{Z}, +)$ is the smallest commutative group including the semigroup $(\mathbb{N}, +)$.

Next, using the operation of multiplication on $\mathbb{N}$ we want to define a binary operation $\odot$, called **multiplication** on $\mathbb{Z}$ which extends the binary operation of multiplication on $\mathbb{N}$, i.e., for, $n, m, k, r \in \mathbb{N}$, we want to have

$$(n - m) \odot (k - r) = (nk + mr) - (mk + nr).$$

This motivates our next theorem.

3.2.14 Theorem (Multiplication on $\mathbb{Z}$). *For $l, s, m, n, k, r \in \mathbb{N}$ let*

$$[m, n] \odot [k, r] := [mk + nr, nk + mr].$$

Then $\odot$ is a well-defined binary operation on $\mathbb{Z}$ and has the following properties:

(i) $[m, n] \odot [k, r] = [k, r] \odot [m, n]$, *i.e.*, $\odot$ *is commutative.*

(ii) $[m, n] \odot ([k, r] \odot [l, s]) = ([m, n] \odot [k, r]) \odot [l, s]$, *i.e.*, $\odot$ *is associative.*

(iii) $[m, n] \odot [k + 1, k] = [m, n] = [k + 1, k] \odot [m, n]$, *i.e.*, $[k + 1, k]$ *is the identity for* $\odot$.

(iv) $[m,n] \odot ([k,r] \oplus [l,s]) = ([m,n] \odot [k,r]) \oplus ([m,n] \odot [l,s])$, i.e., $\odot$ *is distributive over* $\oplus$.

(v) $([k,r] \oplus [l,s]) \odot [m,n] = ([k,r] \odot [m,n]) \oplus ([\ell,s] \odot [m,n])$.

(vi) $[m,n] \odot [k,r] = [1,1]$ *iff either* $[m,n] = [1,1]$ *or* $[k,r] = [1,1]$.

Proof: To prove that $[m,n] \odot [k,r]$ is well-defined, suppose $[m,n] = [m',n']$ and $[k,r] = [k',r']$. Then $m+n' = m'+n$ and $k+r' = k'+r$. We want to show that

$$[m,n] \odot [k,r] = [m',n'] \odot [k',r'],$$

i.e., $[mk+nr, mr+nk] = [m'k'+n'r', m'r'+n'k']$,
i.e., $mk+nr+m'r'+n'k' = m'k'+n'r'+mr+nk$.

The above equality holds iff we have

$$mk+nr+m'r'+n'k'+n'k+m'r+nr'+mk'$$
$$= m'k'+n'r'+mr+nk+m'r+n'k+mk'+nr',$$

i.e.,

$$(m+n')k + (n+m')r + (m'+n)r' + (n'+m)k'$$
$$= m'(k'+r) + n'(r'+k) + m(r+k') + n(k+r'). \quad (3.1)$$

Now using the fact that $m+n' = m'+n$ and $k'+r = k+r'$, equation (3.1) will hold iff

$$(m+n')(k+r+r'+k') = (k'+r)(m'+n'+m+n),$$

i.e.,

$$(m+n')(k'+r)(1+1) = (k'+r)(m+n')(1+1),$$

which is true. This proves that $\odot$ is a well-defined binary operation on $\mathbb{Z}$. Proofs of (i) and (ii) are left as exercises. To prove (iii) we note that $\forall \, k$, using lemma 3.2.3, we have

$$\begin{aligned} [m,n] \odot [k+1, k] &= [m(k+1)+nk, mk+n(k+1)] \\ &= [mk+nk+m, mk+nk+n] \\ &= [m,n]. \end{aligned}$$

3.2 Construction and uniqueness of integers

To check (iv), we note that

$$[m,n] \odot ([k,r] \oplus [\ell,s])$$
$$= [m,n] \odot [k+\ell, r+s]$$
$$= [m(k+\ell) + n(r+s), m(r+s) + n(k+\ell)]$$
$$= [mk + m\ell + nr + ns, mr + ms + nk + n\ell]. \quad (3.2)$$

Also,

$$([m,n] \odot [k,r]) \oplus ([m,n] \odot [\ell,s])$$
$$= [mk + nr, mr + nk] \oplus [m\ell + ns, ms + n\ell]$$
$$= [mk + nr + m\ell + ns, mr + nk + ms + n\ell]. \quad (3.3)$$

From (3.2) and (3.3), the required claim follows. Proof of (v) is similar. Finally to prove (vi), we note that

$$[1,1] \odot [k,r] = [k+r, k+r] = [1,1]$$

and

$$[m,n] \odot [1,1] = [m+n, m+n] = [1,1].$$

Hence the first part of (vi) is true. Conversely, let $[m,n] \odot [k,r] = [1,1]$. Then

$$[mk + nr, mr + nk] = [1,1],$$

i.e.,

$$mk + nr + 1 = mr + nk + 1. \quad (3.4)$$

Suppose that neither $[k,r] = [1,1]$ nor $[m,n] = [1,1]$. Then $k+1 \neq r+1$ and $m+1 \neq n+1$, i.e., $k \neq r$ and $m \neq n$. Suppose $k > r$ and $m > n$. Let $k = r+l, m = n+s$ for some $l, s \in \mathbb{N}$. Then, by (3.4) we have

$$(n+s)k + nr + 1 = mr + n(r+l) + 1,$$

i.e.,

$$n(k+r) + sk + 1 = r(m+n) + nl + 1,$$

i.e.,
$$n(r+l+r) + s(r+l) + 1 = r(n+s+n) + nl + 1,$$
i.e.,
$$sl + 1 = 1,$$
which is not possible. Hence both $k > r, m > n$ are not possible. Similarly, one can show that the other cases, i.e., $k > r, m < n$; $k < r, m < n$; $k \leq r, m > n$ are also not possible. Hence either $[k, r] = [1, 1]$ or $[m, n] = [1, 1]$. ∎

3.2.15 Remarks.

(i) Since $[n+1, 1] \odot [m+1, 1] = [mn+m+n+1+1, m+n+1+1] = [mn+1, 1]$, under the identification of $[k+1, k]$ with $\mathbb{N}$, $\odot$ extends the binary operation of multiplication on $\mathbb{N}$. From now onwards, for $x, y \in \mathbb{Z}$ we write xy for $x \odot y$.

(ii) Theorems 3.2.4 and 3.2.14 tell us that on $\mathbb{Z}$ two binary operations of addition and multiplication are defined respectively with the following properties:

(A) $(\mathbb{Z}, +)$ is a commutative group.

(B) Multiplication is distributive over addition, i.e.,
$$x(y + z) = xy + xz.$$

(C) There exists identity for multiplication, i.e., there exists an element $1 \in \mathbb{Z}$ such that $1x = x1 = x \ \forall \ x \in Z$.

(D) $\mathbb{Z}$ does not have **zero divisors**, i.e., $xy = 0$ iff either $x = 0$ or $y = 0$.

Properties (A), (B) and (C) say that $(\mathbb{Z}, \oplus, \odot)$ is a **Ring** with identity and property (D) says that this ring is in fact an **integral domain** (see section 3.6 for more details).

3.2.16 Exercise. For $x, y, z \in \mathbb{Z}$, show that the following hold:

(i) $x0 = 0x = 0$.

(ii) $-(xy) = (-x)y = x(-y)$.

(iii) $(-x)(-y) = xy$.

3.3 Order on integers

3.3.1 Definitions. Let $m, n \in \mathbb{Z}$. We say m is **larger than** n or n is **smaller than** m, and we write it as $m > n$ or $n < m$, if $m - n \in \mathbb{N}$. We write $m \geq n$ or $n \leq m$ wherever either $m = n$ or $n < m$.

3.3.2 Note. Note that the order on $\mathbb{Z}$ is consistent with the order on $\mathbb{N}$, i.e., for $n, m \in \mathbb{N}, n \geq m$ as natural numbers iff $n \geq m$ as integers.

3.3.3 Theorem (Properties of order on $\mathbb{Z}$). *The order $\leq$ on $\mathbb{Z}$ is a total order and has the following properties:*

(i) For $m, n, k \in \mathbb{Z}$, if $m \leq n$, $m + k \leq n + k$.

(ii) For $m, n, k \in \mathbb{Z}$, if $m \leq n$,

$$mk \leq nk \quad \text{if } k > 0$$

and

$$mk \geq nk \quad \text{if } k < 0.$$

Proof. That $\leq$ is a total order on $\mathbb{Z}$ is easy to check in view of exercise 3.2.12(c). To prove (i), let $m \leq n$. If $m = n$, then obviously $m + k = n + k \ \forall k \in \mathbb{Z}$. So suppose $m < n$. Then $n - m := r \in \mathbb{N}$. Thus $n = m + r$ and we have

$$n + k \ = \ m + r + k = m + k + r.$$

Hence $n+k \geq m+k$, proving (i). Proof of (ii) is similar and is left as an exercise. ∎

3.3.4 Exercise. Prove the following statements:

(i) For $n, m \in \mathbb{N}$, $n + m > 0$ and $nm > 0$.

(ii) For $n, m \in \mathbb{Z}$, $nm > 0$ iff either $n > 0$ and $m > 0$ or $n < 0$ and $m < 0$.

(iii) For every $n \in \mathbb{Z}$, there does not exist any $m \in \mathbb{Z}$ such that $n < m < (n+1)$.

3.3.5 Exercise. Let $n \in \mathbb{Z}$. Show that there is no integer k such that $n < k < n+1$.

3.3.6 Theorem. *For any nonempty subset E of $\mathbb{Z}$ the following hold:*

(i) If E is bounded above, $lub(E)$ exists and $lub(E) \in E$.

(ii) If E is bounded below, $glb(E)$ exists and $glb(E) \in E$.

(iii) E is finite iff E is both bounded above and below.

Statements (i) and (ii) say that $(\mathbb{Z}, \leq)$ is order complete.

Proof. (i) Suppose E is bounded above. Since $E \neq \emptyset$, let $n \in E$. Consider $n, n+1, \ldots$. Then there exists a $k \in \mathbb{N} \cup \{0\}$ such that $n+k \in E$ but $n+k+1 \notin E$, for otherwise, E will not be bounded above. Then $\forall\, m \in E, m \leq n+k \in E$. Hence $(n+k) = glb(E)$.

(ii) Proof is similar to (i) and is left as an exercise.

(iii) Suppose E is finite. Let $n \in E$. Consider the integers $n+1, n+2, \ldots$. Since E is finite, there exists $k \in \mathbb{N}$ such that $n+m \notin E\ \forall\, m \geq k$. Thus $x \leq n+m\ \forall\, x \in E$. Hence E is bounded above. Similarly, E is bounded below. Conversely, suppose E is bounded above and below with $n := glb(E), m := lub(I)$. Then

3.3 Order on integers

$n, m \in E$ and $x \in [n, m] := \{k \in \mathbb{N} | n \leq k \leq m\}$ $\forall\, x \in E$. Thus E is a subset of the finite set $[n, m]$ and hence is itself finite, by example 2.4.6 (ii). ∎

3.3.7 Theorem (Archimedean property). *Let $m, n \in \mathbb{Z}, m > 0$. Then there exists $k \in \mathbb{N}$ such that $mk > n$.*

Proof. Note that if $n \leq 0$, then $m > n$ and hence theorem holds with $k = 1$. So, suppose $m, n \in \mathbb{N}$. Now the required claim is just theorem 2.4.15. ∎

As an application of theorem 3.3.7, we have the following theorem.

3.3.8 Theorem (Division algorithm). *Let $m, n \in \mathbb{Z}, n > 0$. Then there exist unique integers q and r such that $0 \leq r < n$ and $m = nq + r$.*

The integer q is called the **quotient** and r is called the **remainder.**

Proof. Let $A := \{m + nq\,|\,q \in \mathbb{Z}, m + nq \geq 0\}$. Then A is a nonempty subset of $\mathbb{N}$, by theorem 3.3.7 $\exists\, k \in \mathbb{N}$ such that $nk > -m$, i.e., $(m + nk) > 0$. Thus A has a least element by theorem 2.4.10. Let $r := glb(A)$. Then $r \in A$ and hence $m + nq = r$, i.e., $m = n(-q) + r$ for some $q \in \mathbb{Z}$. Clearly $r \geq 0$. Suppose $n < r$. Then $r = n + r_1$ for some $r_1 \in \mathbb{N}$ with $0 < r_1 < r$. But then, $r_1 = r - n = m + nq - n = m + (q - 1)n$. Thus $r_1 < r$ and $r_1 \in A$, which is not possible. Hence $0 \leq r < n$. This proves the existence part of the theorem. To prove the uniqueness, let $nq + r = nq_1 + r_1$ where $0 \leq r, r_1 < n$. Suppose $r < r_1$. Then $0 < r_1 - r < n$ and $r_1 - r = n(q - q_1)$. But $(q - q_1) > 0$ by exercise 2.2.8. Thus $n(q - q_1) = r_1 - r < n$, which is not possible. Hence $r < r_1$ is not true. Similarly $r_1 < r$ is not true. Thus $r = r_1$ and $q = q_1$. ∎

3.3.9 Exercise. Let $n, m \in \mathbb{Z}$ and $k \in \mathbb{N}$. Show that $n - m$ is divisible by k iff n and m have same remainder when divided by k.

3.4 Prime factorization of integers

3.4.1 Definition. An integer $n \neq 0$ is called a **divisor** (or a **factor**) of an integer m (or we say n **divides** m) if there exists an integer q such that $m = nq$.

3.4.2 Theorem. For $m, n \in \mathbb{Z}$, the following hold.

(i) If $n \neq 0$ divides m, there is exactly one integer q such that $m = nq$.

(ii) If $m \neq 0$, m is divisible by $m, -m, 1$ and -1.

(iii) If n divides m and m divides k, n divides k.

Proof. (i) Let n divide m and let there exist q_1 and q_2 such that $m = nq_1 = nq_2$. Then $n(q_1 - q_2) = 0$. Since $n \neq 0$, $q_1 - q_2 = 0$, i.e., $q_1 = q_2$.

(ii) Since $m = m1 = (-m)(-1)$, by exercise 3.2.12, it follows that $m, 1, -m, -1$ are divisors of m.

(iii) Since n divides m and m divides k, $m = nq$ and $k = mp$ for some $q, m \in \mathbb{Z}$. Thus $k = npq$, i.e., n divides k. ∎

3.4.3 Exercise. Let $\mathcal{R}$ be the relation on $\mathbb{N}$ defined by $n\mathcal{R}m$ iff n divides m. Show that $\mathcal{R}$ is not a partial order. Is it an equivalence relation?

3.4.4 Exercise. Prove the following statements:

(i) If n divides m and k, n also divides $m + k$ and $m - k$.

(ii) If n divides $(m + k)$ and n divides m, then n divides k.

3.4.5 Definition. An integer $p > 1$ is called a **prime** if for every $q \in \mathbb{N}, q$ a divisor of p, either $q = 1$ or $q = p$. An integer $p > 1$ which is not a prime is called a **composite** number.

3.4 Prime factorization

3.4.6 Examples. Recall that in our terminology, $\mathbb{N} := \{1, 2, \ldots\}$, where $2 := 1 + 1$ and in general $n := (n-1) + 1$ for $n \geq 2$. Let us look at the natural number 2. If we have $2 = nm$ for $n, m \in \mathbb{N}$, then clearly $n \leq 2$ and $m \leq 2$. Thus either $n = 1$, in which case $m = 2$ or vise versa. Hence 2 is a prime. One can show similarly $3 := 2 + 1$ is a prime. Let $4 := 3 + 1$ and $4 = nm$, where $n, m \in \mathbb{N}$. Then $1 \leq n, m \leq 4$. It is easy to check that the only possible values of n are $1, 2$ and 4. In the second case $4 = 2m$, i.e., $3 + 1 = 2m$, i.e., $3 = 2m - 1$. If $m = 1$, then $3 = 2 - 1$, not true as $3 = 2 + 1$. If $m = 2$, then $2 + 1 = 2(1 + 1) - 1 = 2 + 2 - 1$, which is true. Thus $m = 2$ is possible. Next $m = 3$ is not possible, for then $3 = (1 + 1)3 - 1$ i.e. $3 = 3 + 3 - 1$, i.e. $0 = 2$. Similarly, $m = 4$ is not possible. Thus 4 is not a prime.

3.4.7 Exercise. Show that $n \in \mathbb{N}$ is a composite number iff there exist $m, k \in \mathbb{N}$ such that $1 < m < n, 1 < k < n$ and $n = mk$.

3.4.8 Theorem (Prime factorization). *Every integer $n > 1$ can be expressed as a product of primes.*

Proof. Let $A := \{n \in \mathbb{N} |$ every $k \in \mathbb{N}, 1 < k \leq n$ is a product of primes$\}$. Clearly $A \neq \emptyset$. In fact $2 \in A$. Suppose $n \in A$. Then either $(n + 1)$ is itself a prime and in that case $(n + 1) \in A$. Otherwise, $(n + 1) = n_1 n_2$, where $1 < n_1 < n, 1 < n_2 < n$ (by exercise 3.4.7). By induction hypothesis, both n_1 and n_2 are products of primes. Hence $(n + 1)$ is also a product of primes, i.e., $(n + 1) \in A$. Hence $A = \mathbb{N} \setminus \{1\}$ by induction. ∎

3.4.9 Definition. Let $m, n \in \mathbb{Z}$ be such that at least one of them is nonzero. An integer $d \in \mathbb{Z}$ is called a **common divisor** of n, m if $n = k_1 d$ and $m = k_2 d$ for some $k_1, k_2 \in \mathbb{Z}$. The number $lub\{k \in \mathbb{N} | k$ is a common divisor of n and $m\}$, if it exists is called the **greatest common divisor** of n and m and is denoted by $gcd(n, m)$.

3.4.10 Theorem. *Let $n, m \in \mathbb{Z}$. Then the following hold:*

(i) $gcd(n, m)$, if it exists, is positive and unique.

(ii) If $\alpha > 0$ is a common divisor of n and m, then $\alpha = gcd(n, m)$ if α is divisible by every common divisor of n and m.

(iii) $gcd(n, m)$ exists iff at least one of m, n is not zero.

(iv) If either $n \neq 0$ or $m \neq 0$, then $gcd(n, m)$ exists and can be expressed as $nx + my$ for some $x, y \in \mathbb{Z}$.

(v) If $\alpha = gcd(n, m)$ then $\alpha > 0$ and α is divisible by every common divisor of n and m.

Proof. (i) Since a divisor is a nonzero integer and if d is a divisor then $-d$ is also a divisor, it follows that $gcd(n, m) > 0$ whenever it exists. Let α, β be two gcd's of n and m. Then by definition $\alpha \leq \beta$ and $\beta \leq \alpha$ and hence $\alpha = \beta$.

(ii) Let $\alpha > 0$ be such that every common divisor of n, m also divides α. Then clearly α is the largest of all the common divisors and hence $\alpha = gcd(n, m)$.

(iii) Let $m, n \in \mathbb{Z}$ and $gcd(n, m)$ exist. Then both m, n cannot be zero, for then every integer is a common divisor of n, m and hence the set of divisors is unbounded. Hence either $n \neq 0$ or $m \neq 0$. Conversely, let $n, m \in \mathbb{Z}$ be such that at least one of them is not zero, say $n \neq 0$. Then the set $D := \{k \in \mathbb{N} | k \text{ divides both } n \text{ and } m\}$ is nonempty as $1 \in D$ and is bounded above by n. Hence, by theorem 3.3.6 $lub(D)$ exists and is an element of D.

(iv) Let $n, m \in \mathbb{Z}$ be such that not both are zero, say $n \neq 0$. Consider the set $A := \{nx + my | x, y \in \mathbb{Z}\}$. Choose $y = 0$ and $x := +1$ if $n > 0$, $x := -1$ if $n < 0$. Then $nx + my > 0$, i.e., A has at least one positive integer. Thus $A \cap \mathbb{N}$ is nonempty and hence has the smallest element, say α. Then $\alpha \in A$, i.e., $\alpha = nx_0 + my_0$, for some $x_0, y_0 \in \mathbb{Z}$. Clearly, if $\gamma \in \mathbb{Z}$ divides both n and m, γ divides α. Finally, by division algorithm, let $n = \alpha q + r$ for $q, r \in \mathbb{Z}, 0 \leq$

3.4 Prime factorization

$r < \alpha$. Then $n = q(nx_0+my_0)+r$, i.e., $r = n(1-qx_0)-qmy_0$. Thus, if $r \neq 0$, then $r \in A$ and $r < \alpha$, not possible. Thus $r = 0$, i.e., α divides n. Similarly α also divides m. Hence by (ii) $\alpha = gcd(n,m)$.

(v) Proof of (v) follows obviously from (iv). ■

3.4.11 Definition. Let $n, m \in \mathbb{Z}$. We say n and m are **relatively prime** if $gcd(n, m) = 1$.

3.4.12 Exercise. Let $k, n, m \in \mathbb{Z}$. Prove the following:

(i) $gcd(n, m) = gcd(n, n + m)$.

(ii) $gcd(kn, km) = k\, gcd(n, m)$, if $k > 0$.

(iii) $gcd(n, km) = 1$, if $gcd(n, k) = 1$ and $gcd(n, m) = 1$.

3.4.13 Example. The numbers 3 and 4 are relatively prime, because
$$1 = 3 \times (-1) + 4(+1).$$

3.4.14 Theorem. *Let $n, m, k \in \mathbb{Z}$ be such that $gcd(n, m) = 1$. If n divides mk, n divides k.*

Proof. Since $(n, m) = 1$, $\exists\, x, y \in \mathbb{Z}$ such that $nx + my = 1$. Thus $k = nkx + mky$. Since n divides nkx and n divides mky clearly n divides their sum, i.e., k. ■

3.4.15 Corollary. *Let p be a prime such that p divides the product nm for some $n, m \in \mathbb{Z}$. Then either p divides n or p divides m.*

Proof. Suppose p divides nm but p does not divide n. We claim that $gcd(p, n) = 1$ and hence by theorem 3.4.14, we will have p divide m. Let $d = gcd(p, n)$. Then d divides p, a prime, and hence either $d = 1$ or p. But $d = p$ is not possible for then $p = d$ divides n. Hence $d = 1$. ∎

3.4.16 Exercise. Let p be a prime such that p divides the product $n_1 n_2 \cdots n_k$, where each $n_i \in \mathbb{Z}$. Then p divides at least one n_i, $1 \leq i \leq k$.

3.4.17 Theorem (Fundamental theorem of arithmetic). *Let $n > 1$ be any integer. Then there exist distinct primes $p_1, p_2, \ldots, p_k$ and natural numbers $\alpha_1, \alpha_2, \ldots, \alpha_k$ such that*

$$n = p_1^{\alpha_1} p_2^{\alpha_2} \ldots p_k^{\alpha_k}. \tag{3.5}$$

Further, this representation is unique except for the order of the terms $p_1^{\alpha_1}, \ldots, p_k^{\alpha_k}$ in the above representation. The representation (3.5) is called the **prime decomposition** of n.

Proof. We have already seen in theorem 3.4.8 that every $n > 1$ can be written as a product of a finite number of primes. If $p_1, \ldots, p_k$ are distinct primes appearing in that product and p_i occurs α_i times, $1 \leq i \leq k$, then clearly

$$n = p_1^{\alpha_1} p_2^{\alpha_2} \ldots p_k^{\alpha_k}.$$

We prove the uniqueness by induction on n. (We shall use the general principle of induction as stated in remark 2.4.31.) Suppose for all $1 < i \leq n - 1$, the uniqueness of the representation holds. Suppose if possible

$$n = p_1^{\alpha_1} p_2^{\alpha_2} \ldots p_k^{\alpha_k} = q_1^{\beta_1} q_2^{\beta_2} \ldots q_r^{\beta_r}, \tag{3.6}$$

where p_i's and q_j's are distinct primes. We shall show that $k = r$, $\{p_1, p_2, \ldots, p_r\} = \{q_1, q_2, \ldots, q_r\}$ and $\alpha_i = \beta_j$ whenever $p_i = q_j$. Since p_1 divides n, p_1 divides $q_1^{\beta_1} \ldots q_r^{\beta_r}$ and hence, by exercise 3.4.16, p_1 divides one and only one q_i. Without loss of generality let this be q_1. Since q_1 is a prime, $p_1 = q_1$. Also $\alpha_1 = \beta_1$ for if, say $\alpha_1 < \beta_1$, then $n' := p_2^{\alpha_2} \ldots p_k^{\alpha_k} = q_1^{\beta_1 - \alpha_1} \ldots q_r^{\beta_r}$. But then q_1 divides

3.4 Prime factorization

the right-hand side of the equality but not the middle term, which is not possible. Thus $\alpha_1 < \beta_1$ is not true. Similarly $\alpha_1 > \beta_1$ is not possible. Hence $\alpha_1 = \beta_1$. Thus from (3.6) we will have

$$n' = p_2^{\alpha_2} \ldots p_k^{\alpha_k} = q_2^{\beta_2} \ldots q_r^{\beta_r}.$$

Since $1 < n' \leq n - 1$, by induction hypothesis we have $k = r$ and $p_i = q_i, \alpha_i = \beta_i \ \forall \ i$. ■

3.4.18 Exercise. An integer $n \in \mathbb{Z}$ is said to be **even** if 2 divides n. An integer which is not even is said to be **odd**. For $n \in \mathbb{Z}$ prove the following :

(i) If n is even, $n = 2k$ for some $k \in \mathbb{Z}$.

(ii) If n is odd, $n = 2k + 1$ for some $k \in \mathbb{Z}$.

(iii) n^2 is odd iff n is odd.

(iv) If n^2 is even, n is also even.

We close this section with the following classical result in number theory.

3.4.19 Euclid's Theorem (Infinitude of primes). *Given any finite number of primes $p_1, p_2, \ldots, p_r$, there exists a prime $p \neq p_1, p_2, \ldots, p_r$.*

Proof: Let $n = p_1 p_2 \ldots p_r + 1$ Then either n itself is a prime, in which case $n \neq p_i, 1 \leq i \leq r$. In fact, $n > p_i \ \forall \ 1 \leq i \leq r$. If n is not a prime, let p be a prime that occurs in the prime decomposition of n. Then p divides n and hence p divides $p_1 p_2 \ldots p_r + 1$. But then $p = p_r$ is not possible, as no p_i divides $p_1 p_2 \ldots p_r + 1$. ■

3.5 Representation of integers: Numeration

We had denoted the distinguished element of $\mathbb{N}$ by 1, its successor by 2 and the successor of 2 by 3 and that of 3 by 4. But does this mean that we will have to design a symbol for each natural number (and hence have infinite symbols). The problem arises from the fact that we have infinitely many numbers and we would like to represent them with a finite number of symbols only. In fact, we would like this finite number of symbols to be sufficiently small so that we can remember them and at the same time, the representation of a number with the help of these symbols should not occupy too much space. Both these objectives were achieved by the Hindu Mathematicians in India around 250 B.C. The symbols designed by them were adopted by the Arabs who visited India for trade, and they spread it to Europe and the rest of the world. Using only the symbols 0, 1, 2, 3, 4, 5, 6, 7, 8, 9 and a place value system they gave a method of representing all the numbers. The key to this representation is the following theorem.

3.5.1 Theorem. *Let n and $m \in \mathbb{Z}$ be such that $n > 0$ and $m > 1$. Then there exist unique nonnegative integers $k, r_0, r_1, r_2, \ldots, r_k$ such that*

$$n = m^k r_k + m^{k-1} r_{k-1} + \cdots + m r_1 + r_0,$$

where $\forall\, i = 1, 2, \ldots, k$, $0 \leq r_i < m$ and $r_k \neq 0$.

Proof. By theorem 3.3.8, there exist nonnegative integers q_0, r_0 such that $0 \leq r_0 < m$ and

$$n = mq_0 + r_0.$$

Suppose nonnegative integers $q_{k-1}, \ldots, q_0$ and $r_{k-1}, \ldots, r_0$ have been defined. Define nonnegative integers q_k, r_k such that

$$q_{k-1} = mq_k + r_k, \quad 0 \leq r_k < m,$$

3.5 Representation of integers

as given by theorem 3.3.8. Thus q_k, r_k are defined $\forall\, k \in \mathbb{N}$ and we have

$$\begin{aligned} n &= mq_0 + r_0 \\ &= m(mq_1 + r_1) + r_0 \\ &= m^2 q_1 + mr_1 + r_0 \\ &\cdots \quad \cdots \quad \cdots \\ &= m^{k+1} q_k + m^k r_k \cdots + mr_1 + r_0. \end{aligned}$$

Note that for some k if $q_{k-1} = 0$ then $q_k = 0$. Suppose $q_k \neq 0$. Then $q_k \geq 1$ and hence $q_k(k+1) > k$. Since $m > 1$, by exercise 2.4.8, $m^k > (k+1)$, $\forall\, k \in \mathbb{N}$, and $n \geq m^k q_k \geq (k+1)q_k$. Thus for $k > n$, if $q_k \neq 0$, we will have $n \geq (k+1)q_k > n$, which is a contradiction. Hence $q_k = 0$ for every $k > n$. Hence the set $A := \{k \geq 0 | q_k = 0\}$ is nonempty. Let $k_0 := glb(A)$. Then $q_{k_0} \in A$, i.e., $q_{k_0} = 0$ and $q_k \neq 0$ for $0 \leq k < k_0$. Thus

$$n = m^{k_0} r_{k_0} + \cdots + mr_1 + r_0.$$

Note that $r_{k_0} = mq_{k_0} + r_{k_0} = q_{k_0 - 1} \neq 0$. This proves the existence part of the theorem. To prove the uniqueness, let us suppose there exist nonnegative integers $k'_0, r'_0, r'_1, \ldots, r'_{k'_0}$ with $r'_{k'_0} \neq 0$ such that

$$n = m^{k'_0} r'_{k'_0} + \cdots + mr'_1 + r'_0.$$

Then

$$r'_0 + m(m^{k'_0 - 1} r'_{k'_0} + \cdots + r'_1) = r_0 + m(m^{k_0 - 1} r_{k_0}, + \cdots + r_1)$$

and, by the uniqueness of theorem 3.3.8, we have $r'_0 = r_0$ and $m^{k'_0 - 1} r'_{k'_0} + \cdots + r'_1 = m^{k_0 - 1} r_{k_0} + \cdots + r_1$. Now we can apply induction on n to deduce $r'_i = r_i$ $\forall\, i$ and $k'_0 = k_0$. ∎

3.5.2 Representation in base m. We consider application of theorem 3.5.1 for some special values of m.

(i) **Binary representation.** For $m = 2$ we have only two nonnegative integers less than $m - 2$, i.e., 0 and 1. Let $n \subset \mathbb{N}$ be arbitrary. By theorem 3.5.1 there exist unique nonnegative integers $r_0, r_1, \ldots, r_k$ such that $r_k \neq 0$ and

$$n = 2^k r_k + \ldots + 2r_1 + r_0.$$

We write the representation of n to be $n := r_k \ldots r_2 r_1 r_0$, where each $r_k = 0$ or 1. This is called the representation of numbers in **base 2** or the **binary representation**. For example

$$\begin{aligned} 0 &= 0, & \text{has binary representation: } & 0 \\ 1 &= 1, & \text{has binary representation: } & 1 \\ 2 &= 2^1 \times 1 + 0, & \text{has binary representation: } & 10 \\ 3 &= 2^1 \times 1 + 1, & \text{has binary representation: } & 11, \end{aligned}$$

and so on.

(ii) **Decimal representation.** Recall, we already have the symbols 0 and 1. Let

$$\begin{aligned} 2 &:= 1+1, & \text{called } \mathbf{two}, \\ 3 &:= 2+1, & \text{called } \mathbf{three}, \\ 4 &:= 3+1, & \text{called } \mathbf{four}, \\ 5 &:= 4+1, & \text{called } \mathbf{five}, \\ 6 &:= 5+1, & \text{called } \mathbf{six}, \\ 7 &:= 6+1, & \text{called } \mathbf{seven}, \\ 8 &:= 7+1, & \text{called } \mathbf{eight}, \\ 9 &:= 8+1, & \text{called } \mathbf{nine}. \end{aligned}$$

Let $m := 9+1$ be called **ten**. Let $n \in \mathbb{N}$ be arbitrary and consider unique nonnegative integers $0 \leq r_i \leq 9$, for $0 \leq i \leq k$ such that $r_k \neq 0$

$$n = m^k r_k + \cdots + m r_1 + r_0.$$

Then, we write $n = r_k r_{k-1} \ldots r_1 r_0$ and call it the **decimal representation** of n. For example, for the number ten, i.e., m itself, since $m = m \times 1 + 0$, it has the representation 10, i.e.,

3.5 Representation of integers

$$9+1:=10.$$

The numbers $r_0, r_1, \ldots, r_k$ in the representation of n occupy places in the representation as follows: the place on the extreme right is called the **units place** and represents numbers between 1 and 9. The second place is called **tens place** and these two places together represent numbers between 10 and $10^2 - 1$ and so on. Since, this is the system of representation we have been using since childhood, we assume the usual addition and multiplication tables for them. One can represent numbers in any base. For most scientific purposes, the binary representation is more economical (e.g. in computers) while the representation in decimal is more convenient from the practical point of view.

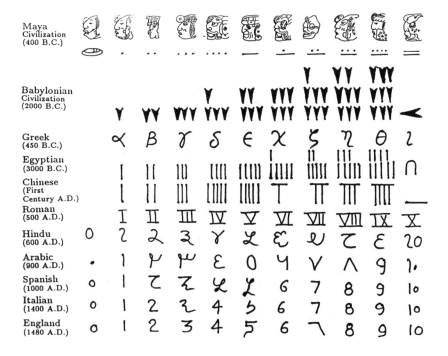

Figure 3.1 : Evolution of number symbols.

3.6 Abstractions

Motivated by the arithmetical properties of the integers, we have the following definition.

3.6.1 Definition. Let X be a set with two binary operations on it, denoted by $\oplus$ and $\odot$. The triple $(X, \oplus, \odot)$ is called a **ring** if the following hold:

(i) $(X, \oplus)$ is a commutative group.

(ii) $(X, \odot)$ is a semigroup.

(iii) $\odot$ is distributive over $\oplus$.

Whenever the binary operations $\oplus$ and $\odot$ are understood from the context, a ring $(X, \oplus, \odot)$ is also denoted by X. If $\odot$ is also commutative, we say that X is a **commutative ring.** We say a ring X is a **ring with identity** if there is an identity for the binary operation $\odot$.

3.6.2 Examples.

(i) The set of integers $\mathbb{Z}$ with addition and multiplication, as defined in section 3.3, forms a commutative ring with identity.

(ii) Consider the set $\mathbb{Z}_e$ of even integers with addition and multiplication of integers. Then $\mathbb{Z}_e$ is a commutative ring without identity.

(iii) Let $C(\mathbb{Z})$ denote the set of all functions from $\mathbb{N}$ into $\mathbb{Z}$. Elements of $C(\mathbb{Z})$ are called **sequences** in $\mathbb{Z}$. For $f, g \in C(\mathbb{Z})$, and $n \in \mathbb{N}$ define

$$(f \oplus g)(n) := f(n) + g(n) \quad \text{and} \quad (f \odot g)(n) := f(n)g(n).$$

Then $C(\mathbb{Z})$ is a commutative ring with identity with the above binary operations $\oplus$ and $\odot$. The identity for $\oplus$ being the constant

3.6 Abstractions

function $f(n) := 0$ $\forall\, n \in \mathbb{N}$ and the identity for $\odot$ being the constant function $g(n) := 1$ $\forall\, n \in \mathbb{N}$.

(iv) Let X be any nonempty set and let $\mathcal{P}(X)$ be the power set of X, i.e., the set of all subsets of X. For $A, B \in \mathcal{P}(X)$, define

$$A \oplus B := A \triangle B = (A \setminus B) \cup (B \setminus A) \quad \text{and} \quad A \odot B = A \cap B.$$

The set $A \triangle B$ is called the **symmetric difference** of sets A and B. It is easy to see that $\mathcal{P}(X)$ becomes a commutative ring with identity under these binary operations, the identity for $\odot$ being the set X.

(v) Let $n > 1$ be any integer. For $x, y \in \mathbb{Z}$, we say x is **congruent modulo** n to y, written as $x \equiv y (mod\ n)$, if $(x - y)$ is divisible by n. In view of exercise 3.3.9, $x \equiv y (mod\ n)$ iff x and y have the same remainders when divided by n. It is easy to show that 'congruent modulo n' is an equivalence relation on $\mathbb{Z}$. Let $\mathbb{Z}_n$ denote the equivalence classes of $\mathbb{Z}$ under this relation. We denote the elements of $\mathbb{Z}_n$ by $[x]$, it is the equivalence class that contains x. In view of the division algorithm, since the possible remainders upon division by n are $0, 1, 2, \ldots, n-1$, we have $\mathbb{Z}_n = \{[0], [1], \ldots, [n-1]\}$. For $[x], [y] \in \mathbb{Z}_n$, define

$$[x] \oplus [y] := [x + y] \quad \text{and} \quad [x] \odot [y] = [xy].$$

Using exercise 3.4.4, it is easy to check that $\oplus$ and $\odot$ are well-defined binary operations on $\mathbb{Z}_n$, i.e., if $[x] = [x_1]$ and $[y] = [y_1]$, then $[x + y] = [x_1 + y_1]$ and $[xy] = [x_1 y_1]$. Further $\mathbb{Z}_n$ becomes a commutative ring with identity under these binary operations, the additive identity being $[0]$, additive inverse of $[x]$ being $[-x]$ and $[1]$ being the identity for $\odot$. The ring $(\mathbb{Z}_n, \oplus, \odot)$ is called the **ring of integers modulo** n.

(vi) Let X be any set. Any ordered collection $x_0, x_1, \ldots$ of elements of X is called a **sequence** in X. We denote this by $\{x_n\}_{n \geq 0}$. (It is more customary to order the members of a sequence by $\mathbb{N}$ instead of $\mathbb{N} \cup \{0\}$, as we had done in example (ii) above.) Let X be any ring and let $\mathcal{P}[X]$ denote the set of all sequences $\{x_n\}_{n \geq 0}$ in X such that $x_n = 0 \ \forall \ n \geq m$, for some $m \geq 0$. For $\underline{x} = \{x_n\}_{n \geq 0}$ and $\underline{y} = \{y_n\}_{n \geq 0}$ in $\mathcal{P}[X]$, we define the binary operations of addition and multiplication as follows:

$$\underline{x} + \underline{y} := \{z_n\}_{n \geq 0}, \quad \text{where } z_n = x_n + y_n \ \forall \ n \geq 0;$$
$$\underline{x}\,\underline{y} := \{w_n\}_{n \geq 0}, \quad \text{where } w_n = \sum_{i=0}^{n} x_i y_{n-i}, n \geq 0.$$

For $\underline{x} = \{x_n\}_{n \geq 0} \in \mathcal{P}[X]$, let $x_n = 0 \ \forall \ n \geq m$, then an informal way of representing $\underline{x}$ is

$$\underline{x} := p(x) := x_0 + x_1 x + x_2 x^2 + \cdots + x_m x^m.$$

The sequence $\underline{x} = p(x)$ is called a **polynomial over** X. For $\underline{x} = \{x_n\}_{n \geq 1}$ if $m \geq 1$ is the largest integer such that $x_m \neq 0$ and $x_n = 0 \ \forall \ n \geq m$, it is called the **degree** of the polynomial. The elements $x_0, \ldots, x_n$ are called the **coefficients** of the polynomials. The coefficient x_0 is called the **constant term** of the polynomial and x_m is called the **leading coefficient** for a polynomial of degree m. Polynomials of degree 1 are called the **linear polynomials** and polynomials of degree 2 are called the **quadratic polynomials**. In the informal representation $p(x)$, x is called a **variable** and x_i is the coefficient of the ith power of the variable. The **addition** (**multiplication**) of two polynomials is the polynomial obtained by adding (multiplying) the informal representations as elements of the ring and gathering the coefficients of similar powers. Thus for

$$p(x) = x_0 + x_1 x + \ldots + x_n x^n$$
$$q(x) = y_0 + y_1 x + \ldots + y_m x^m,$$

if $m \geq n$, we write $x_j = 0 \ \forall \ n+1 \leq j \leq m$ and have

$$p(x) + q(x) = (x_0 + x_1 x + \ldots + x_m x^m)$$
$$+ (y_0 + y_1 x + \cdots + y_n x^n)$$
$$:= (x_0 + y_0) + (x_1 + y_1)x + \ldots + (x_m + y_m)x^m.$$

3.6 Abstractions

Similarly,
$$\begin{aligned} p(x)q(x) &= (x_0 + x_1 x + \cdots + x_n x^n)(y_0 + y_1 x + \cdots + y_m x^m) \\ &= x_0(y_0 + y_1 x + \cdots + y_m x^m) \\ &\quad + x_1 x(y_0 + y_1 x + \cdots + y_m x^m) \\ &\quad + \ldots + x_n x^n(y_0 + y_x + \cdots + y_m x^m) \\ &:= x_0 y_0 + (x_0 y_1 + x_1 y_0)x + \cdots \\ &\quad + (x_0 y_k + x_1 y_{k-1} + \cdots + x_k y_0)x^k + \cdots + x_n y_m x^{n+m}. \end{aligned}$$

A routine verification will tell us that $\mathcal{P}[X]$ is a ring under the addition and multiplication as defined above and it is called the **ring of polynomials** over the ring X. Let $p(x) \in \mathcal{P}[X], p(x) = x_0 + x_1 x + \cdots + x_n x^n$. We say $p(x)$ has a root $z \in X$ if $p(z) := x_0 + x_1 z + \cdots + x_n z^n = 0 \in X$.

3.6.3 Exercise. Let $(X, +, \cdot)$ be a ring with identity 1. Show that $X = \{0\}$ iff $1 = 0$, 0 being the identity for $+$.

We discuss next some rings with additional properties.

3.6.4 Definition. Let $(X, +, \cdot)$ be a commutative ring with identity 1. We say X is an **integral domain** if it has the following property:

if $x, y \in X$ and $xy = 0$ then either $x = 0$ or $y = 0$; 0 being the identity for $+$.

Whenever the binary operations $+$ and $\cdot$ are understood from the context, we denote the integral domain $(X, +, \cdot)$ by X only.

3.6.5 Exercise. Let $(X, +, \cdot)$ be a commutative ring with identity. Show that X is an integral domain iff for $x, y, z \in X$ if $x \neq 0$ and $xy = xz, y = z$. This is called the **cancellation law** for integral domains.

3.6.6 Examples.

(i) The ring of integers with usual addition and multiplication is an integral domain.

(ii) Consider $\mathbb{Z}_n$, the ring of integers modulo n, as defined in example 3.6.2 (v). Since $\mathbb{Z}_n$ is a commutative ring with identity, it will be an integral domain iff $[x][y] = [0] \Rightarrow$ either $[x] = [0]$ or $[y] = [0]$. Suppose $\mathbb{Z}_n$ in an integral domain and if possible $n = mk$ where m, k are positive integers smaller than n. Then $[0] = [n] = [mk] = [m][k]$, but $[m] \neq [0]$ and $[k] \neq [0]$. Hence if $\mathbb{Z}_n$ is an integral domain n must be a prime. Conversely, if n is a prime we show that $\mathbb{Z}_n$ is an integral domain. Let $[x] \neq [0]$ and $[y] \neq [0]$. We claim that $[x][y] \neq [0]$. Since $[x] \neq [0]$ and $[y] \neq [0]$, x is not divisible by n and y is not divisible by n. Since n is a prime, it follows from corollary 3.4.15 that xy is not divisible by n. Hence $[xy] \neq 0$. Thus $\mathbb{Z}_n$ is an integral domain iff n is a prime.

(iii) Consider $\mathbb{Z}_e$, the ring of even integers under usual addition and multiplication of integers. Then $\mathbb{Z}_e$ is not an integral domain for it does not have identity. Note that $\mathbb{Z}_e$ has the property that $xy = 0$ iff either $x = 0$ or $y = 0$.

(iv) For $a, b, c, d \in \mathbb{Z}$, the 2×2 array $\begin{bmatrix} a & b \\ c & d \end{bmatrix}$ is called a 2×2 matrix with integer entries. Let $M(2 \times 2; \mathbb{Z})$ denote all such matrices. For $\begin{bmatrix} a_1 & b_1 \\ c_1 & d_1 \end{bmatrix}$ and $\begin{bmatrix} a_2 & b_2 \\ c_2 & d_2 \end{bmatrix} \in M(2 \times 2, \mathbb{Z})$ we say $\begin{bmatrix} a_1 & b_1 \\ c_1 & d_1 \end{bmatrix} = \begin{bmatrix} a_2 & b_2 \\ c_2 & d_2 \end{bmatrix}$ iff $a_1 = a_2, b_1 = b_2, c_1 = c_2$ and $d_1 = d_2$. Define addition and multiplication on $M(2 \times 2; \mathbb{Z})$ as follows: for $\begin{bmatrix} a_1 & b_1 \\ c_1 & d_1 \end{bmatrix}, \begin{bmatrix} a_2 & b_2 \\ c_2 & d_2 \end{bmatrix} \in M(2 \times 2; \mathbb{Z})$

$$\begin{bmatrix} a_1 & b_1 \\ c_1 & d_1 \end{bmatrix} + \begin{bmatrix} a_2 & b_2 \\ c_2 & d_2 \end{bmatrix} := \begin{bmatrix} a_1 + a_2 & b_1 + b_2 \\ c_1 + c_2 & d_1 + d_2 \end{bmatrix}$$

and

$$\begin{bmatrix} a_1 & b_1 \\ c_1 & d_1 \end{bmatrix} \begin{bmatrix} a_2 & b_2 \\ c_2 & d_2 \end{bmatrix} := \begin{bmatrix} a_1 a_2 + b_1 c_2 & a_1 b_2 + b_1 d_2 \\ c_1 a_2 + d_1 c_2 & c_1 b_2 + d_1 d_2 \end{bmatrix}.$$

3.6 Abstractions

It is easy to verify that $M(2 \times 2; \mathbb{Z})$ becomes a commutative ring with identity under the above binary operations of addition and multiplication. The element $\begin{bmatrix} 0 & 0 \\ 0 & 0 \end{bmatrix}$ is the identity for addition and $\begin{bmatrix} 1 & 0 \\ 0 & 1 \end{bmatrix}$ is the identity for multiplication. However, $M(2 \times 2; \mathbb{Z})$ is not an integral domain. Consider $\begin{bmatrix} 1 & 1 \\ 0 & 1 \end{bmatrix}, \begin{bmatrix} 1 & 0 \\ -1 & 0 \end{bmatrix} \in M(2 \times 2; \mathbb{Z})$. Then both are non-zero elements of $M(2 \times 2; \mathbb{Z})$ but

$$\begin{bmatrix} 1 & 1 \\ 0 & 0 \end{bmatrix} \begin{bmatrix} 1 & 0 \\ -1 & 0 \end{bmatrix} = \begin{bmatrix} 0 & 0 \\ 0 & 0 \end{bmatrix}.$$

3.6.7 Exercise. Show that in an integral domain X for $y \in X, y \neq 0$, the equation $xy = 1$ need not have a solution.

3.6.8 Definition. An integral domain $(X, +, \cdot)$ is called a **field** if $1 \neq 0$ and $\forall y \in X, y \neq 0$, the equation $xy = 1$ has a solution in X, i.e., there exists $x \in X$ such that $xy = 1$.

3.6.9 Remark. The condition $1 \neq 0$ ensures that a field always has at least two elements.

3.6.10 Exercise. Show that a commutative ring with identity $(X, +, \cdot)$ is a field iff $X \setminus \{0\}$ is a commutative group under the binary operation of multiplication.

3.6.11 Examples.

(i) As shown in example 3.6.6 (ii), $\mathbb{Z}_n$ is an integral domain if n is a prime. In fact, $\mathbb{Z}_n$ is a field if n is a prime. For any $[x] \in \mathbb{Z}_n$, consider $[j][x] = [jx], j = 0, 1, \ldots, n-1$. These are n distinct elements of $\mathbb{Z}_n$, for if $[jx] = [kx]$, since $[x] \neq 0$ and $\mathbb{Z}_n$ is an integral domain, we have $[j] = [k]$. Since $[1] \in \mathbb{Z}_n = \{[jx] | j = 0, 1, \ldots, n-1\}$ we have $[jx] = [1]$, for some j, i.e., $\mathbb{Z}_n$ is a field for n prime. (In fact, the above argument shows that any finite integral domain is a field.)

(ii) Consider the set $X := \{0, 1, a, b\}$ with binary operations $\oplus$ and $\odot$ defined on it by the tables 1 and 2 given below. In each table the jth entry of the ith row gives the values of $i \oplus j$ and $i \odot j$, respectively.

$\oplus$	0	1	a	b
0	0	1	a	b
1	1	0	b	a
a	a	b	0	1
b	b	a	1	0

Table 1

$\odot$	0	1	a	b
0	0	0	0	0
1	0	1	a	b
a	0	a	b	1
b	0	b	1	a

Table 2

Figure 3.1 : Field of four elements.

It is easy to check that $(X, \oplus, \odot)$ is a field of four elements.

3.6.12 Note. It can be shown that given any prime p, there exists fields having p^r elements for any $r \geq 1$, see [Art].

We shall see more examples of fields in the chapters to follow. We list below some algebraic properties of a field which are easy to verify.

3.6.13 Theorem. *Let $(X, +, \cdot)$ be a field with 0_X being the identity for addition and 1_X being the identity for multiplication. For $x \in X$, let $-x$ denote the additive inverse and x^{-1} denote the multiplicative inverse of $x \neq 0_X$. Then the following hold for $x, y, z \in X$:*

(i)   *The elements 0_X and 1_X are unique.*

(ii) *The additive inverse and the multiplicative inverse of each element, whenever they exist, are unique.*

(iii) *X has at least two elements and $1_X \neq 0_X$.*

3.6 Abstractions

(iv) If $x + y = x + z, y = z$.

(v)   If $xy = xz$ and $x \neq 0_X, y = z$.

(vi) $x 0_X = 0_X$.

(vii) $(-x)y = x(-y) = -(xy)$.

(viii) $(-x)(-y) = xy$.

Proof. Exercise. ∎

3.6.14 Exercise. Show that in a field $(X, +, .)$ the following equations have unique solutions for arbitrary $x_0, y_0 \in X$:

(i) $x + x_0 = y_0$

(ii) $xx_0 = y_0$.

The existence of order on the integers motivates our next definition.

3.6.15 Definition. Let $(X, +, \cdot)$ be an integral domain. We say X is an **ordered integral domain** if there exists a relation $<$ on X such that for $x, y, z \in X$, the following hold :

(i) The relation $<$ is irreflexive, i.e., $x < x$ is not true for any $x \in X$.

(ii) The relation $<$ is transitive, i.e., $x < y$ and $y < z$ implies $x < z$.

(iii) For every $x, y, z \in X$, one of the following holds : $x < y$, $y < x$, $x = y$.

(iv) If $x > 0$ and $y > 0$, then $x + y > 0$, where 0 is the additive identity on X.

(v) If $x > 0$ and $y > 0$, then $xy > 0$.

We denote the ordered integral domain by $(X, +, \cdot, <)$ or just by X when the binary operation $+, \cdot$ and the relation $<$ are clear from the context.

3.6.16 Examples.

(i) The set of integers $\mathbb{Z}$ with usual addition, multiplication and order form an ordered integral domain.

(ii) Consider the particular case of example 3.6.2(vi) for $X = \mathbb{Z}$, i.e., the ring $\mathcal{P}[\mathbb{Z}]$ of polynomials with integer coefficients. We first show that it is an integral domain. Let $p(x) = a_0 + a_1 x + \cdots + a_m x^m$ and $q(x) = b_0 + b_1 x_1 + \cdots b_n x^n$, where each $a_i, b_i \in \mathbb{Z}$. Then

$$p(x)q(x) = (a_0 b_0) + (a_0 b_1 + a_1 b_0)x + (a_0 b_1 + a_1 b_1 + a_2 b_0)x^2 \\ + \cdots + a_m b_n x^{n+m}.$$

If $p(x) \neq 0$ and $q(x) \neq 0$, say $a_m \neq 0$ and $b_n \neq 0$, then clearly $p(x)q(x) \neq 0$. Hence $\mathcal{P}[\mathbb{Z}]$ is an integral domain. We define a relation $<$ on $\mathcal{P}[\mathbb{Z}]$ as follows: for $p(x), q(x) \in \mathcal{P}[\mathbb{Z}]$, we say $p(x) < q(x)$ if the leading coefficient of $q(x) - p(x)$ is positive. It is easy to check that $\mathcal{P}[\mathbb{Z}]$ becomes an ordered integral domain with this relation.

(iii) Let $(X, +, \cdot, <)$ be an ordered integral domain. Let $\mathcal{L}(X)$ denote the set of all formal expressions of the form:

$$a_{-n} x^{-n} + a_{-n+1} x^{-n+1} + \cdots + a_{-1} x^{-1} + a_0 + a_1 x + \cdots, \quad (3.7)$$

where each $a_i \in X$ and n is a natural number. Such an expression is called a **formal Laurent series** over X. The expression as given in (3.7) can be written formally as

$$\sum_{m=-\infty}^{+\infty} a_m x^m, \quad a_m = 0 \text{ for every } m < -n.$$

Let n be the smallest integer such that $a_n \neq 0$. This a_n is called the **leading coefficient** of the Laurent series $\sum_{n=-\infty}^{+\infty} a_n x^n$. We define

3.6 Abstractions

binary operations $\oplus$ and $\odot$ on $\mathcal{L}(X)$ as follows:

$$\left(\sum_{n=-\infty}^{+\infty} a_n x^n\right) \oplus \left(\sum_{n=-\infty}^{+\infty} b_n x^n\right) := \sum_{n=-\infty}^{+\infty} (a_n + b_n) x^n$$

and

$$\left(\sum_{n=-\infty}^{+\infty} a_n x^n\right) \odot \left(\sum_{n=-\infty}^{+\infty} b_n x^n\right) := \sum_{n=-\infty}^{+\infty} c_n x^n,$$

where $\forall n \in \mathbb{Z}$, $c_n := \sum_{i+j=n} a_i b_j$, the summation being over all those indices i and j such that $i + j = n$. It is easy to check that $(\mathcal{L}(X), \oplus, \odot)$ is an integral domain with the identity element for $\oplus$ being $O_{\mathcal{L}} := \sum_{n=-\infty}^{+\infty} a_n x^n$, $a_n = 0$ $\forall n$ and the identity for $\odot$ being $I_{\mathcal{L}} := \sum_{n=-\infty}^{+\infty} a_n x^n$, where $a_0 = 1$ and $a_n = 0$ $\forall n \in \mathbb{Z}, n \neq 0$. Finally, if $p(x), q(x) \in \mathcal{L}(X)$ with

$$p(x) = a_{-n} x^{-n} + \cdots + a_0 + a_1 x + \ldots, a_{-n} \neq 0$$

and

$$q(x) = a_{-m} x^{-m} + \cdots + b_0 + b_1 x + \cdots, b_{-m} \neq 0.$$

Then

$$p(x) \odot q(x) = a_{-n} b_{-m} x^{-(n+m)} + \cdots + a_0 b_0 + \ldots \neq 0.$$

Thus $p(x) \neq 0, q(x) \neq 0$ imply $p(x) \odot q(x) \neq 0$. Hence $(\mathcal{L}(X), \oplus, \odot)$ becomes an integral domain. Next for $p(x), q(x) \in \mathcal{L}(X)$, we say $p(x)$ is **positive** and write it as $p(x) > 0$ if the leading coefficient of $p(x)$ is positive. For $p(x), q(x) \in \mathcal{L}(X)$, we say $p(x) > q(x)$ if $(p(x) - q(x)) > 0$. For example if $p(x) = x^2, q(x) = x$ and $r(x) = x^{-1}$, then $p(x) < q(x) < r(x)$. In general,

$$0 < \cdots < x^2 < x < 1_{\mathcal{L}} < \ldots n_{\mathcal{L}} < \cdots < x^{-1} < x^{-2} < \cdots,$$

where $n_{\mathcal{L}} := \sum_{m=-\infty}^{+\infty} a_m x^m$ with $a_0 = n$ and $a_m = 0$ $\forall m \neq 0$. We first check that $<$ is a linear order on $\mathcal{L}(X)$. Let

$$\left. \begin{array}{rl} p(x) &= a_{-n} x^{-n} + \cdots + a_{-1} x^{-1} + a_0 + a_1 x + \ldots, \\ \text{and} & \\ q(x) &= b_{-m} x^{-m} + \cdots + b_{-1} x^{-1} + b_0 + b_1 x + \ldots, \end{array} \right\} \quad (3.8)$$

where $a_{-n} \neq 0, b_{-m} \neq 0$. Let $p(x) \neq q(x)$. Suppose $n > m$. Then

$$p(x) - q(x) = a_{-n}x^{-n} + \cdots + a_{-m-1}x^{-m-1} + (a_{-m} - b_{-m})x^{-m} + \cdots.$$

Since $a_{-n} \neq 0$ and X is an ordered integral domain, either $a_{-n} > 0$ or $a_{-n} < 0$. Thus either $p(x) - q(x) > 0$ or $q(x) - p(x) > 0$. Similarly if $n < m$, it can be shown that either $p(x) > q(x)$ or $q(x) > p(x)$. In case $n = m$, as $p(x) \neq q(x)$ there exists some k such that $a_j = b_j \; \forall \, j < k$ and $a_k \neq b_k$. Then $p(x) - q(x)$ will have leading coefficient $a_k - b_k$ and hence $p(x) - q(x) > 0$ if $a_k > b_k$ and $p(x) - q(x) < 0$ if $a_k < b_k$. Thus $<$ is a linear order on $\mathcal{L}(X)$. Finally, let $p(x), q(x) \in \mathcal{L}(\mathbb{R})$ be such that $p(x) > 0$ and $q(x) > 0$. Let $p(x), q(x)$ be as in (3.8) with $a_{-n} > 0$ and $b_{-m} > 0$. Then the leading coefficient of $p(x) + q(x)$ will either be a_{-n} or b_{-m} or $a_{-n} + b_{-m}$ depending upon the fact that $n > m$, or $n = m$, or $n < m$, respectively. In each case, $p(x) + q(x) > 0$. Similarly, since $p(x)q(x)$ has leading coefficient $a_{-n}b_{-m} > 0$, we have $p(x)q(x) > 0$. Hence $(\mathcal{L}(X), \oplus, \odot, <)$ is an ordered integral domain.

3.6.17 Exercise. Let $(X, +, \cdot)$ be a field. Let $\mathcal{L}(X), \oplus$ and $\odot$ be as in example 3.6.16(iii). Show that $(\mathcal{L}(X), \oplus, \odot)$ is a field.

We deduce some general properties of ordered integral domains which will be used later.

3.6.18 Proposition. *Let $(X, +, \cdot, <)$ be an ordered integral domain. Then the following hold:*

(i) *For every $x, y \in X$ one and only one of the following alternatives hold: $x < y$, $x = y$, $y < x$.*

(ii) *For every $x > 0$, $x^2 > 0$; 0 being the additive identity of X.*

(iii) *If $1 \in X$ is the multiplicative identity of X, $-1 < 0 < 1$.*

Proof. (i) We know at least one of the three given alternatives hold. Suppose more than one of the alternatives hold, say $x < y$

3.6 Abstractions

and $y < x$. Then $x < x$, contradicting the irreflexive property. Similarly, no other two alternatives can hold simultaneously.

(ii) For $x > 0$, by the defining property of the integral domain, $x^2 := x, \; x > 0$.

(iii) By (ii), $1 = 1^2 > 0$. Now adding -1 to both sides and using property (ii) again,

$$0 = 1 - 1 < 0 - 1 = -1. \;\blacksquare$$

Other arithmetic rules can be derived similarly, as described in the next exercise.

3.6.19 Exercise. For x, y, z, u in an ordered integral domain $(X, +, \cdot, <)$, prove the following:

(i) If $x > y$ and $z > u$, then $x + z > y + u$.

(ii) If $z > 0$ and $xz < yz$, then $x < y$.

(iii) $x < y$ iff $(y - x) > 0$.

(iv) If $x < 0$ and $y < 0$, then $x + y < 0$.

(v) If $x < 0$ and $y < 0$, then $xy > 0$.

Next we show that every ordered integral domain includes a 'copy' of the natural numbers. For the rest of the section, all our discussions will be with respect to a fixed ordered integral domain $(X, +, \cdot, <)$ with multiplicative identity element $1_X \neq 0_X$, the additive identity of X. For any $x \in X$ and $n \in \mathbb{N}$, we define, nx inductively as follows:

$$nx := \begin{cases} x & \text{if } n = 1, \\ (n-1)x + x & \text{if } n \geq 1. \end{cases}$$

3.6.20 Exercise. Let $(X, +, \cdot, <)$ be an ordered integral domain. For $x \in X$ and $n, k \in \mathbb{N}$, show that

(i) $(n+k)x = nx + kx$.

(ii) $(nk)x = (nx)(k1_X) = (n1_X)(kx)$.

(iii) $nx < kx$ if $n < k$ and $x < 0$.

3.6.21 Exercise. Let $(X, +, \cdot)$ be an integral domain, and k be the smallest $n \in \mathbb{N}$ such that $n1_X = 0$. If $1_X \neq 0_X$, show that n is a prime.

3.6.22 Theorem. *Let $\mathbb{N}_X := \{n1_X | n \in \mathbb{N}\}$. Then $\mathbb{N}_X$ is an infinite subset of X.*

Proof. Let $A := \{n \in \mathbb{N} | 0_X < n1_X\}$. By proposition 3.6.18(iii), $1 \in A$. Suppose $n \in A$. Then $(n+1)1_X = n1_X + 1_X$. Since $n1_X > 0_X$ and $1_X > 0_X$, by the definition of ordered integral domain $(n+1)1_X > 0$. Thus $(n+1) \in A$. By induction $A = \mathbb{N}$, i.e., $n1_X > 0 \; \forall \, n \in \mathbb{N}$. Next for $n, k \in \mathbb{N}$, if $n \neq k$, either $n < k$ or $n > k$. Suppose $n < k$. Then $k = n + r, r > 0$. Thus if $n1_X = k1_X$, then $n1_X = (n+r)1_X = n1_X + r1_X$. Hence $r1_X = 0$, which is not possible. Hence $n1_X \neq k1_X$ if $n < k$. Similarly, $n1_X \neq k1_X$ for $k < n$. Thus $n \longmapsto n1_X$ gives a one-one map from $\mathbb{N}$ onto $\mathbb{N}_X$. Hence $\mathbb{N}_X$ is infinite. ∎

3.6.23 Exercise. A subset A of an ordered integral domain X is said to be **inductive** if $1_X \in A$ and $\forall \, x \in A, \; x + 1_X \in A$. Show that $\mathbb{N}_X$ is the smallest inductive subset of X.

3.6.24 Theorem. *Let $S : \mathbb{N}_X \to \mathbb{N}_X$ be defined by $S(x) := x + 1_X, x \in \mathbb{N}_X$. Then $(\mathbb{N}_X, 1_X, S)$ is a Peano system and hence is isomorphic of $\mathbb{N}$. Elements of $\mathbb{N}_X$ are called **natural numbers** of X.*

Proof. Clearly, the map $S : \mathbb{N}_X \to \mathbb{N}_X$ is a one-one map which is not onto. Since $\mathbb{N}_X$ is the smallest inductive subset of X, for any subset A of $\mathbb{N}_X$, if $x \in A$ implies that $S(x) = x + 1_X \in A$, A will be an inductive subset of X and hence $A = \mathbb{N}_X$. This proves that $(\mathbb{N}_X, 1_X, S)$ is a Peano system. By the uniqueness theorem 2.2.4, $\mathbb{N}_X$ is isomorphic to $\mathbb{N}$. ∎

3.6.25 Note. In theorem 3.6.22 we showed that for an ordered integral domain X, the set $\mathbb{N}_X := \{n1_X | n \in \mathbb{N}\}$ of natural numbers of X is an infinite set. This need not be true if X is just an integral domain. For example, for a prime number n, consider the field $\mathbb{Z}_n$ of example 3.6.2(v). The set of natural numbers $\mathbb{N}_{\mathbb{Z}_n}$ in $\mathbb{Z}_n$ has only n-elements. In general, if for an integral domain $X, n1_X = 0$ for some $n \in \mathbb{N}$, the smallest positive integer n such that $n1_X = 0$ is called the **characteristic** of X. If $n1_X \neq 0$ for every $n \in \mathbb{N}$, one says that X has **characteristic zero**. Clearly, if an integral domain X has characteristic 1 then $X = \{0\}$. If X has characteristic $n, 1 < n < \infty$, then n is a prime (see exercise 3.6.21). If X is an integral domain with characteristic zero, clearly $n1_X \neq m1_X$ for $n \neq m$ and hence the map $n \longmapsto n1_X$ is a one-one map from $\mathbb{N}$ into $\mathbb{N}_X$. Hence $\mathbb{N}_X$ is infinite. Conversely, suppose for an integral domain X, $\mathbb{N}_X$ is infinite (this happens for example when X is an ordered integral domain, see theorem 3.6.22). We claim that X must have characteristic zero. For if X has characteristic p, it is easy to check that $\{n1_X | n \in \mathbb{N}\}$ has exactly p elements. Thus, an integral domain X has characteristic zero iff the set $\mathbb{N}_X$ of its natural numbers is an infinite set.

3.6.26 Definition. Let $\mathbb{Z}_X := \{-x | x \in \mathbb{N}_X\} \cup \{0_X\} \cup \mathbb{N}_X$. Elements of $\mathbb{Z}_X$ are called **integers** of X.

3.6.27 Exercise. For $n \in \mathbb{Z}$ and $x \in X$, let

$$m1_X := \begin{cases} m1_X & \text{if } m \in \mathbb{N}, \\ 0_X & \text{if } m = 0, \\ -((-m)1_X) & \text{if } -m \in \mathbb{N}. \end{cases}$$

Show that $\mathbb{Z}_X = \{m1_X | m \in \mathbb{Z}\}$.

3.6.28 Exercise. For $n, m \in \mathbb{Z}$ show that $(n+m)1_X = n1_X + m1_X$ and $(nm)1_X = (n1_X)(m1_X)$.

3.6.29 Definition. Let $(X, +, \cdot, <)$ be an ordered integral domain and let Y be a subset of X such that $1_X \in Y$ and the restrictions of the binary operations $+, \cdot$ and the relation $<$ to Y are all well-defined. We say Y is a **subdomain** of X if $(Y, +, \cdot, <)$ itself is an ordered integral domain.

3.6.30 Exercise. Show that Y is a subdomain of an ordered integral domain $(X, +, \cdot, <)$ iff $1_X \in Y$ and $x-y, \; xy \in Y \; \forall \; x, y \in Y$.

3.6.31 Theorem. *$\mathbb{Z}_X$ is a subdomain of $(X, +, \cdot, <)$.*

Proof. For $x = m1_X$ and $y = n1_X$, since $x + y = m1_X + n1_X = (m+n)1_X$ and $xy = (m1_X)(n1_X) = (mn)1_X$, addition and multiplication on $\mathbb{Z}_X$ is well-defined. It is easy to check that $\mathbb{Z}_X$ becomes an ordered integral domain under these binary operations and with the order of X restricted to $\mathbb{Z}_X$. ■

3.6.32 Definition.

(i) We say an integral domain (field) $(X, \oplus, \odot)$ is **isomorphic** to an integral domain (field) $(Y, +, \cdot)$ if there exists a bijective map $\phi : X \to Y$ such that $\forall \; x, y \in X, \phi(x \oplus y) = \phi(x) + \phi(y)$ and $\phi(x \odot y) = \phi(x)\phi(y)$. A map ϕ with the above properties is called an **isomorphism** of the integral domains (fields).

(ii) An ordered integral domain (field) $(X, \oplus, \odot, \prec)$ is said to be **isomorphic** to an ordered integral domain (field) $(Y, +, \cdot, <)$ if there exists a map $\phi : X \to Y$ such that $\forall \; x, y \in X, \phi(x \oplus y) = \phi(x) + \phi(y), \phi(x \odot y) = \phi(x)\phi(y)$ and $\phi(x) > \phi(y)$ whenever $x \prec y$. A map ϕ with the above properties is called an **isomorphism** of the ordered integral domains (fields).

3.6.33 Theorem. *The ordered integral domain $\mathbb{Z}_X$ is isomorphic to the integral domain $\mathbb{Z}$.*

3.6 Abstractions

Proof. Define $\phi : \mathbb{Z} \to \mathbb{Z}_X$ by $\phi(n) := n1_X, n \in \mathbb{Z}$. It is easy to see that ϕ has the required properties. ∎

3.6.34 Theorem. *Let $(X, +, \cdot, <)$ be an ordered integral domain. Then X has characteristic zero and $\mathbb{Z}_X$ is the smallest subdomain of X which is isomorphic to the ordered integral domain $\mathbb{Z}$.*

Proof. That X has characteristic zero was shown in note 3.6.25. By theorem 3.6.24, $\mathbb{Z}_X$ is isomorphic to $\mathbb{Z}$. Let Y be any other subdomain of X. Since $1_X \in Y$, we have $\mathbb{Z}_X \subseteq Y$ and hence $\mathbb{Z}_X$ is the smallest subdomain of X isomorphic to $\mathbb{Z}$. ∎

3.6.35 Note. Let $(X, +, \cdot)$ be any integral domain with characteristic zero. Then as noted in 3.6.25, the set $\mathbb{N}_X$ of natural numbers of X is an infinite set. If we define $\mathbb{Z}_X$ as in 3.6.26 and define a relation $<$ on $\mathbb{Z}_X$ by $x < y$ if $x - y \in \mathbb{Z}_x$, it is easy to check that $(\mathbb{Z}_X, +, \cdot, <)$ becomes an ordered integral domain isomorphic to $\mathbb{Z}$. Thus, $\mathbb{Z}$ can also be thought of as the smallest ordered integral domain of characteristic zero.

Chapter 4

RATIONAL NUMBERS

4.1 Historical comments

As we pointed out earlier, numbers and the process of counting were originated due to practical needs. The need to express a part of a whole must have given rise to the notion of fractions. For example, given two numbers $m, n \in \mathbb{N}$, the fraction $\frac{m}{n}$, as we understand it today, was considered by the Greeks (600 B.C.), but they did not regard it as a number or a single entity. For them it was a relationship between the numbers m and n. Thus the fraction $\frac{m}{n}$ or the ratio $m : n$ was regarded more as an ordered pair (m, n), which we still do today to construct the set of rationals from integers - see section 4.2. Around 600 B.C. the Greek mathematicians, the school led by Pythagoras, believed that numbers are the ultimate component of all real material objects. They believed in "All is number". Pythagoreans (members of the Pythagoras school of thought) had developed an elementary theory of ratios and proportions. Two ratios $m : n$ and $p : q$ were said to be proportional if m is same part or parts or multiple of n as p is of q. For example 3 : 6 is proportional to 2 : 4 for 3 is one of the two equal parts of 6 as 2 is one of the two equal parts of 4. More formally, $a : b = c : d$ if $a = mp, b = mq, c = np$ and $d = nq$ for some natural numbers

m, n, p and q. Pythagoreans, motivated by their philosophy that 'All is number', believed that any two line segments are 'commensurable', i.e., they are both multiples of a common unit. With this assumption, the theory of number ratios could be extended to length, areas and volumes. For example given two line segments a and b, their ratio was $m : n$ if $a = mp$ and $b = np$, where p is the common unit. With this definition of proportionality, ratios and proportions were applied to lengths, areas of simple figures and results were obtained. For example, given two rectangles R_1 and R_2 of equal height, it was shown that the ratios of their areas A_1, A_2 is proportional to the ratio of their bases b_1, b_2. To prove such a result, since b_1 and b_2 are commensurable (this is the assumption), we can assume $b_1 = m_1 \ell$ and $b_2 = m_2 \ell$ for some unit ℓ of measurement. Now R_1 and R_2 can be cut into m_1 and m_2, smaller rectangles, respectively to get $A_1 : A_2$ proportional to $b_1 : b_2$, both proportional to $m_1 : m_2$.

All worked very well till a member of the Pythagorean school (believed to be Hippasus of Metapontum in 500 B.C.) discovered that the ratio of the diagonal to the side of a square cannot be commensurable. The proof of this employs the technique specialized by the Greek mathematicians called "proof by contradiction" and is used even today in many proofs in mathematics.

4.1.1 Theorem. *The ratio of the diagonal and the side of a square are not commensurable.*

Proof. Consider a square $ABCD$ with diagonal AC and side AB. From AC cut off $CB_1 = AB$ and draw B_1C_1 such that $\angle CB_1C_1$ is a right angle. Join CC_1. The triangles CBC_1 and CB_1C_1 are congruent and hence $C_1B_1 = C_1B$ and $CB_1 = CB = AB$. It is easy to see that $\angle C_1AB_1 = \angle B_1C_1A$ and hence $AB_1 = C_1B_1 = BC_1$. Thus $AC_1 = AB - AB_1$. Now suppose AB and AC are commensurable, i.e., there exists a line segment AP such that $AB = m(AP)$ and $AC = n(AP)$ for some $m, n \in \mathbb{N}$. We may assume that $n > m$.

4.1 Historical comments

Then
$$AB_1 = AC - CB_1 = AC - AB = (n-m)(AP)$$
and
$$\begin{aligned} AC_1 &= AB - BC_1 \\ &= AB - AB_1 \\ &= AB - AC + CB_1 \\ &= 2AB - AC \\ &= (2m-n)(AP). \end{aligned}$$

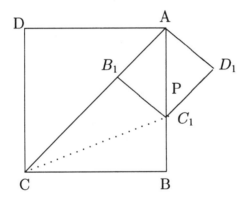

Figure 4.1 : Incommensurability of diagonal to the side.

Thus AC_1 and AB_1 are commensurable with respect to AP. In other words, we have constructed a new square $AB_1C_1D_1$ whose diagonal AC_1 is commensurable with its side AB_1, where $AC_1 : AC = (2m-n) : n$. We claim that $AC_1 : AC < 1 : 2$, i.e., $2(2m-n) < n$, i.e., $4m < 3n$, i.e., $4(AB) < 3(AC)$, i.e., $16(AB)^2 < 9(AC)^2$, i.e., $8(AC)^2 < 9(AC)^2$, as $2(AB)^2 = AC^2$. Since the last statement is true, $AC_1 : AC < 1 : 2$. Thus, we have a square $AB_1C_1D_1$ constructed from the square $ABCD$ with the property that its diagonal AC_1 and its side AB_1 are commensurable with respect to AP and $AC_1 < \frac{1}{2}AC$. By repeating the above process k times, we will get a square $AB_kC_kD_k$ such that its diagonal AC_k and side AB_k are commensurable with respect to AP and $AC_k < \frac{1}{2^k}AC$. Using exercise 1.4.16, choose k sufficiently large so that we have $n < 2^k$. But then $AC_k < \frac{1}{2^k}AC = \frac{n}{2^k}(AP) < AP$, not possible because $AC_k = r(AP)$ for some $r \in \mathbb{N}$. Hence AC and AB are not commensurable. ∎

This discovery of incommensurable lengths, i.e., lengths which are not commensurable, caused a lot of distress among the Pythagoreans. They could not 'Arithmetize' geometry. For example, they wanted to assign to every point on the line (a geometric object) a number. Given the line, any point could be chosen to represent the number 0. By fixing some unit of length, points on the line on the right side of 0 at distance 1 unit from each other could be marked and assigned the numbers 1, 2, ... Similarly, on the left of 0, points could be marked at unit distance apart and assigned numbers $-1, -2, \ldots$.

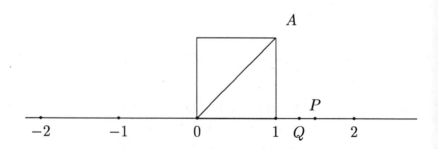

Figure 4.2 : Representation of fractions on the line.

Thus some points on the line get assigned to integers. In fact, one could assign to some more points of the line fractions – if not numbers. For example, point P between 1 and 2 and equidistant

4.1 Historical comments

from both could be assigned the ratio 3 : 2. In other words, if we made our unit smaller (half the size of the earlier unit) then we could assign to P the number 3. So, more and more points of the line could be covered by this assignment. But still, this left gaps in the line, i.e., some points on the line could not be assigned any number, however small the unit of measurement was chosen. For example, consider the point Q on the line such that OQ is equal to the length of the diagonal OA of a square with unit length, i.e., $(OQ)^2 = 2$. To Q no number could be attached. Or with the fixed unit, no ratio could be assigned to Q. In the theory of ratios and proportions, this meant that $\frac{OQ}{OA} \neq \frac{m}{n}$ for any positive integer m, n. We shall give an analytical proof of this fact as given by the Greek mathematician Euclid in section 4.4.

The discovery of such points on the line to which no commensurable ratios could be attached, i.e., the discovery of magnitudes which could be incommensurable caused a major crisis among the Pythagoreans. It shook the logical foundation of their thinking. All geometrical facts were always proved under the assumption that any two line segments are commensurable. For example, the statement that given two triangles of the same height, the ratio of their areas is equal to the ratio of their bases was proved by the Pythagoreans as follows.

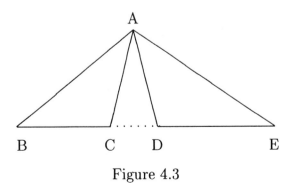

Figure 4.3

Let ABC and ADE be two triangles with common height h, as in figure 4.3. Suppose the bases BC and DE are commensurable. We can divide BC into say m units of length and DE into say n units of length for some unit length say ℓ. Then $\triangle ABC$ will consist of m number of smaller triangles and $\triangle ADE$ will consist of n numbers of smaller triangles where each smaller triangle has the same area, e.g., $\dfrac{\ell h}{2}$ for they have the same height h and equal base ℓ (here we are assuming the fact that triangles with equal base and equal height have equal areas). But then area $\triangle ABC = m(\ell h)$ and area $\triangle ADE = n(\ell h)$. Hence by the Pythagorean definition of equality of ratios, we have

$$BC : DE = m : n = \triangle ABC : \triangle ADE.$$

Thus for any two triangles of the same height, their areas are in the ratio of their bases. But the above proof is no longer valid if BC and DE are not commensurable. This is the crisis the Pythagoreans faced with the discovery of incommensurable lengths. This crisis in the theory of proportion was beautifully resolved by Eudoxus of Cindus (370 B.C.). He gave a new definition for the proportionality of ratios of geometric magnitudes. Let a and b be geometric magnitudes of the same type, for example both are lengths, or areas, or volumes. Also let c, d be another pair of geometric magnitudes of the same type, but not necessarily that of a and b. Eudoxus said "$a : b = c : d$ if for any two positive integers m and n, either of the following hold :

$$\begin{array}{lllll} & na > mb & \text{if} & nc > md \\ \text{or} & na = mb & \text{if} & nc = md \\ \text{or} & na < mb & \text{if} & nc < md." \end{array}$$

Let us first note that this definition extends the Pythagorean definition of proportionality of ratios, i.e., if ratios $a : b$ and $c : d$ are equal in the Pythagorean definition, then they are also equal in the Eudoxus definition. Note that in this definition no mention is made of commensurability. Let us see how the crisis of the Pythagorean school was resolved by this new definition of proportionality. For example let us prove that given two triangles ABC and ADE with

4.1 Historical comments

same height, as shown in figure 4.4, their areas are in the ratio of their bases. Let m, n be any two (positive) numbers. On the line CB produced mark off points $B = B_1, B_2, \ldots, B_m$ such that

$$CB_1 = B_1 B_2 = \cdots = B_{m-1} B_m.$$

Similarly, on the line DE produced mark off points $E = E_1, E_2, \ldots, E_n$ such that
$$DE = E_1 E_2 = \cdots = E_{n-1} E_n.$$

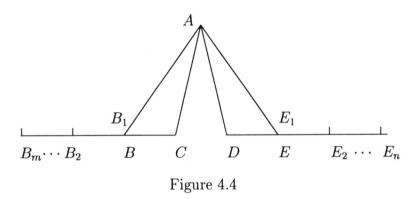

Figure 4.4

Then

$$\begin{aligned} \triangle AB_m C &= m(\triangle ABC), \\ \triangle ABE_n &= n(\triangle ADE), \\ B_m C &= m(BC), \\ \text{and} \quad DE_n &= n(DE). \end{aligned}$$

Since given two triangles with equal height the one with bigger base has bigger area, we get

$$\triangle AB_m C \leq, \quad \text{or} =, \quad \text{or} \geq \triangle ADE_n,$$

whenever $B_m C \leq$, or $=$, or $\geq DE_n$, respectively. This is equivalent to saying that $m(\triangle ABC) \leq$, or $=$, or $\geq n(\triangle ADE)$

whenever $m(BC) \leq$, or $=$, or $\geq n(DE)$, respectively. Hence $\triangle(ABC) : \triangle(ADE) = BC : DE$. This proved that triangles with the same height have areas in the ratio of their bases, without assuming the commensurability of their bases. Thus the new definition of proportionality of ratios enabled the Greeks to work with ratios in much the same way as we do nowadays with real numbers (the ratios of commensurable magnitudes corresponding to the rationals and the ratios of incommensurable magnitudes corresponding to irrationals). Most of the Mathematics developed by the Greeks during 600 B.C.– 300 B.C. was recorded by Euclid (300 B.C.) in his books called 'Elements'. Even though these texts include a theory of numbers and their arithmetic, still a number is not defined in itself – a number is a line segment, product of two numbers is a rectangle and so on. As we mentioned earlier, with the introduction of negative numbers and zero by the Hindus (200–1200 AD) the fractions (positive and negative) were developed and so was their arithmetic. These spread via the Arabs to Europe and to the rest of the world in the following centuries. In nineteenth century efforts were made to develop a theory of rational numbers (the name given to fractions) by Martin Ohm (1792–1872) in 1834, Bernard Bolzano (1781–1848), Hermann Hankel (1839–73) in 1867, Karl Theodore Weierstrass (1815–97) in his lectures in 1860s, and later by G. Peano in 1889. The idea was to construct rational numbers purely through the basic properties of addition, subtraction, multiplication and division. Identification of rational numbers with periodic decimal is due to John Wallis (1616–1703) in 1696. The algebraic definition of field and the abstraction of the construction of fields from integral domains was given by Ernst Steinitz (1871–1928) in 1910.

Eudoxus (408–355 B.C.)

Eudoxus was born in Onidus (Asia Minor) and studied with Archytas in Tarentum (Sicily). Then he moved to Athens where Plato recognized

4.1 Historical comments

his talent and became his friend. Together they travelled to Egypt. Eudoxus then settled down in Cyzicus (Asia Minor) where he taught and spent his last years. His theory of proportions – "the Crown of Greek Mathematics" – resolved the crisis created by the incommensurable ratios. He gave the powerful method of exhaustion, used extensively by the Greek mathematicians in establishing the areas and volumes of curved figures. He also studied medicine and is said to have been a practicing physician. He undertook a serious study of astronomy and was credited with the first model of the movement of the planets.

Karl Theodore Weierstrass (1815–97)

Weierstrass was born in Ostanfeld (Germany). He entered University of Bonn in 1834 to study law but left it after four years. He then turned to mathematics but did not complete his doctoral work. In 1841 he secured a state license to become a high school teacher. Till 1854 he taught school children subjects such as writing and gymnastics. Based on the results he had published during this period, he obtained an honorary degree from the University of Königsberg and in 1856

was able to secure a position in the Industrial Institute at Berlin to teach technical matter. In the same year, he became a lecturer and in 1864 a professor at the University of Berlin. He remained in this post till his death. After Cauchy and Riemann, his efforts brought rigor to analysis. He also made discoveries in algebra, algebraic geometry, calculus of variations, and the theory of complex functions. He did not publish many of his results, most became known to the mathematical world through his lectures at the University of Berlin.

4.2 Construction and uniqueness of rational numbers

Recall that the need to find solutions of linear equations of the type $x + n = m$ for $n, m \in \mathbb{N}$ led to the extension of natural numbers to the integers. To express the difference between two natural numbers m and n, we considered the ordered pair (m, n) and defined the integers $\mathbb{Z}$. Now in order to solve linear equations of the type $nx + m = 0$, for $n, m \in \mathbb{Z}$, we want to consider fractions $\frac{m}{n}$. We represent formally the fraction $\frac{m}{n}$ by (m, n), the ordered

4.2 Construction and uniqueness

pair of integers with $n \neq 0$, and extend the arithmetic of $\mathbb{Z}$ to them.

We start with the set of integers $\mathbb{Z}$ with the usual binary operations of addition and multiplication. In order for a fraction m/n for $m, n \in \mathbb{Z}, n \neq 0$ to be meaningful we consider the set $\mathbb{Z} \times (\mathbb{Z} \setminus \{0\}) := \{(n, m) | n, m \in \mathbb{Z}, m \neq 0\}$. We say two elements $(n, m), (p, q) \in \mathbb{Z} \times (\mathbb{Z} \setminus \{0\})$ are related to each other and write it as $(n, m) \sim (p, q)$ iff $nq = mp$ (here we are guided by the property that for fraction $n/m = p/q$ iff $nq = mp$). It is easy to check that $\sim$ is an equivalence relation on the set $\mathbb{Z} \times (\mathbb{Z} \setminus \{0\})$ and hence will partition it into equivalence classes. We denote the set of equivalence classes by $\mathbb{Q}$. The equivalence class containing the element (n, m) is denoted by $[n, m]$. Elements of $\mathbb{Q}$ are called the **rational numbers**.

4.2.1 Proposition. *For $(n, m), (p, q) \in \mathbb{Q}$, the following hold:*

(i) $(p, q) \sim (kp, kq) \; \forall \; k \in \mathbb{Z}, k \neq 0$.

(ii) *If $(n, m) \sim (p, q)$ and n, m are coprime, then $p = nk$ and $q = mk$ for some $k \in \mathbb{Z}, k \neq 0$.*

(iii) $(n, m) \sim (0, 1)$ *iff* $n = 0$.

(iv) $(n, m) \sim (1, 1)$ *iff* $n = m$.

Proof. (i) Suppose $n = kp$ and $m = kq$ for some $k \in \mathbb{Z}$. Then obviously $nq = kpq = mp$, i.e., $(n, m) \sim (p, q)$.

(ii) Let $(n, m) \sim (p, q)$. If $n = 0$ then $0 = nq = mp$ and hence $p = 0$, for $m \neq 0$. Thus $n = 0 = p$. Similarly if $p = 0$, again $n = 0 = p$. So suppose $n \neq 0, p \neq 0$. Since n and m are coprime and $(n, m) \sim (p, q)$ implies that $nq = mp$, hence n divides p, i.e., $p = nk$ for some $k \in \mathbb{Z}, k \neq 0$. Thus $nq = mnk$ and hence $q = mk$. This proves (ii). Proofs of (iii) and (iv) are left as exercises. ∎

Motivated by the familiar rule of addition of fractions:
$$\frac{n}{m} + \frac{p}{q} = \frac{nq + mp}{mq},$$
we have the following definition.

4.2.2 Definition (Addition on $\mathbb{Q}$). For $x, y \in \mathbb{Q}$, if $x = [n, m]$ and $y = [p, q]$, where $n, m, p, q \in \mathbb{Z}$ and $m \neq 0, q \neq 0$, define
$$x \oplus y := [nq + mp, mq].$$

4.2.3 Theorem (Properties of addition on $\mathbb{Q}$). *The operation $\oplus$ as given above is a well-defined binary operation on $\mathbb{Q}$ and $(\mathbb{Q}, \oplus)$ is a commutative group.*

Proof. Let $x = [n_1, m_1] = [n_2, m_2]$ and $y = [p_1, q_1] = [p_2, q_2]$. Then $n_1 m_2 = m_1 n_2$ and $p_1 q_2 = p_2 q_1$. Thus
$$\begin{aligned}(n_1 q_1 + m_1 p_1) m_2 q_2 &= n_1 q_1 m_2 q_2 + m_1 p_1 m_2 q_2 \\ &= m_1 n_2 q_1 q_2 + p_2 q_1 m_1 m_2 \\ &= (n_2 q_2 + m_2 p_2) m_1 q_1.\end{aligned}$$
Hence $[n_1, m_1] \oplus [p_1, q_1] = [n_2, m_2] \oplus [p_1, q_2]$, i.e., $x \oplus y$ is a well-defined binary operation. It is easy to verify that $\oplus$ is a commutative binary operation. Finally for $(n, m) \in \mathbb{Q}$, $[n, m] \oplus [0, 1] = [n, m]$ and $[n, m] \oplus [-n, m] = [0, 1]$ imply that $(\mathbb{Q}, \oplus)$ is a commutative group with $[0, 1]$ as the identity element for $\oplus$ and $[-n, m]$ as the additive inverse of $[n, m]$. ∎

4.2.4 Definition (Multiplication on $\mathbb{Q}$). For $x = [n, m]$ and $y = [p, q] \in \mathbb{Q}$, define
$$x \odot y := [np, mq].$$
Once again, this is motivated by the familiar multiplication of fractions: $\left(\dfrac{n}{m}\right)\left(\dfrac{p}{q}\right) = \dfrac{np}{mq}.$

4.2 Construction and uniqueness

4.2.5 Theorem (Properties of multiplication). *The operation $\odot$ is a well-defined binary operation and has the following properties:*

(i) $(\mathbb{Q} \setminus \{[0,1]\}, \odot)$ *is a commutative group with identity* $[1,1]$ *and for* $[k,m] \in \mathbb{Q} \setminus \{[0,1]\}$, *the inverse being* $[m,k]$.

(ii) $\odot$ *is distributive over* $\oplus$.

Proof. Let $x = [n_1, m_1] = [n_2, m_2]$ and $y = [p_1, q_1] = [p_2, q_2]$. Then $n_1 m_2 = n_2 m_1$ and $p_1 q_2 = p_2 q_1$. Thus

$$\begin{aligned}(n_1 p_1)(m_2 q_2) &= (n_1 m_2)(p_1 q_2) \\ &= (n_2 m_1)(p_2 q_1) \\ &= (n_2 p_2)(m_1 q_1).\end{aligned}$$

Hence $[n_1, m_1] \odot [p_1, q_1] = [n_2, m_2] \odot [p_2, q_2]$. Thus $x \odot y$ is well-defined. It is easy to check that $(\mathbb{Q} \setminus \{[0,1]\}, \odot)$ is a commutative group with identity $[1,1]$ and for $[n, m] \in \mathbb{Q}\{[0,1]\}$, the inverse being $[m, n]$. To prove that $\odot$ is a distributive over $\oplus$, let $x = [n, m], y = [p, q]$ and $z = [r, s] \in \mathbb{Q}$. Then

$$\begin{aligned}x \odot [y \oplus z] &= [n, m] \odot [ps + rq, qs] \\ &= [n(ps + rq), mqs].\end{aligned}$$

Also, using proposition 4.2.1, we have

$$\begin{aligned}(x \odot y) \oplus (x \odot z) &= [np, mq] \oplus [nr, ms] \\ &= [npms + mqnr, m^2 qs] \\ &= [mn(ps + rq), m(mqs)] \\ &= [n(ps + rq), mqs].\end{aligned}$$

Hence $\odot$ is distributive over $\oplus$. ∎

4.2.6 Theorem. *The triple* $(\mathbb{Q}, \oplus, \odot)$ *is a field, called the field of rational numbers*.

Proof. Follows from theorems 4.2.3 and 4.2.5. ∎

4.2.7 Corollary. *Let $\mathbb{Z}_{\mathbb{Q}} := \{[n,1] \in \mathbb{Q} | n \in \mathbb{Z}\}$. Then the following hold:*

(i) *$(\mathbb{Z}_{\mathbb{Q}}, \oplus, \odot)$ is a subdomain of $\mathbb{Q}$.*

(ii) *For every $x \in \mathbb{Q}$, there exist $y, z \in \mathbb{Z}_{\mathbb{Q}}$ such that $x = y^{-1} \odot z$, where y^{-1} is the multiplicative inverse of y in $\mathbb{Q}$.*

(iii) *$(\mathbb{Z}_{\mathbb{Q}}, \oplus, \odot)$ is isomorphic to the integral domain $(\mathbb{Z}, +, \cdot)$.*

Proof. (i) Since for $x, y \in \mathbb{Z}_{\mathbb{Q}}, x \oplus y$ and $x \odot y \in \mathbb{Z}_{\mathbb{Q}}$, it follows from exercise 3.6.30 that $(\mathbb{Z}_{\mathbb{Q}}, \oplus, \odot)$ is a subdomain of $(\mathbb{Q}, \oplus, \odot)$.

(ii) For $x = [n, m] \in \mathbb{Q}$, $x = [n, m] = [m, 1]^{-1} \odot [n, 1]$. Thus for $y := [m, 1]$ and $z := [n, 1]$, we have $x = y^{-1} \odot z$.

(iii) It is easy to check that the map $[n, 1] \longmapsto n$ is the required isomorphism. ∎

4.2.8 Note. The above theorem tells us that $\mathbb{Q}$, the field of rational numbers, constructed from the integral domain of integers $\mathbb{Z}$, has the property that it includes a copy $\mathbb{Z}_{\mathbb{Q}}$ of $\mathbb{Z}$ such that every element of $\mathbb{Q}$ is a product of an element from $\mathbb{Z}_{\mathbb{Q}}$ with an element of $\mathbb{Z}_{\mathbb{Q}}^{-1} := \{y^{-1} | y \in \mathbb{Z}_{\mathbb{Q}}\}$. We show that this is a characterizing property of the rational numbers.

4.2.9 Theorem (Uniqueness of rationals). *Let $(F, +, \cdot)$ be any field having the following properties:*

(i) *F includes a subdomain $F_{\mathbb{Z}}$ isomorphic to $(\mathbb{Z}, +, \cdot)$.*

(ii) *For every $a \in F$ there exist $b, c \in F_{\mathbb{Z}}$ such that $a = b^{-1}c$.*

Then $(F, +, \cdot)$ is isomorphic to $(\mathbb{Q}, \oplus, \odot)$.

Proof. Let $\phi : (F_{\mathbb{Z}}, +, \cdot) \to (\mathbb{Z}, +, \cdot)$ be the isomorphism of integral domains given by the hypothesis. Define a map $\psi : F \to \mathbb{Q}$ as follows: for $a \in F$, if $a = b^{-1}c$ where $b, c \in F_{\mathbb{Z}}$,

$$\psi(a) := (\phi(b))^{-1}\phi(c).$$

We first show that ψ is well-defined. Let $a = b^{-1}c = e^{-1}f$ for $b, c, e, f \in F_{\mathbb{Z}}$. Then $bf = ec \in F_{\mathbb{Z}}$ and ϕ being an isomorphism, we have

$$\phi(b)\phi(f) = \phi(bf) = \phi(ec) = \phi(e)\phi(c).$$

Thus $(\phi(b))^{-1}\phi(c) = (\phi(e))^{-1}\phi(f)$. Hence ψ is well-defined. It is easy to check that ψ has the required properties. ∎

4.2.10 Definition. A field $(F, +, \cdot)$ is called a **field of rationals** if it includes $F_{\mathbb{Z}}$ an integral domain isomorphic to the integral domain $\mathbb{Z}$ and such that every $a \in F$ can be written as $b^{-1}c$ for $b, c \in F_{\mathbb{Z}}$.

Theorem 4.2.6 tells us that there exists a field of rationals and corollary 4.2.7 tells us that any two fields of rationals are isomorphic. Thus hereafter, we shall call any such field as the **field of rationals** denoted by $\mathbb{Q}$. We shall denote addition and multiplication on $\mathbb{Q}$ by $x + y$ and xy, for $x, y \in \mathbb{Q}$. Also, for $r \in \mathbb{Q}$ and $n \in \mathbb{N}$, we write $(n+1)r := nr + r$ and $r^{n+1} := r^n r$. The field $\mathbb{Q}$, being an integral domain, includes a copy of integers (see theorem 3.6.34) which will be denoted by $\mathbb{Z}$ itself and for any element $n \in \mathbb{Z}, n^{-1}$ will be denoted by $1/n$. Thus we can write $\mathbb{Q} = \{\frac{n}{m} | n, m \in \mathbb{Z}, m \neq 0\}$ with addition and multiplication given by the familiar rules :

$$\frac{n}{m} + \frac{p}{q} = \frac{nq + mp}{mq} \quad \text{and} \quad \frac{n}{m}\frac{p}{q} = \frac{np}{mq}.$$

4.3 Order on rationals

4.3.1 Definition (Order on $\mathbb{Q}$). Let $x = \dfrac{n}{m} \in \mathbb{Q}$. We say that x is **positive** and write it as $x > 0$ if either both $n, m > 0$ or both $n, m < 0$. We say that x is **negative** and write it as $x < 0$ if $-x > 0$. For $x, y \in \mathbb{Q}$, we say that x is **less than** y and write it as $x < y$ if $(y - x) > 0$. We write $x \leq y$ if either $x < y$ or $x = y$.

4.3.2 Theorem (Properties of order). *For $x, y \in \mathbb{Q}$, $x < y$ is a well-defined relation on $\mathbb{Q}$ and has the following properties:*

(i) *If $x > 0$ and $y > 0$, then $x + y > 0$.*

(ii) *If $x > 0$ and $y > 0$, then $xy > 0$.*

(iii) *For $x, y \in \mathbb{Q}$ exactly one of the following alternatives holds: $x > y$; or $x = y$; or $x < y$.*

(iv) *The relation $\leq$ is a total order on $\mathbb{Q}$.*

(v) *For any two positive elements $x, y \in \mathbb{Q}$, there exists $n \in \mathbb{N}$ such that $nx > y$. This is called the **Archimedean** property.*

Proof. Let $x = \dfrac{n_1}{m_1} = \dfrac{n_2}{m_2}$. Then $n_1 m_2 = m_1 n_2$. Suppose both n_1, m_1 are positive integers. Then clearly $n_1 m_2 = m_1 n_2$ implies that either both m_2, n_2 are positive or both are negative integers (by exercise 3.3.4). Thus for $x \in \mathbb{Q}$, the statement $x > 0$ is well-defined and hence the relation $<$ is well-defined on $\mathbb{Q}$.

(i) Let $x = \dfrac{n}{m}$ and $y = \dfrac{p}{q}$ with $x > 0, y > 0$. Suppose n, m, p, q are all positive. Then clearly $nq + pm > 0$ and $mq > 0$ and hence $x + y = \dfrac{nq + pm}{mq} > 0$. Suppose n, m are both positive and p, q are both negative. Then nq, pm and mq are all negative and hence $\dfrac{nq + pm}{mq} > 0$. In other cases similar arguments will also prove that $x + y > 0$ whenever $x > 0, y > 0$. This proves (i).

(ii) Proof is similar to that of (i), and is left as an exercise.

(iii) Let $x \in \mathbb{Q}, x = \dfrac{n}{m}$. We show that exactly one of the alternatives: $x > 0, x = 0, x < 0$ holds. Suppose $x \neq 0$, i.e.,

4.3 Order on rationals

$\frac{n}{m} \neq 0$. Then one and only one of the following cases holds: $n > 0, m > 0$; $n < 0, m < 0$; $n > 0, m < 0$; $n < 0, m > 0$. In the first two cases, we have $x > 0$ and the other two cases give $x < 0$. Now for $x, y \in \mathbb{Q}$, either $x - y > 0$; or $x - y = 0$; or $x - y < 0$. Hence (iii) holds.

(iv) Clearly $x = x$, i.e., $\leq$ is reflexive. Let $x \leq y$ and $y \leq x$. By (iii), $x < y$ and $y \leq x$ are not possible. Similarly, $x \leq y$ and $y < x$ are not possible. Thus, the only possibility is $x = y$. Hence $\leq$ is also symmetric. Finally, if $x \leq y$ and $y \leq z$, then $y - x \geq 0$ and $z - y \geq 0$ and hence $(z - x) = (z - y) + (y - x) \geq 0$. Thus, $\leq$ is also transitive. This proves that $\leq$ is a partial order and in view of (iii) it is also total.

(v) Let $x, y \in \mathbb{Q}$ be such that $x > 0$ and $y > 0$. Choose $k, m, p, q \in \mathbb{N}$ such that $x = \frac{k}{m}$ and $y = \frac{p}{q}$. Then for $n \in \mathbb{N}$,

$$\frac{nk}{m} > \frac{p}{q} \Leftrightarrow nkq > mp.$$

Since we can choose n such that $nkq > mp$ (by theorem 3.3.37) we get $\frac{nk}{m} > \frac{p}{q}$, i.e., $nx > y$. ∎

4.3.3 Note. Theorems 4.2.6, 4.3.2 (i) and 4.3.2 (ii) say that the field $(\mathbb{Q}, +, \cdot)$ is an ordered integral domain (in fact an ordered field, see section 3.6 for definition). Theorem 4.3.2 (v) says that the ordered field $(\mathbb{Q}, +, \cdot)$ is also Archimedean (see section 4.9 for definition). Also, the order on $\mathbb{Q}$ extends the order on $\mathbb{Z}$. Thus we have extended the ordered integral domain $(\mathbb{Z}, +, \cdot, <)$ to an ordered field $(\mathbb{Q}, +, \cdot, <)$. Recall that in the integral domain $\mathbb{Z}$ given $n, m \in \mathbb{Z}$ with $n < m$, there need not exist $k \in \mathbb{Z}$ such that $n < k < m$. In fact, if $m = n + 1$, this will never be true. In $\mathbb{Q}$, the situation is completely different as shown in the next theorem.

4.3.4 Theorem. *For $x, y \in \mathbb{Q}$ with $x < y$, there exists $z \in \mathbb{Q}$ such that $x < z < y$.*

Proof. Given $x, y \in \mathbb{Q}$ with $x < y$, let $z := \frac{1}{2}(x+y)$. Then it is easy to check that $x < z < y$. ∎

4.3.5 Exercise. If $\dfrac{p}{q} < \dfrac{r}{s}$, show that $\dfrac{p}{q} < \dfrac{p+r}{q+s} < \dfrac{r}{s}$.

4.3.6 Theorem. *The set of rationals is not well-ordered.*

Proof. Since $\mathbb{Q}$ includes $\mathbb{Z}$ and $\mathbb{Z}$ is not well-ordered, $\mathbb{Q}$ is also not well-ordered. ∎

4.3.7 Exercise. Show that the set $\{r \in \mathbb{Q} \mid 0 < r < 1\}$ does not have a maximal element and hence deduce that $\mathbb{Q}$ is not well-ordered.

4.3.8 Theorem. *For every rational r, there is a unique integer n such that $n \leq r < (n+1)$.*

This unique n is called the **integral part** of n and is denoted by $[r]$. The rational $r - [r]$ is called the **fractional part** of r.

Proof. Let $A := \{n \in \mathbb{Z} \mid n \leq r\}$. We first show that A is a nonempty set. If $r \geq 0$, then clearly $0 \in A$. If $r < 0$, using the Archimedean property of $\mathbb{Q}$ (theorem 4.3.2 (v)) we can find an integer $m \in \mathbb{N}$ such that $m > -r$. Hence $-m < r$, i.e., $-m \in A$. Since A is a nonempty subset of $\mathbb{Z}$ and is bounded above, $lub(A) := n$ exists by theorem 3.3.6. Also since there exists no integer between n and $n+1$ (see exercise 3.3.4 (iii)), we have $n \leq r < n+1$. To prove the uniqueness of n, suppose $m \in \mathbb{Z}$ is such that $m \leq r < m+1$. If $n < m$, then $r < n+1 \leq m < r$, which is a contradiction. Thus $n < m$ is not true. Similarly $m < n$ is not possible. Hence $n = m$. This proves the theorem completely. ∎

4.4 Order Incompleteness

4.3.9 Exercise. Define $0! := 1, 1! := 1$ and for $n > 1, (n+1)! = (n+1)(n!)$. For every $n \geq 0$ and $0 \leq k \leq n$, let

$$\binom{n}{k} := \frac{n!}{k!(n-k)!}.$$

These are called **Binomial coefficients**. Prove the following statements:

(i) $\binom{n}{k+1} + \binom{n}{k} = \binom{n+1}{k+1} \quad \forall\, n \geq 1, 0 \leq k \leq n-1.$

(ii) For every $x, y \in \mathbb{Q}$

$$(x+y)^n = \sum_{k=0}^{n} \binom{n}{k} x^{n-k} y^k.$$

This is called the **Binomial theorem**.

(iii) For every $x > 0, n \in \mathbb{N}$

$$(1+x)^n \geq 1 + nx.$$

This is called the **Binomial inequality**.

4.3.10 Exercise. Given rationals r and s with $s > 1$, show that there exists a number $k \in \mathbb{N}$ such that $s^k > r$. (Hint: Use 4.3.2 (v) and 4.3.9 (iii).)

4.4 Order incompleteness of $\mathbb{Q}$

We saw in theorem 4.3.2 that the partial order $\leq$ on $\mathbb{Q}$ as defined in 4.3.1 is a total order on $\mathbb{Q}$. The natural question arises: Is the order on $\mathbb{Q}$ a complete order (as was the case with integers)? We show that $\leq$ is not a complete order. This is related to the discovery of incommensurable line segments as stated in section 4.1. This can be stated as the following theorem.

4.4.1 Theorem. *There exists no rational r such that $r^2 = 2$.*

Proof (Euclid). Suppose there exists a rational r such that $r^2 = 2$. We may assume that r is positive, for otherwise we can consider $-r$. Let $m, n \in \mathbb{N}$ be such that $r = \dfrac{m}{n}$. Then

$$\left(\frac{m}{n}\right)^2 = 2, \quad \text{i.e., } m^2 = 2n^2.$$

Since 2 divides $2n^2$, 2 divides m^2 and hence m^2 is even. But then, by exercise 3.4.18, m is even. Let $m = 2k, k \in \mathbb{N}$. Then

$$(2k)^2 = 2n^2, \quad \text{i.e., } 2k^2 = n^2.$$

Since 2 divides $2k^2$, hence 2 divides n^2. Thus 2 divides n, as before. Thus n and m have a common factor 2. But to start with we could have assumed that $gcd(n,m) = 1$ for if $gcd(n,m) = d$ was such that $d > 1$, we could have divided both n, m by d to get n', m' such that $gcd(n', m') = 1$ and $\dfrac{m'}{n'} = \dfrac{m}{n}$. Thus, if we start with $gcd(n,m) = 1$ and $(\dfrac{m}{n})^2 = 2$, we get 2 as a common divisor of n and m, which is a contradiction. Hence our assumption that there exists a rational r with $r^2 = 2$ is not true, i.e., the required claim holds. ∎

Here is a variation of the above proof.

Second proof. Suppose there exists $m, n \in \mathbb{N}$ such that

$$\left(\frac{m}{n}\right)^2 = 2.$$

Let us look at the prime factorizations of both m and n (as given by theorem 3.4.8). Let $m = p_1 p_2 \ldots p_k$ and $n = q_1 q_2 \ldots q_r$, (note that here we are not demanding the primes $p_1, p_2, \ldots, p_k$ to be distinct or $q_1, q_2, \ldots, q_k$ to be distinct). Then m^2 will have $2k$ primes and n^2

4.4 Order Incompleteness

will have $2r$ primes in its factorization. Since $m^2 = 2n^2$, m has $2k$ primes in its prime factorization and $2n^2$ will have $2r + 1$ number of primes in its factorization, not possible. ∎

Here is yet another way of looking at the above proof.

Third proof. Once again suppose $\exists\, m, n \in \mathbb{N}$ such that

$$m^2 = 2n^2.$$

Let us look at the prime factorization of m^2. Since 2 occurs in the prime factorization of $2n^2$, it must occur in the prime factorization of m^2. But it can occur only even number of times, for each factor will be repeated. On the other hand, 2 occurs odd number of times in the prime factorization of $2n^2$ as 2 can occur at most even number of times in the prime factorization of n. This is a contradiction. ∎

4.4.2 Question. What do you think is the difference between the three proofs ? – Look at the next set of exercises and try to prove them by following one of the proofs.

4.4.3 Exercise. Show that there is no rational r such that either of the following holds:

(i) $\quad r^2 = 3$.

(ii) $\quad r^2 = 7$. More generally, there is no rational r such that $r^2 = p$, where p is a prime.

(iii) $\quad r^2 = 6$. More generally, there is no rational r such that $r^2 = k$, where $k \in \mathbb{N}$ is such that its prime factorization has a prime p occurring odd number of times, (i.e., k is not the square of a natural number).

4.4.4 Exercise.

(i) Do you think $\exists\, m \in \mathbb{N}$ such that $(m/7)^2 = 3$?

(ii) Do you think $\exists\, m, n \in \mathbb{N}$ such that $m^3 = 2n^3$?

4.4.5 Exercise.

(i) Let $n, k \in \mathbb{N}$ be such that there is no integer m such that $m^k = n$. Show that there is no rational r such that $r^k = n$.

(ii) Consider the polynomial $p(x) := x^k + a_{k-1}x^{(k-1)} + \cdots + a_1 x + a_0$, where each a_is is an integer. Show that if $p(r) = 0$ for some rational r, then r is an integer.

4.4.6 Theorem. *The partially ordered set $(\mathbb{Q}, \leq)$ does not have the least upper bound property.*

Proof. To prove the required claim, we have to construct a nonempty subset A of $\mathbb{Q}$ which is bounded above but $lub(A)$ does not exist. Consider the set

$$A := \{r \in \mathbb{Q} \mid r > 0 \text{ and } r^2 < 2\}.$$

Clearly A is nonempty, for example $1 \in A$. Also A is bounded above by 2, for if $x \geq 2$, then $x^2 \geq 4$, i.e., $x \notin A$. We shall show that $lub(A)$ does not exist. Suppose $lub(A)$ exists, i.e., $lub(A) := \alpha \in \mathbb{Q}$. Then $\alpha \geq 0$, by theorem 4.4.1, $\alpha^2 \neq 2$. Thus (by theorem 4.3.2 (iii)) only possibilities are $\alpha^2 < 2$ or $\alpha^2 > 2$. Suppose $\alpha^2 < 2$. Then $\alpha \in A$. We try to find some $n \in \mathbb{N}$, such that $\left(\alpha + \dfrac{1}{n}\right) \in A$. Note that

$$(\alpha + \tfrac{1}{n})^2 < 2 \iff \alpha^2 + \frac{2\alpha}{n} + \frac{1}{n^2} < 2$$

$$\iff \frac{2\alpha}{n} + \frac{1}{n^2} < 2 - \alpha^2. \tag{4.1}$$

4.4 Order Incompleteness

Since $\alpha < 2$ and $\dfrac{1}{n^2} < \dfrac{1}{n}$, we have

$$\frac{2\alpha}{n} + \frac{1}{n^2} < \frac{4}{n} + \frac{1}{n} = \frac{5}{n}. \tag{4.2}$$

From (4.1) and (4,2), if $n \in \mathbb{N}$ is such that $\dfrac{5}{n} < 2 - \alpha^2$, then we will have $\left(\alpha + \dfrac{1}{n}\right)^2 < 2$, i.e., $\left(\alpha + \dfrac{1}{n}\right) \in A$. Since $2 - \alpha^2 > 0$, by the Archimedean property of rationals (theorem 4.3.2 (v)) it is possible to choose $n \in \mathbb{N}$ such that $5 < n(2 - \alpha^2)$. For such an $n \in \mathbb{N}$, $0 < \alpha < \alpha + \dfrac{1}{n}$ and $\left(\alpha + \dfrac{1}{n}\right)^2 < 2$. Thus $lub(A) = \alpha < \left(\alpha + \dfrac{1}{n}\right) \in A$, which is a contradiction. Hence $\alpha^2 < 2$ is not possible. Finally, suppose $\alpha^2 > 2$. Since $\alpha = lub(A)$, $\left(\alpha - \dfrac{1}{n}\right)$ cannot be an upper bound for A for any $n \in \mathbb{N}$. Thus there exists $r \in A$ such that $\left(\alpha - \dfrac{1}{n}\right) < r$. Thus

$$(\alpha - \frac{1}{n})^2 < r^2 < 2. \tag{4.3}$$

Also

$$\begin{aligned}(\alpha - \frac{1}{n})^2 &= \alpha^2 - \frac{2\alpha}{n} + \frac{1}{n^2} \\ &> \alpha^2 - \frac{2\alpha}{n}.\end{aligned} \tag{4.4}$$

Thus, if we can choose $n \in \mathbb{N}$ such that $\alpha^2 - \dfrac{2\alpha}{n} > 2$, from (4.3) and (4.4) we will get a contradiction: $2 > (\alpha - \dfrac{1}{n})^2 > 2$, implying that $\alpha^2 > 2$ is also not possible. Now

$$\alpha^2 - \frac{2\alpha}{n} > 2 \quad \text{iff} \quad (\alpha^2 - 2) > \frac{2\alpha}{n} \quad \text{iff} \quad n > \frac{2\alpha}{\alpha^2 - 2}.$$

Since $(\alpha^2 - 2) > 0$, using the Archimedean property of the rationals (theorem 4.3.2 (v)) we can choose $n \in \mathbb{N}$ such that $n(\alpha^2 - 2) > 2\alpha$.

Hence, if $\alpha = lub(A)$ exists then neither of the possibilities: $\alpha^2 > 2, \alpha^2 < 2, \alpha^2 = 2$ is possible, i.e., $lub(A)$ does not exist. ∎

4.4.7 Examples.

(i) Let $S = \{x \in \mathbb{Q} | 0 \le x \le 1\}$. The set S is bounded below by 0 and bounded above by 1. In fact $lub(S) = 1$, $glb(S) = 0$ and both belong to S.

(ii) Let $S_0 = \{x \in \mathbb{Q} | 0 < x < 1\}$. Then S_0 is also bounded above and below with $lub(S_0) = 0, glb(S_0) = 1$ and none of them is an element of S_0.

4.4.8 Exercise.

(i) Let $\alpha \in \mathbb{Q}$ be such that $\alpha^2 > 2$. Let $\beta = \dfrac{\alpha}{2} + \dfrac{1}{\alpha}$. Show that $\beta \in \mathbb{Q}, \beta < \alpha$ and $\beta^2 > 2$. Hence deduce that the set $B := \{\alpha \in \mathbb{Q} | \alpha > 0 \text{ and } \alpha^2 > 2\}$ does not have glb.

(ii) Let $\alpha \in \mathbb{Q}$ and $\alpha^2 < 2$. Consider $\dfrac{2}{\alpha}$ and use (i) above to show that $\exists \beta \in \mathbb{Q}$, such that $\beta > \alpha$ and $\beta^2 < 2$. Hence deduce that $A := \{x \in \mathbb{Q} | x^2 < 2\}$ has no lub.

(iii) Let $\alpha, \beta \in \mathbb{Q}$ with $0 < \alpha < \beta$. Show that the set $S := \{\alpha\delta | 0 < \delta < 1, \delta \in \mathbb{Q}\}$ is bounded above with $lub(S) = \beta$. Deduce that $\exists \delta \in \mathbb{Q}, 0 < \delta < 1$, such that $0 < \alpha < \delta\beta < \beta$.

4.5 Geometric representation of rational numbers

We saw in section 4.1 that one can associate to each natural number a point on the line and that the belief of the Pythagoreans that to every point on the line corresponds to a number, led to the discovery of incommensurable lengths. In any case, the commensurable ratios, i.e., rationals can be assigned to some points on the line.

4.5 Geometric representation

Consider a horizontal line and mark two distinct points O and A on the line, A on the right of O. Choose the segments OA as unit of length. Let O represent the integer zero and A represent the integer 1. Mark points on the line spaced at unit distance apart on the right of A and let them represent positive integers $2, 3, \ldots$, respectively. Similarly, let points on the line spaced at unit distance apart on the left of O represent the negative integers $-1, -2, \ldots$, respectively.

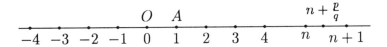

Figure 4.5 : Geometric representation of rationals.

Let $r \in \mathbb{Q}, r > 0$. Let $r = \dfrac{p}{q} + n$, where $n, p, q \in \mathbb{N}$ are unique numbers such that $0 \leq \dfrac{p}{q} < 1$. We divide the segment n to $n+1$ into q equal parts. Let the point corresponding to the right end-point of the pth part be the point on the line which represents r. Negative rationals can be represented similarly. Thus for every rational there is a point on the line and distinct rationals get represented by distinct points. As was noted by Hippasus of Metapontum in 500 B.C., there are points on the line which do not represent any rational (see section 4.1). For example whatever the unit of length be, if we take a right angled triangle with base and height of unit length, then the length of its hypotenuse does not correspond to any rational number. Thus, rationals when put on the line do not cover up the line, even though they are 'dense', i.e., between any two points representing rationals we can find a point in between them which also represents some rational. Let us call the points on the line which do not represent rationals as **irrational points**. They represent 'gaps' on the number line after the rationals have been put on the line. We can define a 'gap' in the line more precisely as follows.

4.5.1 Definition. A pair (A, B) of nonempty subsets of $\mathbb{Q}$ is called a **gap** if the following hold:

(i) $A \cup B = \mathbb{Q}$.

(ii) For every $x \in A$ and $y \in B$, $x < y$.

(iii) A has no maximum and B has no minimum.

4.5.2 Examples.

(i) For $r \in \mathbb{Q}$ fixed, consider the sets $L_r := \{s \in \mathbb{Q} | s < r\}$ and $R_r := \{s \in \mathbb{Q} | s \geq r\}$. Then clearly (L_r, R_r) is not a gap as R_r has a minimum, namely r.

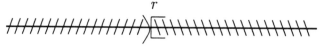

Figure 4.6 : The cut induced r.

(ii) Let $A := \{s \in \mathbb{Q} | s \leq 0\} \cup \{s \in \mathbb{Q} | s > 0 \text{ and } s^2 < 2\}$ and $B := \{s \in \mathbb{Q} | s > 0 \text{ and } s^2 > 2\}$. Then proceeding as in the proof of theorem 4.4.6, it is easy to check that (A, B) is a gap in $\mathbb{Q}$. This gap corresponds to the point x on the number line such that $x^2 = 2$, for example, the point at a distance equal to the diagonal of a square of unit length.

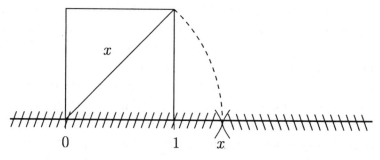

Figure 4.7 : Gap corresponding to x, $x^2 = 2$.

We shall see in the next chapter how these 'gaps' in the line can be 'filled'.

4.6 Decimal representation of rationals

4.6.1 Theorem. *Let $r \in \mathbb{Q}, r \geq 0$. Then there exist unique integers $a_0, a_1, a_2, \ldots$ such that the following hold:*

(i) $a_0 \leq r < a_0 + 1$.

(ii) $0 \leq a_i \leq 9 \ \forall \ i \geq 1$.

(iii) If $r_n := a_0 + \dfrac{a_1}{10} + \ldots + \dfrac{a_n}{10^n}$, then $r_n \leq r < r_n + \dfrac{1}{10^n}$.

Further, there exist positive integers m and j such that $a_n = a_{n+m}$ for all $n \geq j$.

The sequence $a_0, a_1, a_2, \ldots$, is called the **decimal representation** of r and we write it as $r = a_0 \cdot a_1 a_2 \ldots$.

Proof. Let $a_0 := [r]$. Consider $10(r - a_0)$ and let $a_1 := [10(r - a_0)]$. Since $0 \leq 10(r - a_0) < 10, 0 \leq a_1 \leq 9$. Equivalently, a_1 is the largest integer such that

$$a_0 + \frac{a_1}{10} \leq r < a_0 + \frac{a_1 + 1}{10}.$$

In general, if $a_1, a_2, \ldots, a_{n-1}$ with $0 \leq a_i \leq 9$ have been defined, let

$$a_n := [10^n r - (10^n a_0 + 10^{n-1} a_1 + \cdots + 10 a_{n-1})].$$

Equivalently, a_n is the largest integer such that

$$a_0 + \frac{a_1}{10} + \cdots + \frac{a_n}{10^n} \leq r < a_0 + \frac{a_1}{10} + \cdots + \frac{a_n + 1}{10^n}.$$

By induction we get natural numbers $a_0, a_1, a_2, \ldots$ such that, $0 \leq a_n \leq 9$. Let

$$r_n := a_0 + \frac{a_1}{10} + \frac{a_2}{10^2} + \cdots + \frac{a_n}{10^n}.$$

Then clearly,

$$r_n \leq r < r_n + \frac{1}{10^n}.$$

Thus, we get positive integers $a_0, a_1, a_2, \ldots$ and rationals $r_1, r_2, \ldots,$ $r_n, \ldots$ associated with r such that (i), (ii) and (iii) hold. To prove the uniqueness, suppose, if possible there exist positive integers $a'_0, a'_1, a'_2, \ldots$ such that $1 \le a'_i \le 9 \ \forall \ i \ge 1, a'_0 \le r < a'_0 + 1$ and $\forall n \ge 1$,

$$a'_0 + \frac{a'_1}{10} + \cdots + \frac{a'_n}{10^n} \le r < a'_0 + \frac{a'_n}{10} + \cdots + \frac{a'_n + 1}{10^n}. \quad (4.5)$$

Clearly $a_0 = [r] = a'_0$. Suppose $a'_i = a_i$ is true for $0 \le i \le n-1$. Since $a_n = [10^n r - (10^n a_0 + 10^{n-1} a_1 + \ldots + 10 a_{n-1})]$ and (4.5) imply

$$a'_n = [10^n r - (10^n a'_0 + 10^{n-1} a'_1 + \ldots + 10 a'_{n-1})],$$

We have $a_n = a'_n$. Hence by induction, $a_n = a'_n \ \forall \ n$.

To show that there exist positive integers m and j with the required properties, let $R_0 := r - a_0$ and $\forall n \ge 1, R_n := (10^n r - (10^n a_0 + \ldots + 10 a_{n-1})) - a_n$. Then $a_0 = [10 R_0]$ and $\forall n \ge 1, a_n = [10 R_{n-1}]$. Further $10 R_{n-1} = a_n + R_n$, i.e., a_n is the integral part of $10 R_{n-1}$ and R_n is the fractional part of $10 R_{n-1}$. Thus $0 \le R_n < 1 \ \forall \ n$. Since r is a rational, R_0 is also a rational, say $R_0 = \frac{p}{q}$, where $p, q \in \mathbb{N}$ and $0 \le p < q$. Then $R_1 = \frac{10p - a_1 q}{q} := \frac{p_1}{q}$. Similarly $R_n = \frac{p_n}{q}$ where $0 \le p_n < q \ \forall \ n \in \mathbb{N}$. Hence there exists i and j with $i > j$ such that $p_i = p_j$. For example $p_1, \ldots, p_{q+1}$ are $(q+1)$ natural numbers less than q and hence at least two of them are equal. Now if $R_i = R_j$, then $a_{i+1} = [10 R_i] = [10 R_j] = a_{j+1}$. Further $R_{i+1} = 10 R_i - a_{i+1} = 10 R_j - a_{j+1} = R_{j+1}$. Thus if $i = j + m$, we have $R_n = R_{n+m}$ for every $n \ge j$ and hence $a_n = a_{n+m}$ for all $n \ge j + 1$. ∎

4.6.2 Definition. Let $a_0 \in \mathbb{Z}$ and $a_1, a_2, \ldots$ be positive integers such that $1 \le a_i \le 9 \ \forall \ i \ge 1$.

(i) An expression of the form

4.6 Decimal representation

$$a_0 \cdot a_1 a_2 a_3 \ldots$$

is called a **decimal**.

(ii) A decimal $a_0 \cdot a_1 a_2 \ldots$ is called a **periodic decimal** if there exist positive integers m and j such that $a_n = a_{n+m} \; \forall \, n \geq j$.

4.6.3 Example. If $r \in \mathbb{Q}$ and $r < 0$, let $-r$ have the decimal representation $a_0 \cdot a_1 a_2 \ldots$ as given by theorem 4.6.1. Then $-a_0 \cdot a_1 a_2 \ldots$ is called the decimal representation of $-r$. Theorem 4.6.1 tells that to every rational we can associate a periodical decimal, namely its decimal representation. The converse is also true (this will be proved in section 6.4).

We can look at the decimal representation of a rational r from the geometric viewpoint as follows. Divide the number line into disjoint segments $I_n := \{s \in \mathbb{Q} | n \leq s < n+1\}$. Then r will belong to a unique segment, say $[a_0, a_0 + 1)$. Now divide this segment into 10 equal disjoint parts and number them in order from left to right as $0, 1, 2, \ldots, 9$. For each part segment we include the left end-point only. Now locate the segment part in which r lies (there will be one and only one), say it is a_1^{th}. Now divide that segment into 10 equal parts and again number them in order from left to right as $0, 1, 2, \ldots, 9$ and locate the segment part in which r lies, say it is a_2^{th}, and continue this process. Intuitively it is clear that in the decimal representation $a_0 \cdot a_1 a_2 \ldots$ of a rational r there does not exist any m such that $a_n = 9 \; \forall \, n \geq m$. For example $\frac{1}{2} = \cdot 500\ldots$ and not $\cdot 4999\ldots$ in our construction. We prove it in theorem 4.8.17.

4.6.4 Exercise. A rational r is called a **decimal rational** if r equal $\dfrac{p}{10^n}$ for some integer p. Show that given rationals r and $\epsilon > 0$, there exists a decimal rational z such that $r < z < r + \epsilon$.

4.7 Absolute value of rationals

4.7.1 Definition. For $r \in \mathbb{Q}$, we define the absolute value of r, denoted by $|r|$ by

$$|r| := \begin{cases} r & \text{if } r \geq 0, \\ -r & \text{if } r < 0. \end{cases}$$

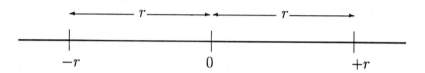

Figure 4.8 : Absolute value of r.

Geometrically, $|r|$ represents the distance between the point O representing 0 and the point P representing r on the number line. Properties of the absolute value function are given in the next theorem.

4.7.2 Theorem. *For $r, s \in \mathbb{Q}$, the following hold:*

(i) $|r| \geq 0$ and $|r| = 0$ iff $r = 0$.

(ii) $|r| = |-r|$.

(iii) $-|r| \leq r \leq |r|$.

(iv) $|r| \leq s$ iff $-s \leq r \leq s$.

(v) $|r + s| \leq |r| + |s|$, *called the* **triangle-inequality.**

Proof. The proofs of (i) and (ii) follow obviously from the definition of absolute value. To prove (iii) we consider two different cases. If $r \geq 0$, then $|r| = r$ and hence $-r \leq 0$. Thus $-|r| \leq r < |r|$. If $r \leq 0, |r| = -r$ and hence $-|r| = r \leq 0 \leq -r = |r|$. To prove (iv) suppose $|r| \leq s$. If $r \geq 0$, then $-s \leq 0 \leq r = |r| \leq s$. If

4.8 Sequences of rationals

$r \leq 0$, then $0 \leq -r = |r| \leq s$. Thus $-s \leq r \leq 0 \leq s$. Conversely, let $-s \leq r \leq s$. If $r \geq 0$, then $|r| = r \leq s$. If $r \leq 0$, then $|r| = -r \leq s$. Thus (iv) holds. To prove (v), note that $-|x| \leq x \leq |x|$ and $-|y| \leq y \leq |y|$. Adding the two we get $-(|x|+|y|) \leq x+y \leq (|x|+|y|)$ and hence by (iv), $|x+y| \leq |x|+|y|$. ∎

4.7.3 Exercise. Show that for $r, s \in \mathbb{Q}$, the following hold:

(i) $|rs| = |r||s|$.

(ii) $||r| - |s|| \leq |r - s|$.

(iii) $|r - s| \leq |r| + |s|$.

(iv) If $b \geq 0$, $|a| \geq b$ iff $a \geq b$ or $-a \leq -b$.

4.8 Sequences of rationals

4.8.1 Definition. An ordered collection $r_1, r_2, \ldots$ of rational numbers is called a **sequence** in $\mathbb{Q}$. In our axiomatic setup, we can define a sequence to be a map $f : \mathbb{N} \to \mathbb{Q}, f(n) := r_n \ \forall \ n \geq 1$, denoted by $\{r_n\}_{n \geq 1}$. The number r_n is called the nth-**term** of the sequence.

Note that a sequence $\{r_n\}_{n \geq 1}$ is not the same as the set $\{r_n | n = 1, 2, \ldots\}$. Sequences of rationals arise in many diverse problems and there is a need to analyze the behaviour of a sequence $\{r_n\}_{n \geq 1}$ for large values of n. We give some examples.

4.8.2 Examples.

(i) Consider a circle with unit radius. We would like to find its area. One way of doing this would be to inscribe a regular polygon of 2^{n+1}-sides inside the circle, compute its area, say A_n, and let the number of sides increase. This process generates a collection of real numbers $A_1, A_2, \ldots, A_n, \ldots$ as shown below:

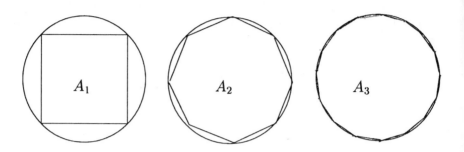

Figure 4.9 : Approximating the area of a circle.

Note that each A_n gives a better approximation for the area of the unit circle. One is interested in knowing what happens to A_n as n increases? Here we are assuming the intuitive concept of 'area' of a region in plane. In fact, it requires advance analysis to define rigorously the concept of 'area'.

(ii) Consider the following gambling game. A coin is tossed. If head appears, the gambler wins a rupee and if tail appears the gambler loses a rupee. Let S_n denote the profit of the gambler at the nth-toss. The average profit of the gambler is $X_n := \dfrac{S_1 + \cdots + S_n}{n}$. The gambler would like to know the behaviour of X_n for large n.

(iii) Consider two cyclists A and B at a distance one km apart moving towards each other at a constant speed of one km per hour. A fly shuttles between A and B at a constant speed of 2 km per hour till the two cyclists meet one another. How far away from the starting point will the fly stop shuttling? How much distance would the fly have covered? The answer to the first question is $\dfrac{1}{2}$ km away from the starting point. This is obvious to someone who knows that A and B are moving towards each other with constant speed. Let us try to work this out from the point of view of the

4.8 Sequences of rationals

fly. Let the fly start from A towards B. Let $P_1, P_2, P_3, \ldots, P_n, \ldots$ denote the consecutive positions of the fly when it turns back from A or B, as shown below:

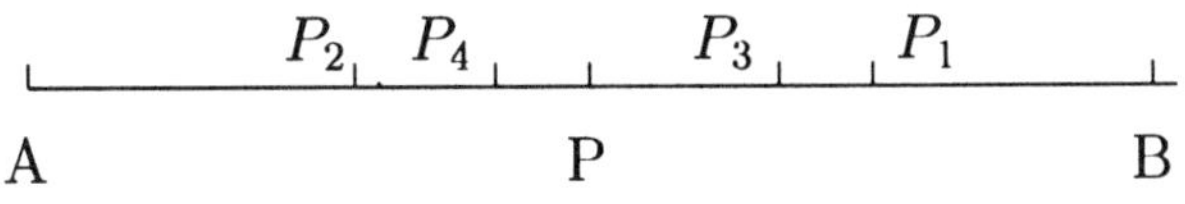

```
|_____P_2_P_4_____|___P_3____P_1___|
A                              P                    B
```

Figure 4.10 : Positions of the fly.

An easy computation shows that

$$x_1 := AP_1 = \frac{2}{3}.$$

$$x_2 := AP_2 = \frac{2}{3} - \frac{2}{9}.$$

$$\ldots \quad \ldots$$

$$x_n := AP_n = \frac{2}{3} - \frac{2}{9} + \ldots + 2(-\frac{1}{3})^{n-1}.$$

As n becomes larger and larger, the points P_n will come closer to P, the point where A and B meet. Thus the final position P is obtained eventually when n becomes larger and larger. To answer the second question, i.e., to find the total distance covered by the fly, again let us proceed inductively. Let y_n denote the distances covered by the fly upto the nth-turning back, i.e.,

$$\begin{aligned} y_n &:= AP_1 + P_1P_2 + \ldots + P_{n-1}P_n \\ &= \frac{2}{3} + \frac{2}{9} + \ldots + 2(\frac{1}{3})^{n-1}. \end{aligned}$$

Each y_n gives a better approximation to the answer (which we know is 1 because total flight time of the fly is 30 minutes, the time taken by A and B to meet, and its speed is 2 km per hour).

In each of the above examples we have a collection of numbers which are labeled as first, second, third, and so on, and we were interested in the behavior of this ordered collection for n large. For example, to calculate the area of the circle, we hope that when n is sufficiently large A_n, the area of the regular polygon of 2^{n+1} sides inscribed in the unit circle will approximate the area of the circle. How good will this approximation be? Intuitively, we should be able to approximate the area of the circle with as much accuracy as we want. To measure this degree of accuracy, the notion of distance or the absolute value comes to our help. For example, if we want to say that the area A_n approximates the area A of the unit circle by error at most $\cdot 01$, we can put it mathematically as

$$|A_n - A| \leq \cdot 01$$

In fact, if $|A_n - A| \leq \cdot 01$, since A_m for $m > n$ is a better approximation as compared to A_n, i.e., $A_n > A_m$ for $n > m$ we should have

$$|A_m - A| \leq \cdot 01 \quad \forall\, m \geq n.$$

Loosely speaking, the sequence $\{A_n\}_{n \geq 1}$ approaches A when n is pushed to its limit. This motivates our next definition.

4.8.3 Definition. Let $\{x_n\}_{n \geq 1}$ be a sequence of rational numbers. We say $\{x_n\}_{n \geq 1}$ has a **limit** $x \in \mathbb{Q}$ if given any rational number $\epsilon > 0$, there exists some $n_0 \in \mathbb{N}$ such that

$$|x_n - x| < \epsilon \quad \forall\, n \geq n_0.$$

In words, the sequence $\{x_n\}_{n \geq 1}$ will have limit x (or approximates x) if given any prescribed accuracy, namely any $\epsilon > 0$, we should be able to find at least one $n_0 \in \mathbb{N}$ such that from the n_0^{th}-term onwards all the terms of $\{x_n\}_{n \geq 1}$ are at most at a distance ϵ away from the number x. Pictorially, imagine yourself standing at the point x on the number line with arms extended on either side by distance $\epsilon > 0$. If you are able to collect all but finitely many terms of the sequence in your arms and if you are able to do this

4.8 Sequences of rationals

for every extension $\epsilon > 0$, then the sequence has limit x.

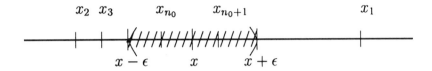

Figure 4.11 : Limit of a sequence.

Another way of viewing a sequence $\{x_n\}_{n\geq 1}$ is by the graph of the function $f : \mathbb{N} \to \mathbb{Q}$, $f(n) := x_n$, as shown in figure 4.12.

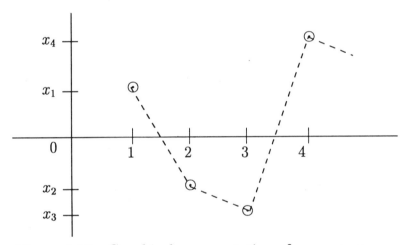

Figure 4.12 : Graphical representation of a sequence.

Saying that a sequence $\{x_n\}_{n\geq 1}$ has a limit x means that if we take a horizontal strip of any given width 2ϵ centered at x, all but finitely many of the terms of $\{x_n\}_{n\geq 1}$ should lie in this strip as shown in figure 4.13.

Whenever the sequence $\{x_n\}_{n\geq 1}$ has a limit x, we say that $\{x_n\}_{n\geq 1}$ is **convergent** to x as $n \to \infty$ (read as n goes to infinity) and we write this as

$$\lim_{n\to\infty} x_n = x \quad \text{or just} \quad x_n \to x.$$

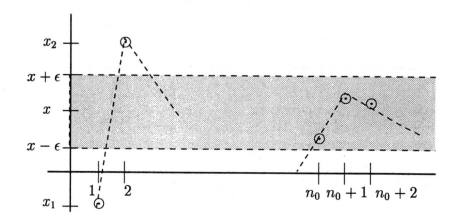

Figure 4.13 : Convergence of a sequence.

Note that ∞ is not a rational number, it is just a symbol (used to indicate that for all large values of n something happens).

4.8.4 Examples.

(i) Consider the sequence $\{x_n\}_{n\geq 1}$ where $x_n := \dfrac{1}{n}, n \in \mathbb{N}$. First few terms of the sequence are $1, 1/2, 1/3, 1/4, \ldots$. Clearly, x_n is becoming smaller and smaller as n becomes larger and larger. In fact, we can make x_n as small as we want, i.e., given a rational $\epsilon > 0$, we can choose $n_0 \in \mathbb{N}$ such that $\dfrac{1}{n_0} < \epsilon$, i.e., $n_0 > \dfrac{1}{\epsilon}$. This is possible because of the Archimedean property of rational numbers. Thus if $n \geq n_0, \dfrac{1}{n} \leq \dfrac{1}{n_0} < \epsilon$. Hence,

$$|x_n - 0| = \dfrac{1}{n} < \epsilon \;\forall\; n \geq n_0, \quad \text{i.e.,} \quad \lim_{n \to \infty} \dfrac{1}{n} = 0.$$

(ii) Consider the sequence $\{x_n\}_{n\geq 1}$, where $x_n = (-1)^n, n \in \mathbb{N}$. All the even terms of the sequence are $+1$ and all the odd terms are -1. Do you think this sequence has a limit?

4.8 Sequences of rationals

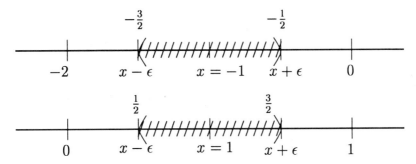

Figure 4.14 : Cases $x = -1; x = 1$.

Since $+1$ and -1 are at a distance 2 apart all x_n's cannot come closer to $+1$ or -1 say by distance 1. For a similar reason it cannot come closer to any real number. So our guess is that this sequence does not have a limit.

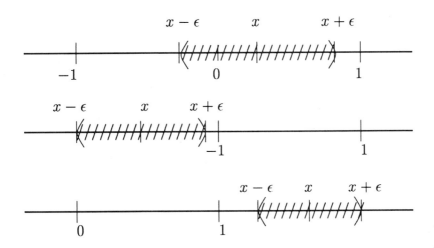

Figure 4.15 : Cases $0 < x < 1;\ x < -1;\ x > 1$.

We can write down the above intuitive idea as follows. Suppose $\lim_{n \to \infty} x_n = x$. If $x = +1$ or -1, for $\epsilon = \frac{1}{2}$, we should have some n_0 such that

$$|x_n - x| < \epsilon \ \forall\, n \geq n_0.$$

In case $x = 1$, then for $n \geq n_0$ and n odd, we have $2 = |x_n - x| < 1/2$, which is a contradiction. Similarly if $x = -1$, for $n > n_0$ and n even, we will have the contradiction $2 = |x_n - x| < 1/2$. Thus $x \neq +1$ or -1 and $|x-1| > 0, |x+1| > 0$. Let $\epsilon = \frac{1}{4}\min\{|x-1|, |x+1|\}$. Then $|x_n - x| < \epsilon$ will not be true for any natural number n.

4.8.5 Remarks.

(i) If $\lim_{n \to \infty} x_n = x$, for a given rational $\epsilon > 0$, the choice of n_0, such that $|x_n - x| < \epsilon \ \forall \ n \geq n_0$, will depend upon ϵ. Normally, the smaller the ϵ, the bigger the n_0 is chosen. For example for the sequence $x_n = \frac{1}{n}, n \in \mathbb{N}$, for $\epsilon = \cdot 01$ any $n_0 \geq 100$ will work. However, if $\epsilon = \cdot 001$, we can select $n_0 \geq 10000$.

(ii) To show that a sequence $\{x_n\}_{n \geq 1}$ does not converge to a rational x, it is enough to show that there exists some rational $\epsilon > 0$ such that $\forall \ n_0, \ \exists$ some $n > n_0$ with the property that $|x_n - x| > \epsilon$. (Have a look at the example 4.8.4 (ii) again.)

(iii) To prove that $\lim_{n \to \infty} x_n = x$, one has to make a guess for x and then prove it.

In view of the above remark, it is natural to ask the following question: can a sequence have more than one limit? Suppose if possible, a sequence $\{x_n\}_{n \geq 1}$ has two different limits x and $y, x \neq y$. Suppose without loss of generality that $x < y$. Since $\{x_n\}_{n \geq 1}$ converges to x and y, given any $\epsilon > 0$, $\exists$ a 'tail' of $\{x_n\}_{n \geq 1}$ inside both $(x - \epsilon, x + \epsilon)$ and $(y - \epsilon, y + \epsilon)$, which is not possible if $x + \epsilon < y - \epsilon$, i.e., $\epsilon < \dfrac{y-x}{2}$. Hence $x = y$.

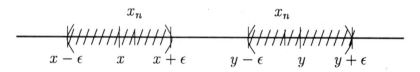

Figure 4.16 : Uniqueness of limit.

4.8 Sequences of rationals

The above argument can be written as follows. Let $\epsilon > 0$ be such that $x + \epsilon < y - \epsilon$. Since $\lim_{n\to\infty} x_n = x$, $\exists\, n_1 \in \mathbb{N}$ such that

$$|x_n - x| < \epsilon \quad \forall\, n \geq n_1.$$

Similarly, $\lim_{n\to\infty} x_0 = y$ implies that $\exists\, n_2 \in \mathbb{N}$ such that

$$|y_n - y| < \epsilon \quad \forall\, n \geq n_2.$$

Let $n_0 = \max\{n_1, n_2\}$. Then $\forall\, n \geq n_0$

$$x - \epsilon < x_n < x + \epsilon < y - \epsilon < x_n < y + \epsilon,$$

a contradiction. Thus, we have proved the following theorem.

4.8.6 Theorem. *Limit of a sequence, if exists, is unique.*

Another property of convergent sequences is given in the next theorem.

4.8.7 Theorem. *Let $\{x_n\}_{n\geq 1}$ be a convergent sequence. Then $\{x_n\}_{n\geq 1}$ is bounded, i.e., $|x_n| \leq M\ \forall\, n \in \mathbb{N}$, for some constant M.*

Proof. Intuitively, the proof is clear. For if $x_n \to x$, say, then there exists a tail $\{x_n | n \geq n_0\}$ of $\{x_n\}_{n\geq 1}$ included in the interval $(x - 1, x+1)$. If we put $m := \inf\{x-1, x_1, \ldots, x_{n_0-1}\}$ and $M := \max\{x+1, x_1, \ldots, x_{n_0-1}\}$, then $m \leq x_n \leq M\ \forall\, n \in \mathbb{N}$. In particular, $|x_n| \leq M$. This proof can be written formally as follows: let $\epsilon = 1$ be given. Then $\exists\, n_0 \in \mathbb{N}$ such that

$$|x_n - x| < 1 \quad \forall\, n \geq n_0.$$

Thus $\forall\, n \geq n_0$,

$$|x_n| \leq |x_n - x| + |x| = |x| + 1.$$

Let

$$M := \max\{|x| + 1, |x_1|, \ldots, |x_{n_0-1}|\}.$$

Then $|x_n| \leq M \ \forall \, n \in \mathbb{N}$. ∎

4.8.8 Exercise. Prove the following:

(i) $\lim_{n \to \infty} [1 + \dfrac{(-1)^n}{n}] = 1$.

(ii) $\lim_{n \to \infty} \dfrac{1}{n^2} = 0$. What can you say about $\lim_{n \to \infty} \dfrac{1}{n^r}$, where r is any positive rational?

(iii) $\lim_{n \to \infty} x_n = x$ iff $\lim_{n \to \infty} x_{n+1} = x$.

(iv) If $\lim_{n \to \infty} x_n = 0$, $\lim_{n \to \infty} |x_n| = |x|$. The converse holds if $x = 0$, but not in general.

(v) Let $\lim_{n \to \infty} x_n = x$ and $x_n < M \ \forall \, n \geq n_1$, for some $n_1 \in \mathbb{N}$. Show that $x \leq M$.

(vi) Let $x_n < y_n \ \forall \, n \geq 1$. If $\lim_{n \to \infty} x_n = x$ and $\lim_{n \to \infty} y_n = y$ then $x \leq y$. Show that $x < y$ need not hold in general.

(vii) Let $\{x_n\}_{n \geq 1}$ be the **constant sequence**, i.e., $x_n = x \ \forall \, n$. Show that $\lim_{n \to \infty} x_n = x$. In this case for every rational $\epsilon > 0$, the choice of n_0 such that $|x_n - x| < \epsilon \ \forall \, n \geq n_0$, does not depend upon ϵ. Suppose $\{y_n\}_{n \geq 1}$ is a sequence such that $\forall \, \epsilon > 0, \ \exists$ the same n_0 such that $|y_n - y| < \epsilon \ \forall \, n \geq n_0$. What can you say about the sequence $\{y_n\}$?

4.8.9 Exercise.

(i) Show that the sequence $\{\dfrac{(-1)^n n}{n+1}\}_{n \geq 1}$ has no limit.

(ii) Does the sequence $\{\dfrac{n}{n+1}\}_{n \geq 1}$ have a limit?

(iii) Let $x_n \geq 0 \ \forall \, n$ and $\lim_{n \to \infty} x_n^2 = 0$. Show that $\lim_{n \to \infty} x_n = 0$.

4.8 Sequences of rationals

In general for a given sequence it is not easy to guess/prove, using the definition alone, that it has a limit. Some useful methods for analyzing the limits of sequences are given in the next theorem.

4.8.10 Theorem. *Let $\{x_n\}_{n\geq 1}, \{y_n\}_{n\geq 1}$ be sequences such that $\lim_{n\to\infty} x_n = x$ and $\lim_{n\to\infty} y_n = y$. Then the following hold:*

(i) The sequence $\{x_n + y_n\}_{n\geq 1}$ is convergent and

$$\lim_{n\to\infty}(x_n + y_n) = x + y.$$

(ii) The sequence $\{x_n y_n\}_{n\geq 1}$ is convergent and

$$\lim_{n\to\infty}(x_n y_n) = xy.$$

(iii) If $y \neq 0$, then $\dfrac{x_n}{y_n}$ is defined for $\forall\, n \geq n_1$, for some $n_1 \in \mathbb{N}$, and the sequence $\left\{\dfrac{x_n}{y_n}\right\}_{n \geq n_1}$, is convergent with

$$\lim_{n\to\infty} \frac{x_n}{y_n} = \frac{x}{y}.$$

Proof. (i) Let a rational $\epsilon > 0$ be given. Choose $n_1, n_2 \in \mathbb{N}$ such that

$$|x_n - x| < \frac{\epsilon}{2} \;\; \forall\, n \geq n_1 \quad \text{and} \quad |y_n - y| < \frac{\epsilon}{2} \;\; \forall\, n \geq n_2.$$

Let $n_0 := \max\{n_1, n_2\}$. Then $\forall\, n \geq n_0$,

$$|x_n + y_n - (x+y)| \leq |x_n - x| + |y_n - y| < \epsilon.$$

This proves (i). To prove (ii), first note that $\{x_n\}_{n\geq 1}$ and $\{y_n\}_{n\geq 1}$ being convergent, are bounded sequences by theorem 4.8.7. Let $|x_n| < M_1$ and $|y_n| < M_2 \;\; \forall\, n$. Then $x \leq M_1$ by exercise 4.8.8 (v). Let a rational $\epsilon > 0$ be given. Choose n_1 and $n_2 \in \mathbb{N}$

such that $|x_n - x| < \dfrac{\epsilon}{2M_2}$ $\forall\, n \geq n_1$ and $|y_n - y| < \dfrac{\epsilon}{2M_1}$ $\forall\, n \geq n_2$ (note that $M_1 > 0$ and $M_2 > 0$). Then $\forall\, n \geq n_0 := \max\{n_1, n_2\}$,

$$\begin{aligned}
|x_n y_n - xy| &\leq |x_n y_n - x y_n| + |x y_n - xy| \\
&\leq |y_n|\,|x_n - x| + |x|\,|y_n - y| \\
&\leq M_2 |x_n - x| + M_1 |y_n - y| \\
&< \frac{\epsilon}{2} + \frac{\epsilon}{2} \\
&= \epsilon.
\end{aligned}$$

Hence, $\lim_{n\to\infty}(x_n y_n) = xy$. This proves (ii). To prove (iii), we have to show that given a rational $\epsilon > 0$, $\exists\, n_0$ such that

$$\left|\frac{x_n}{y_n} - \frac{x}{y}\right| < \epsilon \ \forall\, n \geq n_0.$$

Since $\lim_{n\to\infty} x_n = x$ and $\lim_{n\to\infty} y_n = y$, we have $n_1, n_2 \in \mathbb{N}$ such that

$$\text{and} \qquad \left.\begin{aligned} |x_n - x| < \epsilon \ \forall\, n \geq n_1 \\ |y_n - y| < \epsilon \ \forall\, n \geq n_2. \end{aligned}\right\} \qquad (4.6)$$

Now

$$\begin{aligned}
\left|\frac{x_n}{y_n} - \frac{x}{y}\right| &= \left|\frac{x_n y - x y_n}{y_n y}\right| \\
&= \left|\frac{x_n y - xy + xy - x y_n}{y_n y}\right| \\
&\leq \frac{|x_n - x||y| + |x||y_n - y|}{y_n y}.
\end{aligned}$$

If we take $n_0 := \max\{n_1, n_2\}$, the above inequality will give us

$$\left|\frac{x_n}{y_n} - \frac{x}{y}\right| \leq \frac{|y|\epsilon + |x|\epsilon}{|y_n||y|}. \qquad (4.7)$$

Appearance of y_n in the denominator on the right-hand side is to be removed. That can be done if we can say $\dfrac{1}{|y_n|} < c$, for some

4.8 Sequences of rationals

constant c and for all n large, i.e., $|y_n| > 1/c$ for all large n. This is true since $\lim_{n\to\infty} y_n = y \neq 0$ and we can choose n_3 such that $\forall\, n \geq n_3$.

$$|y_n - y| < \frac{|y|}{2}.$$

Thus $\forall\, n \geq n_3$,

$$\begin{aligned}|y_n| &\geq |y| - |y_n - y| \\ &> |y| - \frac{|y|}{2} \\ &= \frac{|y|}{2} > 0.\end{aligned} \qquad (4.8)$$

Thus for $n > \max\{n_1, n_2, n_3\}$, $\dfrac{x_n}{y_n}$ is defined and we will have from (4.7) and (4.8),

$$\left|\frac{x_n}{y_n} - \frac{x}{y}\right| < 2\left(\frac{|y|\epsilon + |x|\epsilon}{|y||y|}\right) = 2\epsilon\left(\frac{|y| + |x|}{|y|^2}\right).$$

To bring this estimate to the required form, we make changes in (4.6). We choose n_1 such that

$$|x_n - x| < \frac{1}{4}|y|\epsilon \quad \forall\, n \geq n_1,$$

and if $x \neq 0$, we choose n_2 such that

$$|y_n - y| < \frac{1}{4}\frac{|y|^2 \epsilon}{|x|} \quad \forall\, n \geq n_2.$$

Note that both $\dfrac{|y|\epsilon}{4}$ and $\dfrac{|y|^2\epsilon}{4|x|}$ are rationals. Then for $n \geq \max\{n_1, n_2, n_3\}$ we will have

$$\begin{aligned}\left|\frac{x_n}{y_n} - \frac{x}{y}\right| &\leq 2\left(\frac{|x_n - x||y| + |x||y_n - y|}{|y|^2}\right) \\ &< 2\left(\frac{\epsilon}{4} + \frac{\epsilon}{4}\right) = \epsilon.\end{aligned}$$

In case $x = 0$, clearly
$$\left|\frac{x_n}{y_n}\right| \leq \frac{2|x_n|}{|y|} < \epsilon \quad \text{for} \quad n \geq n_1.$$
This proves (iii). ∎

4.8.11 Example. Let us analyze the sequence $\{x_n\}_{n\geq 1}$ where
$$x_n := \frac{n^2 + 3n - 2}{5n^2 - 1}, \quad n \in \mathbb{N}.$$
Since
$$x_n = \frac{1 + \frac{3}{n} - \frac{2}{n^2}}{5 + \frac{1}{n^2}},$$
$\frac{1}{n} \to 0$ and $\frac{2}{n^2} \to 0$, it follows by theorem 4.8.10 that $x_n \to 1/5$.

Another useful result in analyzing the convergence of sequences is the following theorem.

4.8.12 Theorem (Sandwich). Let $\{x_n\}_{n\geq 1}, \{y_n\}_{n\geq 1}$ and $\{z_n\}_{n\geq 1}$ be sequences such that $\forall n \in \mathbb{N}$,
$$y_n \leq x_n \leq z_n,$$
and $\lim_{n\to\infty} y_n = \lim_{n\to\infty} z_n := x$. Then, $\{x_n\}_{n\geq 1}$ is convergent and $\lim_{n\to\infty} x_n = x$.

Proof. Let a rational $\epsilon > 0$ be given. We can choose $n_0 \in \mathbb{N}$ such that $\forall n \geq n_0$
$$|y_n - x| < \epsilon \quad \text{and} \quad |z_n - x| < \epsilon.$$
Then
$$x - \epsilon < y_n \leq x_n \leq z_n < x + \epsilon.$$
Thus
$$|x_n - x_0| < \epsilon \; \forall n \geq n_0.$$

4.8 Sequences of rationals

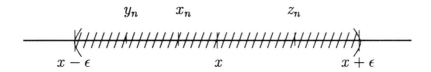

Figure 4.17 : Sandwich theorem.

Hence $\lim\limits_{n\to\infty} x_n = x$. ∎

4.8.13 Examples.

(i) Let
$$x_n := \frac{n}{n^2+1} + \frac{n}{n^2+2} + \ldots + \frac{n}{n^2+n}, \quad n \geq 1.$$
Since $n^2 + n > n^2 + k > n^2 + 1 \;\forall\; 1 \leq k \leq n$, we have
$$\frac{n^2}{n^2+n} \leq x_n \leq \frac{n^2}{n^2+1}.$$
Since
$$\lim_{n\to\infty} \frac{n^2}{n^2+n} = 1 = \lim_{n\to\infty} \frac{n^2}{n^2+1},$$
by sandwich theorem, $\lim\limits_{n\to\infty} x_n = 1$.

(ii) Let $x \in \mathbb{Q}$ be such that $|x| < 1$. Consider the sequence $\{x^n\}_{n\geq 1}$. Since $|x| < 1$, we can write $|x| = \dfrac{1}{1+h}, h > 0$. Now using Binomial theorem we have
$$0 < |x^n| = |x|^n = \frac{1}{(1+h)^n} = \frac{1}{1+nh+\frac{n}{2}(n-1)h^2+\cdots+h^n}$$
$$< \frac{1}{nh}.$$
Taking $y_n = 0$ and $z_n = \dfrac{1}{n} \;\forall\; n \in \mathbb{N}$, by the sandwich theorem we have
$$\lim_{n\to\infty} x^n = 0.$$

4.8.14 Exercise. Prove the following:

(i) $\lim\limits_{n\to\infty} \left(\sqrt{n+1} - \sqrt{n}\right) = 0.$

(ii) Show that $\lim\limits_{n\to\infty} x^{\frac{1}{n}} = 1.$

(Hint: for $x > 1$, write $x^{\frac{1}{n}} = 1 + y_n$, $y_n > 0$, and use Binomial theorem to show that $0 < y_n \leq \dfrac{x-1}{n}$ $\forall\ n$.)

(iii) $\lim\limits_{n\to\infty} n^{1/n} = 1.$

(iv) Show that $\lim\limits_{n\to\infty} y_n = 1$, where y_n is as in example 4.8.2 (iii).

(Hint: Use theorem 4.8.16)

4.8.15 Exercise. Let $\{x_n\}_{n\geq 1}$ be a sequence of nonnegative rationals such that
$$0 < \frac{x_{n+1}}{x_n} = r < 1 \ \forall\ n \geq 1.$$
Show that $x_n \to 0$. In fact $x_n \to 0$ if $\dfrac{x_{n+1}}{x_n} \to r$, $0 < r < 1$.

4.8.16 Theorem. *Let $r \in \mathbb{Q}$ be such that $|r| < 1$ and let $x_n := 1 + r + r^2 \ldots + r^{n-1}$. Then the following hold:*

(i) $x_n = \dfrac{1 - r^{n+1}}{1 - r}.$

(ii) $\lim\limits_{n\to\infty} x_n = \dfrac{1}{1-r}.$

Proof. (i) As $x_n = 1 + r + \ldots + r^n$ and $rx_n = r + r^2 + \ldots + r^{n+1}$, we have $x_n(1 - r) = 1 - r^{n+1}$, i.e., $x_n = \dfrac{1 - r^{n+1}}{1 - r}$. This proves (i).
(ii) Follows from (i) and example 4.8.13 (ii). ∎

We show next that in the decimal representation of a rational, as given in 4.6.3, all the terms cannot be equal to the number 9 after some stage.

4.8 Sequences of rationals

4.8.17 Theorem. *Let $a_0 \cdot a_1 a_2 \ldots$ be the decimal representation of a rational r as given by theorem 4.6.1. Then there does not exist any m such that $a_n = 9 \ \forall \ n \geq m$.*

Proof. Recall that $a_0 = [r]$ and $\forall \ n \geq 1 \ a_n := [10^n r - (10^n a_0 + \ldots + 10 a_{n-1})]$. Further, if $r_n := a_0 + \dfrac{a_1}{10} + \ldots + \dfrac{a_n}{10^n}$, then $r_n \leq r < r_n + \dfrac{1}{10^n}$. Since $\dfrac{1}{10^n} \to 0$ as $n \to \infty$, by example 4.8.13 (ii) we have $r = \lim\limits_{n \to \infty} r_n$. Now suppose $\exists \ m$ such that $a_n = 9 \ \forall \ n \geq m$. Then $\forall \ n > m$,

$$\begin{aligned} r_n &= a_0 + \frac{a_1}{10} + \ldots + \frac{a_{m-1}}{10^{m-1}} + \frac{9}{10^m} + \ldots + \frac{9}{10^n} \\ &= a_0 + \frac{a_1}{10} + \ldots + \frac{a_{m-1}}{10^{m-1}} + \frac{9}{10^m}\left[1 + \frac{1}{10} + \ldots + \frac{1}{10^{n-m}}\right]. \end{aligned}$$

Thus, using theorem 4.8.16,

$$\begin{aligned} r &= \lim_{n \to \infty} r_n \\ &= a_0 + \frac{a_1}{10} + \ldots + \frac{a_{m-1}}{10^{m-1}} \\ &\quad + \frac{9}{10^m}\left[\lim_{n \to \infty}\left(1 + \frac{1}{10} + \ldots + \frac{1}{10^{n-m}}\right)\right] \\ &= a_0 + \frac{a_1}{10} + \ldots + \frac{a_{m-1}}{10^{m-1}} + \frac{9}{10^m} \times \frac{10}{9} \\ &= a_0 + \frac{a_1}{10} + \ldots + \frac{a_{m-1}+1}{10^{m-1}}. \end{aligned}$$

But then $a_{m-1} = [10^{m-1}r - (10^{m-1}a_0 + \ldots + 10 a_{m-1})] = a_{m-1} + 1$, a contradiction. Hence there does not exist any m such that $a_n = 9 \ \forall \ n \geq m$. ∎

Let $\{x_n\}_{n\geq 1}$ be a sequence in $\mathbb{Q}$. We have analysed properties of convergent sequences. However, the question arises: how to decide whether $\{x_n\}_{n\geq 1}$ is convergent or not? The definition of convergence, i.e., $\lim\limits_{n \to \infty} x_n = x$ required the knowledge (or a guess at least) of a point $x \in \mathbb{Q}$. Can we have a criterion for convergence which is intrinsic, i.e., it requires the knowledge about the sequence only? For example, we showed in theorem 4.8.7 that it will be bounded. Thus, boundedness of a sequence is a necessary condition for convergence, (i.e., if a sequence is not bounded, it cannot converge) and it is an intrinsic property. Another such condition is given by the following theorem which says that if a sequence is convergent, then its terms are arbitrarily close to each other from some stage onwards.

4.8.18 Theorem. Let $\{x_n\}_{n\geq 1}$ be a convergent sequence and $\epsilon > 0$ be given. Then $\exists\ n_0 \in \mathbb{N}$ such that

$$|x_n - x_m| < \epsilon \quad \forall\ n, m \geq n_0.$$

Proof. Intuitively this is obvious, for if $\lim\limits_{n \to \infty} x_n = x$, the elements of the sequence x_n will be arbitrarily close to x and hence to each other after some stage.

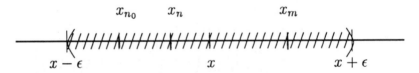

Figure 4.18 : Convergent implies Cauchy.

Formally, let a rational $\epsilon > 0$ be given. Then $\exists\ n_0 \in \mathbb{N}$ such that

$$|x_n - x| < \epsilon/2 \quad \forall\ n \geq n_0.$$

Thus for $n, m \geq n_0$,

$$|x_n - x_m| \leq |x_n - x| + |x_n - x| = \epsilon/2 + \epsilon/2 = \epsilon. \ \blacksquare$$

4.8.19 Definition. A sequence $\{x_n\}_{n\geq 1}$ is called a **Cauchy sequence** if $\forall$ rational $\epsilon > 0$, $\exists\ n_0 \in \mathbb{N}$ such that

$$|x_n - x_n| < \epsilon \quad \forall\ n, m \geq n_0.$$

4.8 Sequences of rationals

4.8.20 Remark. Theorem 4.8.18 says that every convergent sequence is Cauchy. Note that saying that a sequence is Cauchy requires knowledge of the elements of the sequence alone. Thus it is an 'intrinsic' property of the sequence. Also if a sequence of rationals is convergent, it is necessarily Cauchy. Unfortunately, the converse need not be true as shown in the next theorem.

4.8.21 Exercise. Let $C(\mathbb{Q})$ denote the set of all Cauchy sequences in $\mathbb{Q}$. For $\{r_n\}_{n\geq 1}, \{s_n\}_{n\geq 1} \in C(\mathbb{Q})$, prove the following:

(i) Every sequence $\{r_n\}_{n\geq 1} \in C(\mathbb{Q})$ is bounded.

(ii) $\{r_n + s_n\}_{n\geq 1}$ is a Cauchy sequence.

(iii) $\{\alpha r_n\}_{n\geq 1}$ is a Cauchy sequence $\forall \, \alpha \in \mathbb{Q}$.

(iv) $\{r_n s_n\}_{n\geq 1}$ is a Cauchy sequence.

4.8.22 Exercise. Let $\{r_n\}_{n\geq 1}$ be a sequence in $\mathbb{Q}$, such that $\{r_n^2\}_{n\geq 1}$ is a Cauchy sequence and there exists a positive rational α such that $r_n \geq \alpha > 0 \ \forall \, n \geq 1$. Show that $\{r_n\}_{n\geq 1}$ is also a Cauchy sequence.

4.8.23 Theorem (Cauchy incompleteness of $\mathbb{Q}$). *There exists sequences $\{r_n\}_{n\geq 1}$ in $\mathbb{Q}$ which are Cauchy but not convergent.*

Proof. Let $r_0 = 1$ and for every $n \geq 0$, let

$$r_{n+1} := \frac{r_n}{2} + \frac{1}{r_n}.$$

Then $\{r_n\}_{n\geq 1}$ is a sequence of nonnegative rationals. Using induction, it is easy to show that $r_n^2 > 2$ and $r_n < r_{n+1} \ \forall \, n \geq 1$. Thus

$$r_{n+1}^2 - 2 = \left[\frac{1}{2}\left(r_n + \frac{2}{r_n}\right)\right]^2 - 2$$

$$
\begin{aligned}
&= \frac{1}{4}\left[r_n^2 + 4 + \frac{4}{r_n^2}\right] - 2 \\
&= \frac{r_n^4 + 4r_n^2 + 4 - 8r_n^2}{4r_n^2} \\
&= \frac{(r_n^2 - 2)^2}{4r_n^2} \\
&\leq \frac{1}{8}(r_n^2 - 2)^2.
\end{aligned}
$$

Hence by induction, we have for every $n \geq 1$,

$$r_{n+1}^2 - 2 \leq \frac{(r_0^2 - 2)^{2^n}}{(8)^{n-1}} = \frac{1}{8^{n-1}}.$$

Thus $\{r_n^2\}_{n\geq 1}$ converges to 2 and hence $\{r_n^2\}_{n\geq 1}$ is a Cauchy sequence by theorem 4.8.18. It follows from exercise 4.8.22 that $\{r_n\}_{n\geq 1}$ is also a Cauchy sequence. Suppose if possible, there exists a rational r such that $\lim_{n\to\infty} r_n = r$. Then it follows from exercise 4.8.8 (iii) and 4.8.10 that

$$r = \lim_{n\to\infty} r_{n+1} = \lim_{n\to\infty}\left(\frac{r_n}{2} + \frac{1}{r_n}\right) = \frac{r}{2} + \frac{1}{r}.$$

Thus $r^2 = 2$, which is not true. Hence $\{r_n\}_{n\geq 1}$ is a Cauchy sequence which is not convergent. ∎

4.8.24 Remark. See corollary 5.5.4 for another proof of theorem 4.8.23.

4.8.25 Note. In fact the sequence $\{r_n\}_{n\geq 1}$ as defined in theorem 4.8.23 can only 'converge' to $\sqrt{2}$, which is yet to be defined. Since the terms r_n come closer to $\sqrt{2}$, it can be used to get rational approximations to $\sqrt{2}$. This is the method of finding $\sqrt{n}$ which is used even today in computers and was proposed by the Greek mathematician Heron of Alexandria, around 250 B.C.. By a similar (sequential) method Archimedes had approximated the values of π, the area of the unit circle (see example 6.3.7 and note 6.3.10 (ii)).

4.9 Abstractions

Motivated by the note 4.3.3 we have the following definition.

4.9.1 Definition. Let $(F, \oplus, \odot)$ be a field and let $<$ be a relation on F. We say $(F, \oplus, \odot, <)$ is an **ordered field** if $(F, \oplus, \odot, <)$ is an ordered integral domain.

4.9.2 Examples.

(i) The field of rationals, as constructed in section 4.2 and with order as defined in section 4.3, is an ordered field. In fact, this construction can be repeated with $\mathbb{Z}$ replaced by any ordered integral domain. The resulting field is called the ordered **field of fractions** of the given integral domain.

(ii) Let $(X, +, \cdot, <)$ be an ordered field and let $(\mathcal{L}(X), \oplus, \odot, <)$ be the ordered integral domain as constructed in example 3.6.16 (iii). It follows from exercise 3.6.10 that $(\mathcal{L}(X), \oplus, \odot, <)$ is an ordered field.

4.9.3 Note. Every field $(F, \oplus, \odot)$ has at least two distinct elements, the additive identity 0_F and the multiplicative identity 1_F. In general, it can be a finite set. For example, the field $\mathbb{Z}_n$ of example 3.6.11 is a finite field. Every ordered field $(F, \oplus, \odot, <)$ being an ordered integral domain, includes a copy of the ordered integral domain $(\mathbb{Z}, +, \cdot, <)$; we call it the **integers** of the ordered field F and denote it by $\mathbb{Z}_F$, as in theorem 3.6.34. We show next that every ordered field includes a copy of the ordered field of rationals $(\mathbb{Q}, +, \cdot, <)$, as constructed in section 4.3.

4.9.4 Definition. Let $(F, \oplus, \odot)$ be a field and X be a subset of F. We say X is a **subfield** of F if under the restriction of the binary operations $\oplus$ and $\odot$ to X, $(X, \oplus, \odot)$ is itself a field. If $(F, \oplus, \odot, <)$ is an ordered field, a subfield $(X, \oplus, \odot)$ of F is called an **ordered subfield** of F if under the restriction of the order $<$ to X, $(X, \oplus, \odot, <)$

itself becomes an ordered field.

4.9.5 Exercise. Let $(F, \oplus, \odot)$ be a field and X be a subset of F. Then X is a subfield of F iff $0_F, 1_F \in X$ and $\forall\, x, y \in X, x - y \in X$ and $xy^{-1} \in X$.

4.9.6 Example. Let $(F, \oplus, \odot)$ be any field having characteristic $p > 1$. Then p is a prime (see exercise 3.6.21) and $\mathbb{N}_F$, the set of natural numbers of F, is a finite set given by $\{0, 1_F, 2\,1_F, \ldots, (p-1)1_F\}$. Clearly $\mathbb{N}_F$ is a subfield of F, isomorphic to $\mathbb{Z}_p$.

4.9.7 Theorem. Let $(F, \oplus, \odot)$ be any field and let $\mathbb{Q}_F := \{x \odot y^{-1} | x, y \in \mathbb{N}_F\}$. Then $(\mathbb{Q}_F, \oplus, \odot)$ is a subfield of every subfield of F. Further if F has characteristic 0 then $(\mathbb{Q}_F, \oplus, \odot)$ is isomorphic to the field of rationals $(\mathbb{Q}, +, \cdot)$.

Proof. It is easy to check that $\mathbb{Q}_F$ is a subfield by exercise 4.9.5. Let X be any subfield of F. Then $\mathbb{N}_F \subseteq X$ and hence $\mathbb{Q}_F \subseteq X$. Thus $\mathbb{Q}_F$ is a subfield of X. Suppose F has characteristic zero. Consider the map $\Phi : \mathbb{Q} \to \mathbb{Q}_F$ defined by

$$\Phi\left(\frac{n}{m}\right) = n 1_F \odot (m 1_F)^{-1}, \quad \frac{n}{m} \in \mathbb{Q}.$$

This is a well-defined map, for if $\frac{n}{m} = \frac{n_1}{m_1}$, $(nm_1)1_F = (n_1 m)1_F$, i.e., $(n 1_F) \odot (m_1 1_F) = (n_1 1_F) \odot (m 1_F)$ and hence $n 1_F \odot (m 1_F)^{-1} = n_1 1_F \odot (m_1 1_F)^{-1}$. It is easy to check that Φ is a bijective map and has the following properties:

$$\Phi\left(\frac{n}{m} + \frac{p}{q}\right) = \Phi\left(\frac{n}{m}\right) \oplus \Phi\left(\frac{p}{q}\right),$$

$$\Phi\left(\frac{n}{m} \times \frac{p}{q}\right) = \Phi\left(\frac{n}{m}\right) \odot \Phi\left(\frac{p}{q}\right).$$

Hence $(\mathbb{Q}, +, \cdot)$ is isomorphic to $(\mathbb{Q}_F, +, \cdot)$. ∎

4.9 Abstractions

We can now state another uniqueness of the field of rationals.

4.9.8 Corollary. *Let $(F, \oplus, \odot)$ be a field with characteristic zero such that it has no proper subfield. Then $(F, \oplus, \odot)$ is isomorphic to $(\mathbb{Q}, +, \cdot)$.*

Proof. The proof follows from theorem 4.9.7. ∎

Thus $(\mathbb{Q}, +, \cdot)$ is the smallest field of characteristic zero.

4.9.9 Corollary. *Let $(F, \oplus, \odot, <)$ be an ordered field. Then $(\mathbb{Q}_F, \oplus, \odot, <)$ is an ordered subfield of F and is isomorphic to the ordered field of rationals $(\mathbb{Q}, +, \cdot, <)$.*

Proof. The field $(\mathbb{Q}_F, \oplus, \cdot, <)$ is an ordered subfield of F is obvious. Since F is an ordered field, it has characteristic zero (see section 3.6.34) and hence by theorem 4.9.7, $(\mathbb{Q}_F, \oplus, \odot)$ is isomorphic to $(\mathbb{Q}, +, \cdot)$. Let Φ be the isomorphism as constructed in theorem 4.9.7. We show that Φ is also an isomorphism between the ordered fields $(\mathbb{Q}, +, \cdot, <)$ and $(\mathbb{Q}_F, \oplus, \odot, <)$. For that we have to show that Φ is order preserving. Let $\frac{n}{m}, \frac{p}{q} \in \mathbb{Q}$ and $\frac{n}{m} < \frac{p}{q}$. Then $nq < mp$ and using exercise 3.6.20, we have

$$(nq) \odot 1_F < (mp)1_F,$$

i.e.,

$$(n1_F) \odot (q1_F) < (m1_F) \odot (p1_F),$$

i.e.,

$$(n1_F) \odot (m1_F)^{-1} < (p1_F) \odot (q1_F)^{-1},$$

i.e.,

$$\Phi(\frac{n}{m}) < \Phi(\frac{p}{q}). \quad \blacksquare$$

The Archimedean property of the integral domain of integers and the field of rationals motivates the following definition.

4.9.10 Definition. Let $(F, \oplus, \odot, <)$ be an ordered integral domain. We say F is an **Archimedean** ordered integral domain if $\forall\ x, y \in F, x > 0_F$ there exists $n \in \mathbb{N}$ such that $nx > y$.

4.9.11 Examples.

(i) As shown in theorem 4.3.2(v) $(\mathbb{Q}, +, \cdot, <)$ the ordered field of rational numbers is Archimedean.

(ii) Consider the ordered field $(\mathcal{L}(X), \oplus, \odot, <)$ as constructed in example 4.9.2 (ii). We show that it is not Archimedean. As noted in 3.6.16 (iii), for every $n \in \mathbb{N}, n1_\mathcal{L} < \dfrac{1}{x}$. Thus, given $1_\mathcal{L}, \dfrac{1}{x} \in \mathcal{L}(X)$ there does not exist any $n \in \mathbb{N}$ such that $n1_\mathcal{L} > \dfrac{1}{x}$. Hence $\mathcal{L}(X)$ is not Archimedean.

Corresponding to theorem 4.3.8, we have the following theorem.

4.9.12 Theorem. *Let $(F, +, \cdot, <)$ be an Archimedean ordered field. Then for every $x \in F$ there exists a unique $n \in \mathbb{Z}$ such that $(n1_F) \leq x < (n+1)1_F$.*

Proof. Let $x \in F$ and $S := \{n \in \mathbb{Z} | n1_F \leq x\}$. If $x \geq 0_F$, then clearly $0_F \in S$. If $x < 0_F$, using the Archimedean property of F we can choose $n \in \mathbb{N}$ such that $n1_F > -x$. Thus $(-n)1_x < x$, i.e., $-n \in S$. Thus S is a nonempty set. Again, using the Archimedean property we can find $m \in \mathbb{N}$ such that $x < m1_F$. Thus, S is bounded above, and it follows from theorem 3.3.6 that $n_0 := lub(S)$ exists and $n_0 \in S$, i.e., $n_0 1_F \leq x$. Clearly $(n+1)1_F > x$, for otherwise $(n_0 + 1)1_F$ will be the least upper bound of S. Thus $n_0 1_F \leq x < (n_0 + 1)1_F$. To prove the uniqueness of n_0, suppose $\exists\ m \in \mathbb{Z}$ such that $m1_F \leq x < (m+1)1_F$. Suppose $m < n$. Then $m < (m_0 + 1) \leq n$ and using exercise 3.6.20, we have $x < (m+1)1_F \leq n_0 1_F \leq x$, which is a contradiction. Thus $m < n_0$ is not possible. Similarly, $n_0 < m$ is also not possible. Hence $m = n_0$.

4.9 Abstractions

This completes the proof. ∎

The property of an ordered field to be Archimedean can be described in various equivalent ways, as shown in the next theorem.

4.9.13 Theorem. *Let $(F, +, \cdot, <)$ be an ordered field. Then the following are equivalent:*

(i) F is Archimedean.

(ii) $\mathbb{N}_F$, the set of natural numbers of F, is not bounded above.

(iii) For every $\epsilon \in F, \epsilon > 0_F$, there exists $n \in \mathbb{N}_F$ such that $(n1_F)^{-1} < \epsilon$.

(iv) For every $x, y \in F$ with $x < y$, there exists $z \in \mathbb{Q}_F$ such that $x < z < y$.

Proof. (i) $\Rightarrow$ (ii). Let F be Archimedean but $\mathbb{N}_F$ be bounded above, i.e., there exists $x \in F$ such that $n1_F \leq x \ \forall \ n \in \mathbb{N}$. By the Archimedean property, $\exists \ m \in \mathbb{N}$ such that $m1_F > x$, which is a contradiction. Thus $\mathbb{N}_F$ is not bounded above.

(ii) $\Rightarrow$ (iii). Let $\mathbb{N}_F$ be not bounded above and let $\epsilon \in F, \epsilon > 0_F$ be given. Suppose $\forall \ n \in \mathbb{N}, (n1_F)^{-1} \geq \epsilon$. Then $n1_F \leq \epsilon^{-1} \ \forall \ n \in \mathbb{N}$, which is not true. Hence there exists some $n \in \mathbb{N}$ such that $(n1_F)^{-1} < \epsilon$, i.e., (iii) holds.

(iii) $\Rightarrow$ (iv). Let $x, y \in F, 0_F < x < y$. Consider $(y - x) > 0_F$. By (iii), there exists $n \in \mathbb{N}_F$ such that $(n1_F)^{-1} < (y - x)$. Let $S := \{k \in \mathbb{N} | nx < k1_F\}$. Since $x > 0, (nx)^{-1} > 0_F$ and by (iii) there exists k such that $(k1_F)^{-1} < (nx)^{-1}$, i.e., $k1_F > nx$. Thus $S \neq \emptyset$. Thus S has the smallest element, say $m := glb(S)$. Then $nx < m1_F$ and $(m-1)1_F \leq nx < m1_F$. Thus

$$\begin{aligned} x < (m1_F)(n1_F)^{-1} &= ((m-1)1_F)(n1_F)^{-1} + (n1_F)^{-1} \\ &\leq x + (n1_F)^{-1} \\ &< x + (y - x) = y. \end{aligned}$$

The required claim follows from $z := (m1_F)(n1_F)^{-1}$. In case $x = 0_F$, then $y > 0_F$ and we have $n \in \mathbb{N}$ such that $0_F < (n1_F)^{-1} < y$. Then $z := (n1_F)^{-1}$ will do the job. Next, if $x < 0_F < y$, then $z = 0_F$ is the required element of $\mathbb{Q}_F$. For $x \leq y = 0_F$, we can consider $-x \geq y \geq 0_F$ and apply the earlier cases.

(iv) $\Rightarrow$ (i) Let $x, y \in F$ with $x > 0_F$. If $y \leq 0_F$, $nx > y$ with $n = 1$. So, let $y > 0_F$. Again if $x \geq y$, then $nx > y$ with $n = 2$. Thus, let $y > x$. Then $0_F < xy^{-1} < 1_F$, and by (iv) there exist $m, n \in \mathbb{N}$ such that

$$0_F < (m1_F)(n1_F)^{-1} < xy^{-1} < 1_F, \text{ i.e., } y < (m1_F)^{-1}nx < nx.$$

Hence (i) holds. ∎

4.9.14 Note. Given an ordered field $(F, +, \cdot, <)$, one can analyze whether it is **order complete**: does every nonempty subset A of F which is bounded above has lub? Also using the order $<$, we can define the notion of absolute value for elements of F: if $x \in F$ and $x \geq 0_F$, $|x| := x$, and $|x| := -x$ if $x \leq 0_F$. As was proved for the field of rationals, one can show that the absolute value function has properties similar to that of theorem 4.7.2. We can also consider sequences of elements of F and analyze their convergence and the property of being Cauchy with respect to this absolute value. We call the ordered field F to be **Cauchy complete** if every Cauchy sequence of elements of F converges to an element in F. We saw that the ordered field $\mathbb{Q}$ is neither order complete nor Cauchy complete. In the next chapter we shall consider extension of the ordered field $\mathbb{Q}$ to ordered fields which have these properties.

Chapter 5

REAL NUMBERS: CONSTRUCTION AND UNIQUENESS

5.1 Historical comments

The origin of arithmetic dates back to the Babylonian period of 2100 B.C. It seems that the Babylonians had the knowledge of irrational (incommensurable) ratios. They worked with these incommensurable ratios through rational approximations. However, the discovery of incommensurable ratios is attributed to a Greek mathematician Hippasus of Metapontum around 500 B.C. We also saw in section 3.1 that the crisis created by this discovery was resolved brilliantly in 370 B.C. by Eudoxus of Cindus, another Greek mathematician. After that Greek mathematicians didn't do much about arithmetical calculations, the emphasis being more on geometry. Later, Hindu and Arab mathematicians developed arithmetical rules which worked for incommensurable magnitudes as well. Gradually irrational numbers began to be treated like rational numbers, as far as arithmetical operations were concerned. But still, there was some uncertainty. For example M. Stifel (1544) in his book 'Arithmetica Integra' writes: "Just as an infinite number is no number, so an irrational number is no number, because it is so to speak concealed under a fog of infinity". The interest in the number concept was renewed after the discovery of calculus in the 17th and 18th centuries. That was also the time when mathematicians were concerned about the looseness in concepts and proofs in many branches of analysis. The concept of a function was not clear, convergence of series had no rigorous foundation and the basic concepts of calculus – derivative and integral, were only understood geometrically. Analysis still had its foundation based on geometry. A need was felt to instill rigor in analysis and to achieve this it was felt that results of analysis should be deduced purely on the basis of arithmetical concepts. This process of 'Arithmetization of Analysis' was started by Bernard Bolzano (1781–1848), Augustin Louis Cauchy (1789–1857), Niels Hernik Abel (1802–29), Peter Gustav Lejeune Dirichlet (1805–59) and Karl Weierstrass (1815–97). Through the efforts of these mathematicians, the need to give the 'real number system' a logical foundation was realized. The 'real number system' was interpreted as points on the number line. Another reason (i.e., other than proving the results of analysis in a rigorous fashion) for the logical foundation of real number system was the following: because of the discovery of non-Euclidean geometry, geometry had lost its status as truth. But at the same time it was felt that the mathematics built on the ordinary arithmetic must be an unquestionable reality. The problem was to interpret the geometric property of the line: 'a continuum of points' to be phrased in terms of arithmetic rules. This was achieved by Weierstrass in 1880 (formally done by P. Bachmann in 1892), Georg Cantor in 1883, and by Richard Dedekind in 1872. In 4.1.1 and 4.1.2 we briefly described the constructions of real numbers given by Dedekind and Cantor. The details are given in sections 4.2 and 4.3. Otto Stolz (1842–1905) in 1886 showed that

5.1 Historical comments

irrational numbers can be represented as non-periodic decimals and that this can be taken as a defining property.

Richard Dedekind (1831–1916)

Dedekind was born in Brunswick, the birthplace of Gauss. At the age of 19 he entered the University of Göttingen and at the age of 21 he received his doctorate. In 1854 Dedekind was appointed as a lecturer at Göttingen and in 1857 as a professor at the Zürich Polytechnic. In 1862 he returned to Brunswick as a professor at the technical high school, where he ended his career. Though he gave no early evidence of being a mathematical genius, in the words of Edmund Landau "Dedekind was one of the wholly great in the history of mathematics". He (along with Kummer and Kronecker) was the creator of the theory of algebraic numbers (1857); he (in collaboration with H. Weber) gave a completely algebraic treatment of algebraic curves (1882); he gave the construction of integers (1888) and the construction of real numbers (1872). He remained fresh of mind and robust of body till his death in 1916.

5.1.1 Dedekind's method. Dedekind looked at the continuity of the line as follows: Consider points on the line and call a point x smaller than y, written as $x < y$, if x is strictly on the left of y. Now given any point x on the line, let L_x be the set of points on the line such that $y < x$ and let R_x be the set of points on the line such that $x \leq y$. These sets have the following properties:

(i) $L_x \neq \emptyset, R_x \neq \emptyset$.

(ii) Every point on the line belongs to either L_x or R_x.

(iii) If $y \in L_x$ and $z \in R_x$ then $y < z$.

(iv) L_x has no largest element.

We can think of the pair (L_x, R_x) as a division or a **cut** of the line produced by the point x.

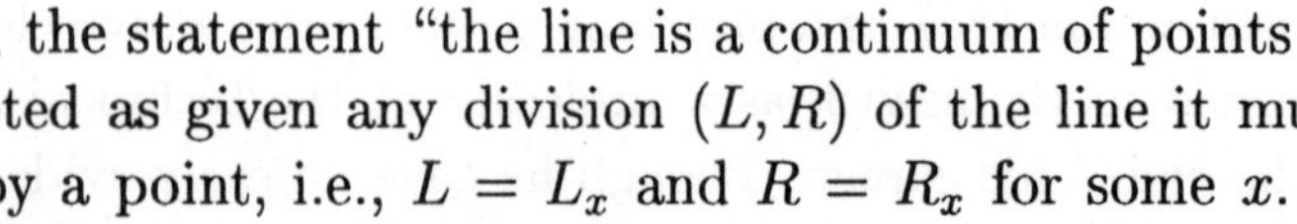

Figure 5.1 : Cut produced by the point x.

Conversely, the statement "the line is a continuum of points" can be interpreted as given any division (L, R) of the line it must be produced by a point, i.e., $L = L_x$ and $R = R_x$ for some x. Since the rationals on the line were well understood, Dedekind defined a real number to be a **cut** of the 'rationals number line', i.e., an ordered pair (α, β) of two subsets of $\mathbb{Q}$ with the properties:

(i) Neither α nor β is empty.

5.1 Historical comments

(ii) Every rational belongs to either α or β.

(iii) Every element of α is less than every element of β.

(iv) α has no largest element.

On $\mathbb{R}$, the collection of all cuts, order can be defined in a very natural way. Addition and multiplication can also be defined so that $\mathbb{R}$ becomes a complete ordered field.

At this point it is worth recalling the definition of proportionality of ratios given by Eudoxus in 370 B.C. (see section 3.1). As a consequence, we can say that a ratio a/b is bigger than a ratio c/d, if $\exists\, m, n$ such that $ma > nb$ but $mc \leq nd$. Let us write this as $a/b > c/d$. Since the commensurable ratios are nothing but rationals, by the above definition of Eudoxus each incommensurable ratio gives us a division of rationals into two classes: If $x = a/b$ is an incommensurable ratio, consider

$$L_x := \left\{\frac{c}{d} \,\Big|\, \frac{c}{d} \text{ is a commensurable ratio, } \frac{c}{d} < \frac{a}{b}\right\}$$

and

$$R_x := \left\{\frac{c}{d} \,\Big|\, \frac{c}{d} \text{ is a commensurable ratio } \frac{c}{d} \geq \frac{a}{b}\right\}.$$

Then (L_x, R_x) is a cut as defined above. So even though the idea of a real number was there, to treat the incommensurable ratios a new viewpoint was needed.

We note that some cuts are produced by rational points on the line, i.e., given a rational number r, consider $L_r := \{s \in \mathbb{Q} \,|\, s < r\}$, $R_r := \{s \in \mathbb{Q} \,|\, r \geq s\}$. On the other hand not every cut need to be of this form, for example if L_x and R_x both do not have any least and biggest element, respectively (see example 5.2.4(ii)). These are the cuts which correspond to non-rational points on the line. We can think of them as gaps in the number line. Thus cuts of the latter type fill the gaps on the number line. Dedekind's method will be described in detail in section 5.2.

5.1.2 Cantor's method. To understand Cantor's approach, let us consider the number line again with the intuitive notion of 'distance' between two points on the line: if these are rational points, say r and s, then $|r-s|$ is the distance between them. Let $\{x_n\}_{n\geq 1}$ be a Cauchy sequence of rational points on the line, i.e., given ϵ, any small unit of length, $\exists\ m_0$ such that $\forall\ n, m \geq m_0$ the distance between x_n and x_m is at most ϵ. In a sense, the elements of a sequence are coming closer to each other. We expect that they will come close to a point x on the line. So, we should expect every Cauchy sequence on the line to 'converge' to a point on the line, i.e., the distance of the terms of the sequence from a point becomes smaller and smaller. Equivalently, to locate a point on the line, we can consider any sequence of rational points on the line that 'converges' to x. In fact, given any point on the line, we can find a sequence $\{r_n\}_{n\geq 1}$ of rational points which locates x. To see this, for a given point x choose rational points r, s such that $r < x < s$ (which is possible, for example there exists integer points r and s such that $r < x < s$). Consider the segment S_0 from r to s, only right end-point included. Divide it into two equal parts and consider the two smaller segments (each including its right-end point only). Then x will be in only one of these segments, say S_1. Let r_1 be the right-end point of the segment S_1. Now divide S_1 into two equal parts and look at the part that contains x. Put r_2 to be right-end point of that segment and so on. This will give a sequence of rationals $\{r_n\}_{n\geq 1}$ 'converging' to x.

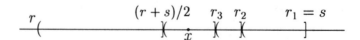

Figure 5.2 : Sequence of rationals approaching x.

Now suppose that there are two sequences of rationals $\{r_n\}_{n\geq 1}$ and $\{s_n\}_{n\geq 1}$, both 'converging' to x. Then clearly, $\{r_n - s_n\}_{n\geq 1}$ is a sequence of rationals converging to 0. Conversely, if $\{r_n\}_{n\geq 1}$ and

$\{s_n\}_{n\geq 1}$ are Cauchy sequences such that $\{r_n - s_n\}_{n\geq 1}$ converges to 0, clearly they will locate the same point on the line. Consider the set $\mathcal{C}$ of all Cauchy sequences of rationals and call two sequences $\{r_n\}_{n\geq 1}$ and $\{s_n\}_{n\geq 1}$ equivalent if they locates the same point on the line, i.e., $\{r_n - s_n\}_{n\geq 1}$ converges to 0. It is easy to see that this is an equivalence relation on $\mathcal{C}$. We can say that all the elements of an equivalence class will locate the same point on the line. So, Cantor defined $\mathbb{R}$, the set of real numbers, to be the set of equivalence class of $\mathcal{C}$ under the above relation. On $\mathbb{R}$, addition and multiplication are defined naturally and an order can also be defined such that $\mathbb{R}$ becomes a complete ordered field. We describe this construction in detail in section 5.3.

Both constructions assume the existence of the ordered field of rationals and construct an ordered field which is both order and Cauchy complete. Further this new ordered field includes a copy of $\mathbb{Q}$, the field of rationals. In a sense we can say that both methods 'extend' the field of rationals to a complete ordered field. We shall show that both constructions lead to the same object.

5.2 Construction of real numbers by Dedekind's method

We start with $\mathbb{Q}$, the ordered field of rationals, as constructed in chapter 4. Our discussion of 5.1.1 motivates the following definition.

5.2.1 Definition. A subset α of rational numbers is called a **cut** if it has the following properties:

(i) $\alpha \neq \emptyset, \alpha \neq \mathbb{Q}$.

(ii) For every $r \in \alpha$ and $s \in \mathbb{Q} \setminus \alpha$, $r < s$.

(iii) α has no greatest element.

We denote the set of all cuts by $\mathbb{R}$.

5.2.2 Exercise. Show that condition (ii) in definition 5.2.1 is equivalent to the condition that $\forall\, r \in \alpha$, if $s \in \mathbb{Q}$ is such that $s < r, s \in \mathbb{Q}$ then $s \in \alpha$.

5.2.3 Definition. Let α and β be two cuts. We say $\alpha = \beta$ if α and β are equal as sets.

5.2.4 Examples.

(i) Let r be any rational number and let
$$\alpha_r := \{s \in \mathbb{Q} \mid s < r\}.$$
Since $r - 1 < r$ and $r - 1 \in \mathbb{Q}$, $\alpha_r \neq \emptyset$. Also $\alpha_r \neq \mathbb{Q}$, for $r \notin \alpha$. Finally if $s \in \alpha_r$, then $s < \dfrac{s+r}{2} < r$ and thus $\dfrac{s+r}{2} \in \alpha_r$. Thus α_r has no greatest element and hence α_r is a cut. The cut α_r is called a **rational cut** or the **cut induced** by the rational r. The subset of all rational cuts, $\{\alpha_r \mid r \in \mathbb{Q}\}$ will correspond to a 'copy' of $\mathbb{Q}$ in $\mathbb{R}$.

(ii) Consider $\alpha := \{s \in \mathbb{Q} \mid s^2 < 2\}$. It follows from 4.5.2(ii) that α is a cut but is not a rational cut.

5.2.5 Exercise.

(i) Show that a cut α is a rational cut iff $\mathbb{Q} \setminus \alpha$ has a smallest element. In that case $\alpha = \alpha_r$, r being the smallest element of $\mathbb{Q} \setminus \alpha$.

(ii) Let $r, s \in \mathbb{Q}$ and $r \neq s$. Show that $\alpha_r \neq \alpha_s$.

5.2.6 Proposition. *Let $\alpha \in \mathbb{R}$. Then for every positive rational ϵ there exists $r \in \alpha$ and $s \notin \alpha$ such that $(s - r) < \epsilon$.*

Proof. By the definition of α, $\exists$ rationals $r' \in \alpha$ and $s' \notin \alpha$. By property (ii) of a cut, $r' < s'$. For each positive integer n consider

5.2 Construction by Dedekind's method

the points $r', r' + \frac{1}{n}(s'-r'), r' + \frac{2}{n}(s'-r'), \ldots, r' + \frac{n-1}{n}(s'-r')$, s'. Since $r' \in \alpha$ and $s' \notin \alpha$, there exists $1 \leq k \leq n$ such that $r' + \frac{k-1}{n}(s'-r') \in \alpha$ and $r' + \frac{k}{n}(s'-r') \notin \alpha$. Let

$$r := r' + \frac{k-1}{n}(s'-r')$$

and

$$s := r' + \frac{k}{n}(s'-r').$$

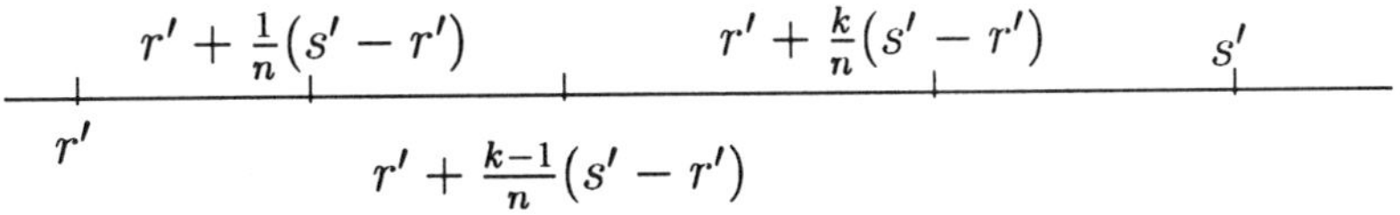

```
        r' + 1/n (s' − r')              r' + k/n (s' − r')              s'
────────┼───────────────────┼───────────────────────┼──────────────────
   r'                    r' + (k−1)/n (s' − r')
```

Figure 5.3

Then $(s - r) = \frac{s' - r'}{n}$. By the Archimedean property of $\mathbb{Q}$, we can choose n sufficiently large such that $\frac{s'-r'}{n} < \epsilon$. ∎

Next we want to define the binary operation of addition on $\mathbb{R}$ such that for the rational cuts α_r and α_s, the addition of α_r and α_s should give us the rational cut α_{r+s} defined by $r+s$. Since

$$\alpha_{r+s} = \{t \in \mathbb{Q} \mid t < r+s\},$$

for $t \in \alpha_{r+s}$, let $r_1 := r - \frac{1}{2}(r+s-t)$ and $s_1 := s - \frac{1}{2}(r+s-t)$. Then $r_1 < r$ and $s_1 < s$. Thus $t = r_1 + s_1$, where $r_1 \in \alpha_r$ and $s_1 \in \alpha_s$. Conversely, if $r_1 \in \alpha_r$ and $s_1 \in \alpha_s$, then $t := r_1 + s_1 < r + s$ and hence $t \in \alpha_{r+s}$. Thus $\alpha_{r+s} = \{r_1 + s_1 \mid r_1 \in \alpha_r, s_1 \in \alpha_s\}$. This motivates our next theorem.

5.2.7 Theorem (Addition on $\mathbb{R}$). *For every $\alpha, \beta \in \mathbb{R}$, let $\alpha \oplus \beta := \{r + s \mid r \in \alpha, s \in \beta\}$. Then $\alpha \oplus \beta \in \mathbb{R}$ and is called the*

sum *or the* **addition** *of α with β. The binary operation $\oplus$ has the following properties:* $\forall\, \alpha, \beta, \gamma \in \mathbb{R}$,

(i) $\alpha \oplus \beta = \beta \oplus \alpha$, *i.e., $\oplus$ is commutative.*

(ii) $(\alpha \oplus \beta) \oplus \gamma = \alpha \oplus (\beta \oplus \gamma)$, *i.e., $\oplus$ is associative.*

(iii) $\alpha \oplus \alpha_0 = \alpha_0 \oplus \alpha = \alpha$, *where $\alpha_0 := \{r \in \mathbb{Q} \mid r < 0\}$. Thus α_0 is the identity for $\oplus$.*

(iv) $\alpha_r \oplus \alpha_s = \alpha_{r+s}$ *for every $r, s \in \mathbb{Q}$.*

Proof. For $\alpha, \beta \in \mathbb{R}$ since $\alpha \neq \emptyset$ and $\beta \neq \emptyset$,

$$\alpha \oplus \beta := \{r + s \mid r \in \alpha, s \in \beta\} \neq \emptyset.$$

Let $r \in \alpha$ and $s \in \beta$ be arbitrary. Since $\exists\, r' \in \alpha$ and $s' \in \beta$ such that $r' > r$ and $s' > s$, we have $r' + s' > r + s$, i.e., $\alpha \oplus \beta$ has no largest element. Also, by proposition 5.2.6, we can find $r \in \alpha, s \in \beta$ such that $(r + 1/2) \notin \alpha$ and $(s + 1/2) \notin \beta$. Thus $(r + s) \in \alpha \oplus \beta$ and $(r + s + 1) \notin \alpha \oplus \beta$, for if $r + s + 1 \in (\alpha \oplus \beta)$, then $r + s + 1 = r' + s'$ for some $r' \in \alpha$ and $s' \in \beta$. But $r' < (r + 1/2)$, and $s' < (s + 1/2)$ as $(r + 1/2) \notin \alpha$ and $(s + 1/2) \notin \beta$. Thus $r' + s' < r + s + 1$, a contradiction. Hence $(r + s + 1) \notin (\alpha + \beta)$. Finally, let $t \in (\alpha \oplus \beta)$ and $v \notin (\alpha \oplus \beta)$. We shall show $t < v$. Let $t = r + s, r \in \alpha, s \in \beta$. If $t \geq v$, then $t = v + k, k \geq 0$. Thus $r + s = v + k$ and we have $(r - k) + s = v$. Since $r - k \leq r$, $(r - k) \in \alpha$ because $r \in \alpha$. Thus $v = (r - k) + s \in \alpha \oplus \beta$, not possible as $v \notin (\alpha + \beta)$. Hence $t \geq v$ is not true, i.e., $t < v$. This proves that $(\alpha \oplus \beta) \in \mathbb{R}$. Proofs of (i), (ii) and (iii) are easy and are left as exercises. We have already proved (iv). ∎

Next, for every $\alpha \in \mathbb{R}$ we want to define its additive inverse. For a rational cut $\alpha_r, r \in \mathbb{Q}$, $\alpha_r \oplus \alpha_{-r} = \alpha_0$. Further

$$\alpha_{-r} \;=\; \{s \in \mathbb{Q} \mid s < -r\}$$

5.2 Construction by Dedekind's method

$$\begin{aligned}
&= \{s \in \mathbb{Q} \mid -s > r\} \\
&= \{-t \in \mathbb{Q} \mid t > r\} \\
&= \{-t \in \mathbb{Q} \mid t \notin \alpha_r, t \neq r\} \\
&= \{-t \in \mathbb{Q} \mid t \notin \alpha_r, t \neq \inf(\mathbb{Q} \setminus \alpha_r)\}.
\end{aligned}$$

This motivates the following theorem.

5.2.8 Theorem. *For any $\alpha \in \mathbb{R}$, let*

$$-\alpha := \{-s \in \mathbb{Q} \mid s \notin \alpha, s \neq \inf(\mathbb{Q} \setminus \alpha)\}.$$

Then $(-\alpha) \in \mathbb{R}$ is the unique cut such that

$$\alpha \oplus (-\alpha) = \alpha_0 = (-\alpha) \oplus \alpha.$$

The element $-\alpha$ is called the **additive inverse** of α.

Proof. We first show that $-\alpha$ is indeed a cut. Since $\alpha \neq \emptyset$ and $\alpha \neq \mathbb{Q}$, we have $-\alpha \neq \emptyset$ and $-\alpha \neq \mathbb{Q}$. Next, let $s \in (-\alpha)$ and $t \notin (-\alpha)$. Then $-s \notin \alpha$, $-t \in \alpha$ and hence $-t < -s$, i.e., $s < t$. Clearly $-\alpha$ has no largest element, for $\mathbb{Q} \setminus \alpha$ may have a smallest element, say r_0, but then $-r_0 \notin (-\alpha)$. Hence $(-\alpha) \in \mathbb{R}$. To check that α has the property $\alpha \oplus (-\alpha) = \alpha_0$, note that $\forall\, r \in \alpha$ and $s \in -\alpha$ we have $s = -t, t \notin \alpha$. Thus $r < t$ and hence $r + s = r - t = -(t - r) < 0$. Thus $(r + s) \in \alpha_0$. Conversely if $t < 0$, then $\forall\, r \in \alpha, t = r + (t - r)$. Let $s = t - r$. We show that $s \in (-\alpha)$, i.e., $-s \notin \alpha$. For that we only note that $-s = (r - t) > r$ and $r \in \alpha$, thus $-s = (r - t) \notin \alpha$. This proves that $\alpha \oplus (-\alpha) = \alpha_0$. Since $(-\alpha) \oplus \alpha = \alpha \oplus (-\alpha)$, by theorem 5.2.7, we also have $(-\alpha) \oplus \alpha = \alpha_0$. To prove the uniqueness of $(-\alpha)$, suppose $\beta \in \mathbb{R}$ also has the property that $\alpha \oplus \beta = \beta \oplus \alpha = \alpha$. Then using theorem 5.2.7, we have $\beta = \beta \oplus \alpha_0 = \beta \oplus (\alpha \oplus (-\alpha)) = (\beta \oplus \alpha) \oplus (-\alpha) = \alpha_0 \oplus (-\alpha) = -\alpha$. This proves the theorem completely. ■

5.2.9 Definition. A real number $\alpha \in \mathbb{R}$ is said to be **positive** if $0 \in \alpha$ and α is said to be **negative** if $0 \notin \alpha$ and $\alpha \neq \alpha_0$. We write $\alpha > 0$ if α is positive and $\alpha < 0$ if α is negative.

Note that $\alpha \in \mathbb{R}$ is positive iff it contains some positive rationals and α will be negative iff $\mathbb{Q} \setminus \alpha$ contains some negative rationals.

5.2.10 Exercise. For $\alpha, \beta \in \mathbb{R}$, prove the following:

(i) One and only one of the alternatives holds: $\alpha = \alpha_0$; α is positive; α is negative.

(ii) $-\alpha_0 = \alpha_0$.

(iii) If $\alpha > 0$ and $\beta > 0$ then $(\alpha \oplus \beta) > 0$.

(iv) $-(-\alpha) = \alpha$.

(v) $-(\alpha \oplus \beta) = (-\alpha) \oplus (-\beta)$.

(vi) If $\alpha \neq \alpha_0$, α is positive iff $-\alpha$ is negative.

Next, we would like to define the binary operation of multiplication, denoted by $\odot$, on $\mathbb{R}$ in such a way that it extends the binary operation of multiplication of rationals. Thus for $r, s \in \mathbb{Q}$, we would like to define $\alpha_r \odot \alpha_s$ in such a way that $\alpha_r \odot \alpha_s = \alpha_{rs}$. We consider the case when both r and s are nonnegative rationals. Since $\alpha_{rs} = \{t \in \mathbb{Q} | t < rs\}$, for $t \in \alpha_{rs}$ either $t \leq 0$ or $t > 0$. If $t > 0$, using exercise 4.3.4 we can choose a $\delta \in \mathbb{Q}$ such that $0 < \delta < 1$ and $t < \delta rs$. Let $r' := r\delta$ and $s' := t/r'$. Then $r' < r$ and $s' < s$. Hence $r' \in \alpha_r$ and $s' \in \alpha_s$ with $t = r's'$. Thus $\alpha_{rs} \subseteq \{t \in \mathbb{Q} | t \leq 0\} \cup \{t = r's' | r' > 0, s' > 0, r' \in \alpha_r \text{ and } s' \in \alpha_s\}$. Clearly, $\{t \in \mathbb{Q} | t \leq 0\} \cup \{t = r's' | r' \in \alpha_r, s' \in \alpha_s, r' > 0, s' > 0\} \subseteq \alpha_{rs}$. Hence, for $r > 0, s > 0$ we have

$$\alpha_{rs} = \{t \in \mathbb{Q} | t \leq 0\} \cup \{t = r's' | r' \in \alpha_r, s' \in \alpha_s, r' > 0, s' > 0\}.$$

This motivates our next definition.

5.2 Construction by Dedekind's method

5.2.11 Definition. Let $\alpha, \beta \in \mathbb{R}$. Define $\alpha \odot \beta$, called the **product** of α with β, as follows:
If $\alpha > 0, \beta > 0$, let

$$\alpha \odot \beta := \{r \in \mathbb{Q} | r \leq 0\} \cup \{r = st | s \in \alpha, t \in \beta, s > 0, t > 0\}.$$

In other cases, let

$$\alpha \odot \beta := \begin{cases} \alpha_0 & \text{if either } \alpha = 0 \text{ or } \beta = 0, \\ (-\alpha) \odot (-\beta) & \text{if } \alpha < 0 \text{ and } \beta < 0, \\ -((-\alpha) \odot \beta) & \text{if } \alpha < 0 \text{ and } \beta > 0, \\ -(\alpha \odot (-\beta)) & \text{if } \alpha > 0, \beta < 0. \end{cases}$$

The product $\alpha \odot \beta$ is well-defined for every $\alpha, \beta \in \mathbb{R}$ and has the properties expected of it, is proved in the next theorem.

5.2.12 Theorem (Multiplication on $\mathbb{R}$). *For every $\alpha, \beta, \gamma \in \mathbb{R}$ the following holds:*

(i) $\alpha \odot \beta \in \mathbb{R}$, *i.e.*, $\odot$ *is a well-defined binary operation on $\mathbb{R}$ and $\alpha \odot \beta > 0$ if either $\alpha > 0$ and $\beta > 0$, or if $\alpha < 0$ and $\beta < 0$.*

(ii) $\alpha \odot \beta = \beta \odot \alpha$, *i.e.*, $\odot$ *is commutative.*

(iii) $\alpha \odot (-\beta) = (-\alpha) \odot \beta = -(\alpha \odot \beta)$.

(iv) $(\alpha \odot \beta) \odot \gamma = \alpha \odot (\beta \odot \gamma)$, *i.e.*, $\odot$ *is associative.*

(v) $\alpha \odot \alpha_0 = \alpha_0 \odot \alpha = \alpha_0$ *and* $\alpha \odot \alpha_1 = \alpha_1 \odot \alpha = \alpha$, *where α_1 is the cut induced by the rational $1 \in \mathbb{Q}$. Thus α_1 is an identity for $\odot$ on $\mathbb{R}$.*

(vi) *For $\alpha > 0$, let $\alpha^{-1} := \{r \in \mathbb{Q} | r \leq 0\} \cup \{r = 1/s | s \notin \alpha, s > 0$ and $s \neq s_0\}$, where s_0 is the least element of $\mathbb{Q} \setminus \alpha$, if any. For $\alpha < 0$, let $\alpha^{-1} := -(-\alpha)^{-1}$. Then $\alpha \odot \alpha^{-1} = \alpha^{-1} \odot \alpha = \alpha_1 \ \forall \ \alpha \in \mathbb{R}, \alpha \neq \alpha_0$. Thus for the binary operation $\odot$ on $\mathbb{R} \setminus \{\alpha_0\}$, every element has an inverse with respect to the identity α_1.*

(vii) $\alpha \odot (\beta \oplus \gamma) = (\alpha \odot \beta) \oplus (\alpha \odot \gamma)$ and $(\beta \oplus \gamma) \odot \alpha = (\beta \odot \alpha) \oplus (\gamma \odot \alpha)$ $\forall\, \alpha, \beta, \gamma \in \mathbb{R}$, *i.e.*, $\odot$ *distributes over* $\oplus$.

(viii) $\alpha_r \odot \alpha_s = \alpha_{rs}$ *for every* $r, s \in \mathbb{Q}$.

Proof. To prove (i), we only have to check that $\alpha \odot \beta \in \mathbb{R}$ for $\alpha > 0, \beta > 0$ where

$$\alpha \odot \beta := \{r \in \mathbb{Q} \mid r \leq 0\} \cup \{r = st \mid s \in \alpha, t \in \beta, s > 0, t > 0\}.$$

Obviously $\alpha \odot \beta \neq \emptyset$. Since $\alpha > 0, \beta > 0$, using proposition 5.2.6 we can find $s > 0, t > 0$ such that $s \in \alpha$ and $t \in \beta$ with $s + 1 \notin \alpha$, $t + 1 \notin \beta$. We claim that $st + s + t + 1 \notin \alpha \odot \beta$ and hence $\alpha \odot \beta \neq \mathbb{Q}$. For if not, as $(st + s + t + 1) > 0$, we will have $st + s + t + 1 = s't'$ for some $s' \in \alpha, t' \in \beta$ with $s' > 0, t' > 0$. Since $s' \in \alpha$, and $s + 1 \notin \alpha$, we have $s' < s + 1$. Similarly $t' < t + 1$ and hence $s't' < st + s + t + 1$, a contradiction. Thus $st + s + t + 1 \notin \alpha \odot \beta$. Next, we show that $\forall\, r \in \alpha \odot \beta$ and $s \notin \alpha \odot \beta, r < s$. Without loss of generality, we may assume that $r > 0$, for $s \notin \alpha\beta$ implies $s > 0$ and hence $r < s$ if $r \leq 0$. Suppose $r > s$. Then $r = st$ for some $t > 1$. Since $r > 0$ and $r \in \alpha \odot \beta, r = r's'$ for $r' \in \alpha, s' \in \beta$ with $r' > 0, s' > 0$. Thus $st = r's'$ and hence $s = \dfrac{r's'}{t}$. Since $\dfrac{r'}{t} < r', \dfrac{r'}{t} \in \alpha$ and $s' \in \beta$. Thus $s \in \alpha \odot \beta$, a contradiction. Hence $r < s$. Finally, we show that $\alpha \odot \beta$ has not greatest element. Let $r \in \alpha$. Without loss of generality let $r > 0$ and $r = st$ where $s \in \alpha, t \in \beta$ and $s > 0, t > 0$. Since α has no greatest element, $\exists\, s_1 \in \alpha$ with $s_1 > r$. Then $r_1 = s_1 t \in \alpha \odot \beta$ and $r_1 = s_1 t > st = r$. Hence $\alpha \odot \beta$ has no greatest element. Thus $\alpha \odot \beta \in \mathbb{R}$. That $\alpha \odot \beta > 0$ if either $\alpha > 0$ and $\beta > 0$ or if $\alpha < 0$ and $\beta < 0$, is obvious because in that case $\alpha \odot \beta$ will have some positive rationals. This proves (i).

To prove (ii), (iii), (iv) and (v) it is enough to prove them for the case $\alpha > 0, \beta > 0$ and $\gamma > 0$. For example to prove (ii), let $\alpha > 0, \beta > 0$. It follows clearly from the definition that $\alpha \odot \beta = \beta \odot \alpha$. In case $\alpha = \alpha_0$ or $\beta = \alpha_0$, clearly $\alpha_0 \odot \beta = \beta \odot \alpha_0$. If $\alpha < 0$ and

5.2 Construction by Dedekind's method

$\beta > 0$, then by definition

$$\begin{aligned} \alpha \odot \beta &= -((-\alpha) \odot \beta) \\ &= -(\beta \odot (-\alpha)) \\ &= \beta \odot \alpha. \end{aligned}$$

This proves (ii) completely. Proofs of (iii), (iv) and (v) can be supplied on similar lines.

We prove (vi) next. Let $\alpha > 0$ and

$$\alpha^{-1} := \{r \in \mathbb{Q} | r \leq 0\} \cup \{r = 1/s | s \notin \alpha, s > 0 \text{ and } s \neq s_0\},$$

where s_0 is the least element of $\mathbb{Q} \setminus \alpha$, if any. We first check that α^{-1} is a cut. Clearly $\alpha^{-1} \neq \emptyset$ and since $\alpha > 0$, $\exists\, s > 0, s \in \alpha$, i.e., $1/s \notin \alpha^{-1}$. Thus $\alpha^{-1} \neq \mathbb{Q}$. Let $r \in \alpha^{-1}$ and $s \notin \alpha^{-1}$. Then either $r \leq 0$ in which case $s > 0 \geq r$. Or $r > 0$ and $1/r \notin \alpha$. Since $s \notin \alpha^{-1}, 1/s \in \alpha$ and hence $1/r > 1/s$, i.e., $s > r$. Finally, let $r \in \alpha^{-1}, r > 0$. Since $r = 1/s$ for some $s \notin \alpha$ with $s > 0$ and s being not the least element of $\mathbb{Q} \setminus \alpha$, we can find $s_1 \notin \alpha$ such that $0 < s_1 < s$. But then $r_1 := \dfrac{1}{s_1} > \dfrac{1}{s} = r$ and hence α^{-1} has no largest element. This shows that α^{-1} is a cut. We show next that

$$\alpha \odot \alpha^{-1} = \{r \in \mathbb{Q} | r \leq 0\} \cup \{rs | r \in \alpha, s \in \alpha^{-1}, r > 0, s > 0\} = \alpha_1,$$

where α_1 is the cut induced by the rational 1. If $t \in \alpha \odot \alpha^{-1}$ and $t > 0$, then $t = rs$ for some $r \in \alpha, s \in \alpha^{-1}$ with $r > 0, s > 0$. Since $s \in \alpha^{-1}, s = 1/t'$ for some $t' \notin \alpha$ and t' not the least element of $\mathbb{Q} \setminus \alpha$. Thus $r < t'$ and hence $rs = r/t' < 1$, i.e., $t < 1$. Thus $\alpha\alpha^{-1} \subseteq \alpha_1$. Conversely if $t \in \alpha_1$, i.e., $t < 1$, choose any $r \in \alpha, r > 0$ and put $s = t/r$. We show that $s \in \alpha^{-1}$, i.e., $s = 1/k$ for some $k \notin \alpha$ and k not the least element of $\mathbb{Q} \setminus \alpha$. As $s = t/r$ we take $k = r/t$ and note $r/t > r$ and hence $r/t \notin \alpha$. Also r/t cannot be the least element of $\mathbb{Q} \setminus \alpha$, for α has no largest element. This completes the proof that $\alpha \odot \alpha^{-1} = \alpha_1$, when $\alpha > 0$. For $\alpha < 0$ we use (iii) and the above argument to deduce $\alpha \odot \alpha^{-1} = \alpha_1$. This proves (vi).

Finally, to prove (vii), we again consider different cases. For example assume first that $\alpha > 0, \beta > 0$ and $\gamma > 0$. Then $\alpha \odot (\beta \oplus \gamma) > 0$ by exercise 4.2.7. To show that $\alpha\odot(\beta\oplus\gamma) = (\alpha\odot\beta)\oplus(\beta\odot\gamma)$, let $r \in \alpha \odot (\beta \oplus \gamma)$. If $r \leq 0$, clearly r belongs to both $\alpha \odot \beta$ and $\alpha \odot \gamma$ and hence $r \in (\alpha \odot \beta) \oplus (\alpha \odot \gamma)$ also. So suppose $r > 0$. Then $r = s(t+k)$ where $s \in \alpha, t \in \beta, k \in \alpha$ and $(t+k) > 0, s > 0$. Thus $r = st + sk$. If $t \leq 0$, then $k > 0$ and $st < 0$. Hence $st \in \alpha \odot \beta$ and $sk \in \alpha \odot \gamma$ showing that $r \in (\alpha \odot \beta) \oplus (\alpha \odot \gamma)$. Similarly if $k \leq 0$ and $t > 0$, then again $r \in (\alpha \odot \beta) \oplus (\alpha \odot \gamma)$. Finally if both $t > 0$ and $k > 0$, we have $r \in (\alpha \odot \beta) \oplus (\alpha \odot \gamma)$. Conversely, let $r \in (\alpha \odot \beta) \oplus (\alpha \odot \gamma)$. If $r \leq 0$, then clearly $r \in \alpha \odot (\beta \oplus \gamma)$. Let $r > 0$ and $r = st + s'k$, where $s, s' \in \alpha, t \in \beta, k \in \gamma$ and all are positive. If $s = s'$, then clearly $r = s(t+k) \in \alpha \odot (\beta \oplus \gamma)$. Thus we may assume without loss of generality that $s > s'$, i.e., $s'/s < 1$. Then

$$r = st + s'k = s(t + \frac{s'k}{s}).$$

Since $ks'/s < k$ and $k \in \gamma$, it follows that $k(s'/s) \in \gamma$ and hence $r \in \alpha \odot (\beta \oplus \gamma)$. This proves that $\alpha \odot (\beta \oplus \gamma) = (\alpha \odot \beta) \oplus (\alpha \odot \gamma)$, where $\alpha > 0, \beta > 0, \gamma > 0$. The proofs for the other cases can be given using this case and the definition of the product. For example, let $\alpha < 0, \beta > 0$ and $\gamma > 0$. Then $(\beta \oplus \gamma) > 0$ and using (iii) above we have

$$\begin{aligned} \alpha \odot (\beta \oplus \gamma) &= -((-\alpha) \odot (\beta \oplus \gamma)) \\ &= -(((-\alpha) \odot \beta) \oplus ((-\alpha) \odot \gamma)) \\ &= -((-(\alpha \odot \beta)) \oplus (-(\alpha \odot \gamma))) \\ &= (\alpha \odot \beta) \oplus (\alpha \odot \gamma). \end{aligned}$$

We leave details of the other cases as exercises. This proves (vii). The statement (viii) has already been proved just before definition 5.2.11. ∎

5.2 Construction by Dedekind's method

5.2.13 Definition. For $\alpha, \beta \in \mathbb{R}$ we say that α is **bigger** than β (or β is **smaller** than α) written as $\alpha > \beta$ (or $\beta < \alpha$), if $\beta \subsetneq \alpha$, i.e., β is a proper subset of α. We write $\alpha \geq \beta$ if either $\alpha > \beta$ or $\alpha = \beta$.

5.2.14 Theorem (Properties of order). *For $\alpha, \beta, \gamma \in \mathbb{R}$, the following hold:*

(i) $\alpha > 0$ iff $\alpha > \alpha_0$.

(ii) *If* $\alpha > \beta$, $\alpha \oplus \gamma > \beta \oplus \gamma$, *and* $\alpha \odot \gamma > \beta \odot \gamma$ *if* $\gamma > 0$ *also.*

(iii) $\alpha > \beta$ *iff* $\alpha \oplus (-\beta) > \alpha_0$. *We denote* $\alpha \oplus (-\beta)$ *by* $\alpha \ominus \beta$.

(iv) *One and only one of the following holds:*

  (a) $\alpha = \beta$.

  (b) $\alpha < \beta$.

  (c) $\alpha > \beta$.

(v) *The relation $<$ is irreflexive (i.e., $\alpha < \alpha$ does not hold for $\alpha \in \mathbb{R}$), antisymmetric (i.e., $\alpha < \beta$ and $\beta < \alpha$ imply $\alpha = \beta$) and transitive (i.e., $\alpha < \beta$ and $\beta < \gamma$ imply $\alpha < \gamma$).*

(vi) *If $\alpha > 0$ and $\beta > 0$, $\alpha \oplus \beta > 0$ and $\alpha \odot \beta > 0$.*

(vii) *For $r, s \in \mathbb{Q}$, $\alpha_r < \alpha_s$ iff $r < s$.*

Proof. (i) Let $\alpha > 0$. Then $0 \in \alpha$ and hence, by definition of a cut, $s \in \alpha$ if $s < 0$. Thus $\alpha_0 \subsetneq \alpha$, i.e., $\alpha > \alpha_0$. Conversely, let $\alpha > \alpha_0$. Then $\exists \, s \in \mathbb{Q}$ such that $s \in \alpha$ but $s \notin \alpha_0$. Again by the definition of the cut, this implies $s > 0$ and hence $0 \in \alpha$. Thus $\alpha > 0$.

(ii) Let $\alpha > \beta$. Let $r \in \alpha$ be such that $r \notin \beta$. Suppose for every $s \in \gamma$, $r + s \in \beta \oplus \gamma$. Then $r = (r+s) - s \in (\beta \oplus \gamma) \oplus (-\gamma) = \beta \oplus \alpha_0 = \beta$, not true. Thus there exists $s \in \gamma$ such that $r + s \notin \beta \oplus \gamma$. Clearly $r + s \in \alpha + \gamma$. Hence $\alpha \oplus \gamma > \beta \oplus \gamma$. Similar arguments will show that $\alpha \odot \gamma > \beta \odot \gamma$ if $\gamma > 0$, proving (ii).

To prove (iii) let $\alpha > \beta$. Then $\beta \subsetneq \alpha$ and $\exists\, r \in \alpha$ such that $r \notin \beta$. Choose $s > r, s \subset \alpha$ (which is possible since α has no largest element). Then $(s - r) > 0$ and $(s - r) \in (\alpha \ominus \beta)$. Thus $\alpha \ominus \beta$ has positive rationals and hence $(\alpha - \beta) > \alpha_0$. Conversely, if $(\alpha \ominus \beta) > \alpha_0$, then $\exists\, r \in (\alpha \ominus \beta)$ such that $r > 0$. Since $r \in (\alpha \ominus \beta)$, $\exists\, t \in \alpha$ and $s \notin \beta$ such that $r = t - s$ and s is not the least element of $\mathbb{Q} \setminus \beta$. Since $t > s, s \in \alpha$. Hence $\exists\, s \in \alpha$ but $s \notin \beta$. To complete the proof we claim that $\forall\, s \in \beta$, $\exists\, r \in \alpha$ such that $s < r$ and hence $s \in \alpha$. For if not, then $\exists\, s_0 \in \beta$ such that $s_0 \geq r$, $\forall\, r \in \alpha$. But then $r \in \beta$ and hence $\alpha \subseteq \beta$. Either $\alpha = \beta$ or $\alpha < \beta$. If $\alpha = \beta$, then $\alpha \ominus \beta = \alpha_0$, not possible as $(\alpha \ominus \beta) > \alpha_0$. If $\alpha < \beta$, then by earlier part $(\beta \ominus \alpha) > \alpha_0$, i.e., $-(\alpha \ominus \beta) > \alpha_0$. Now by (ii), we have $\alpha_0 > \alpha_0$ which is not true. Hence $\beta < \alpha$. This proves (iii).

(iv) Suppose $\alpha \neq \beta$. Consider $\alpha \ominus \beta$. Either $\alpha \ominus \beta$ will have positive rationals, in which case $\alpha \ominus \beta > \alpha_0$ and hence, by (iii), $\alpha > \beta$. Or $\mathbb{Q} \setminus \alpha \ominus \beta$ will include some negative rationals, in which case $\alpha \ominus \beta < \alpha_0$. Thus either $\alpha > \beta$ or $\beta > \alpha$. Clearly both of the given alternatives cannot hold together.

(v) Follows from (iv) and the definition of the order $<$.

(vi) Follows from (i), exercise 5.2.10(iii) and theorem 5.2.12(i).

(vii) Let $r, s \in \mathbb{Q}$ be such that $r < s$. Note that $\forall\, r' < r < s$, $r' \in \alpha_s$ but $r \notin \alpha_r$. Thus $\alpha_r < \alpha_s$. Conversely, if $\alpha_r < \alpha_s$ then $\exists\, p \in \alpha_s$ such that $p \notin \alpha_r$. Thus $r \leq p < s$, i.e., $r < s$. ∎

Theorems 5.2.7, 5.2.12 and 5.2.14 tell us that under the binary operations of the addition, multiplication and with the order defined in them, $\mathbb{R}$ becomes an ordered field. We show that $\mathbb{R}$ is also order complete. Note that if $S \subseteq \mathbb{R}$ is any finite nonempty set, then $lub(S)$ exists because of theorem 5.2.14(iii) and in fact $lub(S) \in S$. Thus every finite nonempty subset S of $\mathbb{R}$ has a largest element. Similarly, every finite nonempty subset of $\mathbb{R}$ has a smallest element.

5.2 Construction by Dedekind's method

5.2.15 Theorem (Order completeness). *The ordered field $\mathbb{R}$ has least upper bound property, i.e., for every nonempty subset S of $\mathbb{R}$ that is bounded above, the $lub(S)$ exists.*

Proof. Let $\beta \in \mathbb{R}$ be an upper bound of S, i.e., $\alpha \leq \beta \; \forall \; \alpha \in S$. Let $\gamma := \bigcup_{\alpha \in S} \{r \in \mathbb{Q} | r \in \alpha \text{ for some } \alpha \in S\}$. We first show that $\gamma \in \mathbb{R}$. Clearly $\gamma \neq \emptyset$. Since $\alpha \leq \beta \; \forall \; \alpha \in S$, clearly $\gamma \subseteq \beta$ and since $\beta \neq \mathbb{Q}$, we have $\gamma \neq \mathbb{Q}$. If $r \in \gamma$ and $s \notin \gamma$, then $r \in \alpha$ for some $\alpha \in S$ and $s \notin \alpha$. Thus $r < s$. Finally γ has no largest element, for if $r \in \gamma$ then $r \in \alpha$ for some α and hence $\exists \; r' \in \alpha$ with $r < r'$. Clearly $r' \in \gamma$. This shows that $\gamma \in \mathbb{R}$. Clearly $\gamma \geq \alpha \; \forall \; \alpha \in S$ and $\gamma \leq \beta$. Hence $\gamma = lub(S)$. ∎

5.2.16 Exercise. Let $\emptyset \neq S \subset \mathbb{R}$ be bounded below. Show that $glb(S)$ exists.

As a consequence of theorem 5.2.15 we have the following theorem. For $r \in \mathbb{Q}$ and $\alpha \in \mathbb{R}$, we write $s\alpha$ for $\alpha_s \odot \alpha$.

5.2.17 Theorem (Archimedean property). *Let $\alpha, \beta \in \mathbb{R}$ be such that $\alpha > 0$. Then there exists a positive integer n such that $n\alpha > \beta$.*

Proof. Let $S := \{n\alpha | n \in \mathbb{N}\}$. Suppose the required claim is not true. Then $\forall \; n \in \mathbb{N}, n\alpha \leq \beta$. Thus S is bounded above and hence by theorem 5.2.15, $\tilde{\alpha} := lub(S)$ exists. Since $\alpha > 0, \tilde{\alpha} - \alpha < \tilde{\alpha}$ and hence $(\tilde{\alpha} - \alpha)$ is not an upper bound for S, as $\tilde{\alpha}$ is the least upper bound. Thus $\exists \; m \in \mathbb{N}$, such that $\tilde{\alpha} - \alpha < m\alpha$, i.e., $\tilde{\alpha} < (m+1)\alpha$. This is not possible since $(m+1)\alpha \in S$ and $\tilde{\alpha} = lub(S)$. Hence our assumption is wrong, i.e., the required claim holds. ∎

In the above discussion we have constructed a set $\mathbb{R}$; defined binary operations of addition and multiplication on $\mathbb{R}$ so that $\mathbb{R}$ becomes a field; and finally defined an order on $\mathbb{R}$ so that $\mathbb{R}$ becomes an ordered field which is order complete. We shall see the consequences of this completeness later in chapter 6. Presently, we just look at it from the geometric viewpoint. First note that $\mathbb{R}$ includes a copy of $\mathbb{Q}$. To see this recall that some elements of $\mathbb{R}$ can be constructed using elements of $\mathbb{Q}$, for example for every $r \in \mathbb{Q}$ we had $\alpha_r \in \mathbb{R}$, the cut induced by r. We saw in theorem 5.2.7 (iv), theorem 5.2.12 (viii) and 5.2.14 (v) that the association $r \longmapsto \alpha_r$ from $\mathbb{Q}$ to $\mathbb{R}$ is a map which has the following properties:

(i) $\quad \alpha_r \oplus \alpha_s = \alpha_{r+s} \ \forall \ r, s \in \mathbb{Q}$.

(ii) $\quad \alpha_{rs} = \alpha_r \odot \alpha_s \ \forall \ r, s \in \mathbb{Q}$.

(iii) $\quad \alpha_r < \alpha_s$ iff $r < s$ for some $r, s \in \mathbb{Q}$.

In other words, if we identify the points α_r of $\mathbb{R}$ with $r \in \mathbb{Q}$, then this identification respects the addition, multiplication and order of $\mathbb{Q}$. Thus we can regard the subset $\{\alpha_r : r \in \mathbb{Q}\}$ of $\mathbb{R}$ as a copy of $\mathbb{Q}$ and call them the **rational numbers**. One says rationals form a subfield of $\mathbb{R}$ or that $\mathbb{R}$ is a field extension of $\mathbb{Q}$. In view of this, we shall denote $\alpha_r \in \mathbb{R}$ by r itself. Also we shall write $\alpha + \beta$ for $\alpha \oplus \beta$ and $\alpha\beta$ for $\alpha \odot \beta$ from now onwards. The points $\alpha \in \mathbb{R}$ which are not of the form α_r for any $r \in \mathbb{Q}$, are called the **irrational numbers**. The 'gaps' which existed on the rational number line are filled by the irrational points and the horizontal line represents $\mathbb{R}$. Now we can say that the dream of Pythagoras is fulfilled: to every point on the line corresponds a number (and it took about 2000 years to realize that dream). We had seen that rationals were dense, i.e., between any two distinct rationals there existed another rational different from the given ones. It is easy to see that between any two distinct real numbers there are 'infinite' numbers of distinct real numbers. Geometrically this seems obvious. It is also not difficult to see this with the present definition of $\mathbb{R}$. Let $\alpha, \beta \in \mathbb{R}$ and $\alpha < \beta$. Then $(\alpha + \alpha) < (\alpha + \beta)$, i.e., $\alpha(\alpha_1 + \alpha_1) < (\alpha + \beta)$, recall that α_1 is the cut induced by the rational 1. Thus $\alpha < (\alpha + \beta)(\alpha_1 + \alpha_1)^{-1}$. Let $\gamma := (\alpha_1 + \alpha_1)^{-1}$. Then $\alpha < (\alpha+\beta)\gamma$. Similarly, $(\alpha+\beta) < (\beta+\beta) = \beta(\alpha_1+\alpha_1)$. Thus

5.2 Construction by Dedekind's method

$(\alpha+\beta)(\alpha_1+\alpha_1)^{-1} < \beta$, i.e., $(\alpha+\beta)\gamma < \beta$. Hence $\alpha < (\alpha+\beta)\gamma < \beta$. Thus, we have proved the following theorem.

5.2.18 Theorem. *If $\alpha, \beta \in \mathbb{R}, \alpha < \beta$, then $\exists \, \delta \in \mathbb{R}$ such that $\alpha < \delta < \beta$.*

In fact we have the following theorem:

5.2.19 Theorem (Denseness of rationals and irrationals). *Let $\alpha, \beta \in \mathbb{R}, \alpha < \beta$. Then $\exists$ a rational δ and an irrational γ such that $\alpha < \delta < \beta$ and $\alpha < \gamma < \beta$.*

Proof. Since $\alpha < \beta, \alpha$ is a proper subset of β, i.e., $\exists \, r \in \beta$ such that $r \notin \alpha$. Then clearly $\alpha < \alpha_r < \beta$, for if $t \in \alpha$, then $t < r$ (as $r \notin \alpha$) and hence $t \in \alpha_r$. Similarly, if $s \in \alpha_r$ then $s < r \in \beta$ and hence $s \in \beta$. Thus the rational $\delta := \alpha_r$ satisfies the required property. To show the existence of an irrational, let γ' be any irrational number (they exist as shown in example 5.2.4(ii)). Since $\alpha < \beta$, we have $\alpha - \gamma' < \beta - \gamma'$. Choose any rational α_r such that $(\alpha - \gamma') < \alpha_r < \beta - \gamma'$. Then $\alpha < \alpha_r + \gamma' < \beta$. Put $\gamma := \alpha_r + \gamma'$. We show that γ is an irrational number. For if not, i.e., γ is rational then $\gamma + \alpha_{-r} = \alpha_r + \alpha_{-r} + \gamma' = \alpha_0 + \gamma' = \gamma'$ will also be rational, which is not true. ∎

Next we look at the Cauchy completeness of the ordered field $\mathbb{R}$. One can introduce the notion of **absolute value** for elements of $\mathbb{R}$ (as was done on $\mathbb{Q}$ in section 4.7) using the order on $\mathbb{R}$. For $\alpha \in \mathbb{R}$ define $|\alpha| := \alpha$ if $\alpha \geq \alpha_0$ and $|\alpha| = -\alpha$ if $\alpha \leq \alpha_0$. It is easy to see (as in theorem 4.7.2) that $|\alpha+\beta| \leq |\alpha|+|\beta| \; \forall \; \alpha, \beta \in \mathbb{R}$. This is called the **triangle inequality**. Note that for $\alpha \in \mathbb{Q}$, its absolute value as defined here coincides with its absolute value as defined in section 4.7. We can also consider sequences of real numbers, analyze their Cauchyness and convergence using this absolute value in the same manner as we did for sequences in $\mathbb{Q}$ in section 4.8. We shall show that every Cauchy sequence in $\mathbb{R}$ converges, i.e., $\mathbb{R}$

is Cauchy complete. We first give some elementary facts about sequences. These facts will be useful later on also.

5.2.20 Proposition. *For every sequence $\{\alpha_n\}_{n\geq 1}$ in $\mathbb{R}$ the following hold:*

(i) *If $\{\alpha_n\}_{n\geq 1}$ is a Cauchy sequence then it is **bounded**, i.e., $|\alpha_n| \leq \beta$ for some $\beta \in \mathbb{R}$ and $\forall\, n \in \mathbb{N}$.*

(ii) *If $\{\alpha_n\}_{n\geq 1}$ converges, then it is also Cauchy.*

(iii) *Let $\lim_{n\to\infty} \alpha_n = \alpha \neq 0$. Then there exists $n_0 \in \mathbb{N}$ and a positive rational K such that $|\alpha_n| \geq K\ \forall\, n \geq n_0$.*

Proof. (i) Let $\{\alpha_n\}_{n\geq 1}$ be a Cauchy sequence in $\mathbb{R}$ and let $\epsilon > 0$ be any real number. Choose $n_0 \in \mathbb{N}$ such that
$$|\alpha_n - \alpha_m| \leq \epsilon \quad \forall\, n, m \geq n_0.$$
In particular,
$$|\alpha_n - \alpha_{n_0}| \leq \epsilon \quad \forall\, n \geq n_0.$$
Thus $\forall\, n \geq n_0$, using triangle inequality we have
$$|\alpha_n| \leq |\alpha_n - \alpha_{n_0}| + |\alpha_{n_0}| \leq \epsilon + |\alpha_{n_0}|.$$
Let $\beta := \max\{|\alpha_1|, |\alpha_2|, \ldots, |\alpha_{n_0-1}|, \epsilon + |\alpha_{n_0}|\}$, or choose β any real number bigger than $|\alpha_1|, |\alpha_2|, \ldots, |\alpha_{n_0-1}|$ and $\epsilon + |\alpha_{n_0}|$, for example the sum of all these will do. Then $|\alpha_n| \leq \beta\ \forall\, n$. This proves (i).

To prove (ii), let $\{\alpha_n\}_{n\geq 1}$ converge to α and let $\epsilon > 0$ be given. Choose $n_0 \in \mathbb{N}$ such that $|\alpha_n - \alpha| \leq \epsilon/2\ \forall\, n \geq n_0$. Then $\forall\, n, m \geq n_0$, we have
$$|\alpha_n - \alpha_m| \leq |\alpha_n - \alpha| + |\alpha_m - \alpha| \leq \epsilon/2 + \epsilon/2 = \epsilon.$$
Hence $\{\alpha_n\}_{n\geq 1}$ is also Cauchy.

(iii) Suppose the required claim is not true. Then given any arbitrary rational $\delta > 0$, and any positive integer n we can choose

5.2 Construction by Dedekind's method

a positive integer $n_1 > n$ such that $|\alpha_{n_1}| < \delta$. Since $\{\alpha_n\}_{n\geq 1}$ being Cauchy, given any $\epsilon > 0$, $\exists$ a positive integer n_0 such that

$$|\alpha_n - \alpha_m| < \frac{\epsilon}{2} \quad \forall\, n, m \geq n_0.$$

Apply the above argument for $\delta = \frac{\epsilon}{2}$ and $n = n_0$ to get n_1 such that $|\alpha_{n_1}| < \frac{\epsilon}{2}$. Then $\forall\, n \geq n_2 = \max\{n_0, n_1\}$, we have

$$\begin{aligned}
|\alpha_n| &= |\alpha_n - \alpha_{n_1} + \alpha_{n_1}| \\
&\leq |\alpha_n - \alpha_{n_1}| + |\alpha_{n_1}| \\
&< \epsilon/2 + \epsilon/2 = \epsilon,
\end{aligned}$$

i.e., $\{\alpha_n\}_{n\geq 1}$ converges to 0, a contradiction. Hence the required claim holds. ∎

5.2.21 Definition.

(i) Given a sequence $\{\alpha_n\}_{n\geq 1}$, consider a sequence $\{n_k\}_{n\geq 1}$ of positive integers such that $n_1 < n_2 < \ldots$. Then $\{\alpha_{n_k}\}_{k\geq 1}$ is called a **subsequence** of $\{\alpha_n\}_{n\geq 1}$.

(ii) A sequence $\{\alpha_n\}_{n\geq 1}$ is said to be **monotonically increasing** if $\alpha_n \leq \alpha_{n+1}\ \forall\, n \geq 1$ and $\{\alpha_n\}_{n\geq 1}$ is said to be **monotonically decreasing** if $\alpha_n \geq \alpha_{n+1}\ \forall\, n \geq 1$.

5.2.22 Theorem. *Let $\{\alpha_n\}_{n\geq 1}$ be a sequence in $\mathbb{R}$. Then the following hold:*

(i) *$\{\alpha_n\}_{n\geq 1}$ can converge to at most one point.*

(ii) *$\{\alpha_n\}_{n\geq 1}$ has a subsequence $\{\alpha_{n_k}\}_{k\geq 1}$ which is either monotonically increasing or monotonically decreasing.*

(iii) *If $\{\alpha_n\}_{n\geq 1}$ is monotonically increasing and bounded above, it is convergent.*

(iv) If $\{\alpha_n\}_{n\geq 1}$ is monotonically decreasing and is bounded below, it is convergent.

(v) If $\{\alpha_n\}_{n\geq 1}$ is Cauchy and has a subsequence $\{\alpha_{n_k}\}_{k\geq 1}$, which is convergent, say to $\alpha \in \mathbb{R}$, $\{\alpha_n\}_{n\geq 1}$ itself is convergent to α.

(vi) If $\{\alpha_n\}_{n\geq 1}$ is bounded, $\exists$ a subsequence $\{\alpha_{n_k}\}_{k\geq 1}$ of $\{\alpha_n\}_{n\geq 1}$ which is convergent.

This is called the **Bolzano-Weierstrass property**.

(vii) If $\{\alpha_n\}_{n\geq 1}$ is Cauchy, it is convergent.

This is called the **Cauchy completeness** of $\mathbb{R}$.

Proof. (i) The proof is a repetition of the arguments as in the proof of theorem 4.8.6. Suppose if possible, $\{\alpha_n\}_{n\geq 1}$ converge to α and β. Then given $\epsilon > 0$ we can find positive integers n_1 and n_2 such that

$$|\alpha_n - \alpha| < \epsilon \; \forall \, n \geq n_1 \quad \text{and} \quad |\alpha_n - \beta| < \epsilon \; \forall \, n \geq n_2.$$

These two inequalities imply that $\forall \, n \geq n_0 := \max\{n_1, n_2\}$,

$$\alpha - \epsilon \leq \alpha_n \leq \alpha + \epsilon \quad \text{and} \quad \beta - \epsilon \leq \beta_n < \beta + \epsilon.$$

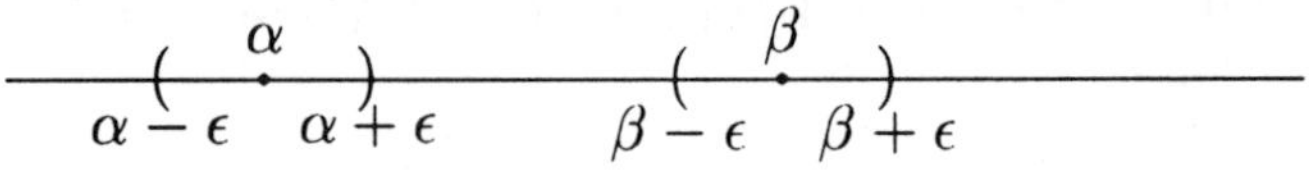

Figure 5.4 : Uniqueness of limit.

Assume $\alpha < \beta$. Then, we could start our proof with an $\epsilon > 0$ such that $\alpha + \epsilon < \beta - \epsilon$, i.e., $(\beta - \alpha) > 2\epsilon$. In that case $\forall \, n \geq n_0$,

$$\alpha - \epsilon < \alpha_n < \alpha + \epsilon < \beta - \epsilon < \alpha_n < \beta + \epsilon,$$

5.2 Construction by Dedekind's method 203

which is obviously absurd. Hence $\alpha < \beta$ is not possible. Similarly, $\beta < \alpha$ is also not possible and hence $\alpha = \beta$ by theorem 5.2.14.

(ii) Let $\{\alpha_n\}_{n\geq 1}$ be a sequence. Consider the set $C := \{n \in \mathbb{N} \,|\, \alpha_m \geq \alpha_n \text{ if } m > n\}$. In case $C = \emptyset, \alpha_n \geq \alpha_m \,\forall\, m > n$. In particular $\alpha_n \geq \alpha_{n+1} \,\forall\, n$, and hence $\{\alpha_n\}_{n\geq 1}$ is monotonically decreasing. Suppose $C \neq \emptyset$ and has finite number of elements. Let n_0 be the largest of these elements. Then $\forall\, n, m > n_0$, if $m > n$ then $\alpha_n \geq \alpha_m$. Thus the sequence $\{\alpha_n\}_{n\geq 1}$ is monotonically decreasing from $(n_0 + 1)^{\text{th}}$ term onwards. Let $n_1 := n_0 + 1, n_2 := n_0 + 2, \ldots$, and get the subsequence $\{\alpha_{n_k}\}_{k\geq 1}$ which is monotonically decreasing. The last possibility is that C is an infinite subset of $\mathbb{N}$. By theorem 2.4.26, $\exists\, n_1 < n_2 < n_3 < \ldots$ in $\mathbb{N}$ such that $n_i \in C \,\forall\, i$. Then clearly by definition $\{\alpha_{n_i}\}_{i\geq 1}$, is a monotonically increasing subsequence of $\{\alpha_n\}_{n\geq 1}$.

(iii) Suppose $\{\alpha_n\}_{n\geq 1}$ is monotonically increasing and $\alpha_n \leq \beta \,\forall\, n \geq 1$. Consider the set $S := \{\alpha_1, \alpha_2, \ldots\}$. This is a nonempty subset of $\mathbb{R}$ and is bounded above. Thus by theorem 5.2.15, $lub(S)$ exists. Let $\alpha := lub(S)$. We claim that $\{\alpha_n\}_{n\geq 1}$ converges to α. To see this, let $\epsilon > 0$ be any real number. Then $\exists\, n_0$ such that $\alpha - \alpha_{n_0} < \epsilon$, for if not then $\alpha - \alpha_n \geq \epsilon \,\forall\, n$, i.e., $\alpha - \epsilon \geq \alpha_n \,\forall\, n$. Hence $\alpha - \epsilon$ will be an upper bound for S, which is not possible as $\alpha - \epsilon < \alpha$. Thus $\alpha - \alpha_{n_0} < \epsilon$, which implies that $\alpha - \alpha_n < \epsilon \,\forall\, n \geq n_0$ as $\alpha_n \geq \alpha_{n_0} \,\forall\, n \geq n_0$. Thus $\{\alpha_n\}$ converges to α. This proves (iii).

(iv) The proof is similar to that of (iii) (use exercise 5.2.13).

(v) Let $\{\alpha_n\}_{n\geq 1}$ be a Cauchy subsequence and let $\{\alpha_{n_k}\}_{k\geq 1}$ be a subsequence of $\{\alpha_n\}_{n\geq 1}$ such that $\{\alpha_{n_k}\}_{k\geq 1}$ converges, say to α. Then after some stage all the elements of $\{\alpha_{n_k}\}_{k\geq 1}$ will be close to α. But as $\{\alpha_n\}_{n\geq 1}$ is Cauchy, its elements are close to each other. Thus, we can see that elements of $\{\alpha_n\}_{n\geq 1}$ will be close to α. More precisely, let $\epsilon > 0$ be given. We can choose positive integers n_0, k_0 such that

$$|\alpha_n - \alpha_m| < \epsilon/2 \quad \forall\, n \geq n_0$$

and

$$|\alpha_{n_k} - \alpha| < \epsilon/2, \quad \forall\, k \geq k_0.$$

Let k be chosen such that n_k greater than both n_0 and n_{k_0}. Then $\forall\, n \geq n_k$

$$|\alpha_n - \alpha| \leq |\alpha_n - \alpha_{n_k}| + |\alpha_{n_k} - \alpha| \leq \epsilon/2 + \epsilon/2 = \epsilon.$$

Hence $\{\alpha_n\}_{n\geq 1}$ converges to α.

(vi) Let $\{\alpha_n\}_{n\geq 1}$ be a sequence which is bounded, i.e, $\exists\, \beta \in \mathbb{R}$ such that $-\beta \leq \alpha_n \leq \beta\ \forall\, n$. By (ii), $\exists$ a subsequence $\{\alpha_{n_k}\}_{k\geq 1}$ of $\{\alpha_n\}_{n\geq 1}$ which is either monotonically increasing or monotonically decreasing. Since $\{\alpha_{n_k}\}_{k\geq 1}$ is also bounded, by (iii) and (iv) $\{\alpha_{n_k}\}_{k\geq 1}$ is convergent.

(vii) Let $\{\alpha_n\}_{n\geq 1}$ be a Cauchy sequence. By proposition 5.2.20, $\{\alpha_n\}_{n\geq 1}$ is bounded. By (vi) it has a convergent subsequence and now by (v) $\{\alpha_n\}_{n\geq 1}$ is convergent. ■

5.2.23 Note. A careful scrutiny of the proofs of theorems 5.2.17, 5.2.20 and 5.2.22 show that we did not used the special nature of elements of $\mathbb{R}$. The above proofs work in any ordered field which is order complete, i.e., these arguments imply that every ordered field which is order complete is also Archimedean and is Cauchy complete.

We close our discussion of the Dedekind method with an answer to the following question. To construct $\mathbb{R}$ from $\mathbb{Q}$ via cuts, we used only the fact that $\mathbb{Q}$ is an ordered field and this process gave us an ordered field which is 'bigger' than $\mathbb{Q}$ and is order complete. Supposing we repeat this construction for the ordered field $\mathbb{R}$, i.e., form cuts of $\mathbb{R}$ and construct the set of all cuts of $\mathbb{R}$. Will we get something 'bigger' than $\mathbb{R}$? The answer is no as is proved in the next theorem.

5.2.24 Theorem (Dedekind). *Let A be any subsets of $\mathbb{R}$ such that*

(i) $\quad A \neq \emptyset, A \neq \mathbb{R}$.

(ii) $\quad \forall\, a \in A\ $ and $\ b \notin A, a < b$.

Then there exists one and only one $x \in \mathbb{R}$ such that either $A = \{y \in \mathbb{R} | y < x\}$ or $A = \{y \in \mathbb{R} \,|\, y \leq x\}$.

Proof. In case the set A has the greatest element, say $x := lub(A) \in A$, then clearly $A \subseteq \{y \in \mathbb{R} | y \leq x\}$ and since $\forall \, y \leq x, y \in A$ (because if $y \notin A$ then $x < y$) we have $A = \{y \in \mathbb{R} | y \leq x\}$. So suppose A does not have the greatest element. Let us consider $\mathbb{R} \setminus A = \{y \in \mathbb{R} | y \notin A\}$. We shall show that $\mathbb{R} \setminus A$ has the smallest element, say x, and $A = \{y \in \mathbb{R} | y < x\}$. Let $x := glb(\mathbb{R} \setminus A)$, which exist for $\mathbb{R} \setminus A$ is bounded below. Clearly $x \notin A$, for otherwise x will be the greatest element of A. Thus $x \in \mathbb{R} \setminus A$, i.e., $\mathbb{R} \setminus A$ has x as its smallest element. Then $\forall \, y \in A, y < x$, as $x \notin A$, and thus $A \subseteq \{y \in \mathbb{R} | y < x\}$. Also, if $y < x$ then $y \in A$ by the given property of A. Hence we will have $A = \{y \in \mathbb{R} | y < x\}$. To prove the uniqueness of x, let $x_1 \in \mathbb{R}$ be such that $y \leq x_1 \leq z \, \forall \, y \in A$ and $z \notin A$. Then x_1 is an upper bound of A and hence $x \leq x_1$. If $x < x_1$, then $t = \dfrac{x + x_1}{2}$ is such that $x < t < x_1$. Either $t \in A$ or $t \notin A$. But $t \in A$ contradicts the fact that $y \leq z \, \forall \, y \in A$, and $t \notin A$ contradicts the fact that $x_1 \leq z \, \forall \, z \notin A$. Hence $x < x_1$ is not true, i.e., $x = x_1$. ∎

5.3 Construction of real numbers by Cantor's method

Once again we start with $\mathbb{Q}$, the ordered field of rational numbers as constructed in chapter 4. As we saw in 5.1.2, geometrically locating a point on the line is equivalent to considering Cauchy sequences of rational numbers converging to that point. The fact that the elements of Cauchy sequences, converging to the same point, are ultimately close to each other motivates the next definition. We denote by $\mathcal{C}$ the set of all Cauchy sequences in $\mathbb{Q}$.

5.3.1 Definition. We say elements $\{r_n\}_{n\geq 1}$ and $\{s_n\}_{n\geq 1}$ of $\mathcal{C}$ are **equivalent** and denote this as $\{r_n\} \sim \{s_n\}$, if given any rational $\epsilon > 0$, there exists a positive integer n_0 such that

$$|r_n - s_n| < \epsilon \quad \forall\, n \geq n_0.$$

5.3.2 Proposition. *The relation $\sim$ has the following properties:*

(i) Let $\{r_n\}_{n\geq 1}$ and $\{s_n\}_{n\geq 1}$ be elements of $\mathcal{C}$ such that both converge to the same point $r \in \mathbb{Q}$. Then $\{r_n\}_{n\geq 1} \sim \{s_n\}_{n\geq 1}$.

(ii) Let $\{r_n\}_{n\geq 1}$ and $\{s_n\}_{n\geq 1}$ be elements of $\mathcal{C}$ such that $\{r_n\}_{n\geq 1} \sim \{s_n\}_{n\geq 1}$. If $\{r_n\}_{n\geq 1}$ converges to some point $r \in \mathbb{Q}$, then $\{s_n\}_{n\geq 1}$ also converges to the same point $r \in \mathbb{Q}$.

(iii) The relation $\sim$ is an equivalence relation on $\mathcal{C}$.

Proof. (i) Let $\{s_n\}_{n\geq 1}, \{r_n\}_{n\geq 1}$ elements of $\mathcal{C}$ be such that both converge to the same point $r \in \mathbb{Q}$. Then, given any rational $\epsilon > 0$, $\exists\, n_1$ and n_2 such that

$$|r_n - r| < \epsilon/2 \quad \forall\, n \geq n_1 \quad \text{and} \quad |s_n - r| < \epsilon/2 \quad \forall\, n \geq n_2.$$

Thus $\forall\, n \geq n_0 := \max\{n_1, n_2\}$,

$$\begin{aligned}
|r_n - s_n| &= |r_n - r + r - s_n| \\
&\leq |r_n - r| + |s_n - r| \\
&< \epsilon/2 + \epsilon/2 = \epsilon.
\end{aligned}$$

Hence $\{r_n\}_{n\geq 1} \sim \{s_n\}_{n\geq 1}$.

(ii) Let $\{r_n\}_{n\geq 1}$ and $\{s_n\}_{n\geq 1}$ be elements of $\mathcal{C}$ such that $\{r_n\}_{n\geq 1} \sim \{s_n\}_{n\geq 1}$ and $\{r_n\}_{n\geq 1}$ converge to $r \in \mathbb{Q}$. Let a rational $\epsilon > 0$ be given. Then $\exists$ positive integers n_1 and n_2 such that

$$|r_n - s_n| < \epsilon/2 \quad \forall\, n \geq n_1 \quad \text{and} \quad |r_n - r| < \epsilon/2 \quad \forall\, n \geq n_2.$$

5.3 Construction by Cantor's method

Then $\forall\, n \geq n_0 := \max\{n_1, n_2\}$, we have

$$\begin{aligned} |s_n - r| &= |s_n - r_n + r_n - r| \\ &\leq |s_n - r_n| + |r_n - r| \\ &< \epsilon/2 + \epsilon/2 = \epsilon. \end{aligned}$$

Hence $\{s_n\}_{n\geq 1}$ converges to $r \in \mathbb{Q}$.

(iii) Showing that the relation $\sim$ is reflexive and symmetric is easy. To show that the relation $\sim$ is also transitive, let $\{r_n\}_{n\geq 1}, \{s_n\}_{n\geq 1}$ and $\{t_n\}_{n\geq 1}$ be elements of $\mathcal{C}$ such that $\{r_n\}_{n\geq 1} \sim \{s_n\}_{n\geq 1}$ and $\{s_n\}_{n\geq 1} \sim \{t_n\}_{n\geq 1}$. Let $\epsilon > 0$ be any rational. Choose positive integers n_1 and n_2 such that

$$|r_n - s_n| < \epsilon/2 \quad \forall\, n \geq n_1 \quad \text{and} \quad |s_n - t_n| < \epsilon/2 \quad \forall\, n \geq n_2.$$

Then $\forall\, n \geq n_0 := \max\{n_1, n_2\}$,

$$\begin{aligned} |r_n - t_n| &= |r_n - s_n + s_n - t_n| \\ &\leq |r_n - s_n| + |s_n - t_n| \\ &< \epsilon/2 + \epsilon/2 = \epsilon. \end{aligned}$$

Hence $\{r_n\}_{n\geq 1} \sim \{t_n\}_{n\geq 1}$. ∎

5.3.3 Exercise. A sequence $\{r_n\}_{n\geq 1}$ of rationals is called a **null sequence** if $\{r_n\}_{n\geq 1}$ converges to zero. Show that for $\{r_n\}_{n\geq 1}$ and $\{s_n\}_{n\geq 1} \in \mathcal{C}, \{r_n\}_{n\geq 1} \sim \{s_n\}_{n\geq 1}$ iff $\{r_n - s_n\}_{n\geq 1}$ is a null sequence.

The equivalence relation $\sim$ partitions the set $\mathcal{C}$ into equivalence classes. Each sequence in a given equivalence class locates the same point on the line. The equivalence class containing the Cauchy sequence $\{r_n\}_{n\geq 1}$ is denoted by $[\{r_n\}]$ and we call the sequence $\{r_n\}_{n\geq 1}$ a representative of this equivalence class. We denote by $\mathbb{R}$ the set of all equivalence classes of $\mathcal{C}$ under the above relation $\sim$ and call it the set of **real numbers**. Note that for $\alpha, \beta \in \mathbb{R}, \alpha = \beta$ iff $\forall$ Cauchy sequence $\{r_n\}_{n\geq 1} \in \alpha$ and

$\{s_n\}_{n\geq 1} \in \beta, \{r_n\}_{n\geq 1} \sim \{s_n\}_{n\geq 1}$. For any rational $r \in \mathbb{Q}$, consider the sequence $\{r_n\}_{n\geq 1}, r_n := r \ \forall \ n \geq 1$. This is called the **constant sequence**. Obviously, the constant sequences are Cauchy, in fact, convergent. The equivalence class containing the constant sequence $\{r_n\}_{n\geq 1}, r_n = r \ \forall \ n \geq 1$, will be denoted by $[r]$. Note that if $\{r_n\}_{n\geq 1} \in [r]$ then $r_n = r$ for all $n \geq n_0$, for some n_0 and hence $\{r_n\}_{n\geq 1}$ converge to r. Thus $\{r_n\}_{n\geq 1}$ locates the point $r \in \mathbb{Q}$ on the line. Our next aim is to define binary operations of addition, $\oplus$, and multiplication, $\odot$, on $\mathbb{R}$. In case $[\{r_n\}]$ locates a point $r \in \mathbb{Q}$ and $[\{s_n\}]$ locates a point $s \in \mathbb{Q}$, we would like $[\{r_n\}] \oplus [\{s_n\}]$ to locate the point $r + s \in \mathbb{Q}$. This forces us to have $[\{r_n\}] \oplus [\{s_n\}] := [\{r_n + s_n\}]$. Thus in general for $[\{r_n\}], [\{s_n\}]$ in $\mathbb{R}$, we may define $[\{r_n\}] \oplus [\{s_n\}] := [\{r_n + s_n\}]$. But $[\{r_n + s_n\}]$ is defined by choosing a representative $\{r_n\}_{n\geq 1} \in [\{r_n\}]$ and a representative $\{s_n\}_{n\geq 1} \in [\{s_n\}]$. To have $[\{r_n + s_n\}]$ uniquely defined, we should show that if $\{r'_n\}_{n\geq 1} \in [\{r_n\}]$ and $\{s'_n\}_{n\geq 1} \in [\{s_n\}]$, then $[\{r'_n + s'_n\}] = [\{r_n + s_n\}]$. We show this in theorem 5.3.4 and also give other properties of addition. Similarly, to locate the product of $[\{r_n\}]$ and $[\{s_n\}]$, it is natural to consider the point located by the product $\{r_n s_n\}_{n\geq 1}$, i.e., define $[\{r_n\}] \odot [\{s_n\}] := [\{r_n s_n\}]$. That, this is well-defined and the other properties of multiplication so defined are proved in theorem 5.3.5. We prove in theorem 5.3.6 that $\mathbb{R}$ forms a field under this addition and multiplication. To compare two points $[\{r_n\}], [\{s_n\}]$, on the line, if $[\{r_n\}]$ is on the left of $[\{s_n\}]$, it is natural to expect that elements of the locating sequence $\{r_n\}_{n\geq 1}$, lie on the left of the elements of the locating sequence $\{s_n\}_{n\geq 1}$ and away from each other by a finite distance after some stage onwards. For example this happens when $[\{r_n\}]$ and $[\{s_n\}]$ determine rationals. This definition of order is made precise in definition 5.3.9 and that the order so defined makes $\mathbb{R}$ an ordered field is proved in theorem 5.3.12. The Cauchy completeness and the order completeness of the ordered field $\mathbb{R}$ so constructed are proved in theorems 5.3.18 and 5.3.20.

5.3 Construction by Cantor's method

5.3.4 Theorem (Addition on $\mathbb{R}$).

(i) Let $\{r_n\}_{n\geq 1}$ and $\{s_n\}_{n\geq 1}$ be elements of C. Then $\{(r_n+s_n)\}_{n\geq 1}$ is also an element of C.

(ii) For $\{r_n\}_{n\geq 1}, \{r'_n\}_{n\geq 1}, \{s_n\}_{n\geq 1}, \{s'_n\}_{n\geq 1}$ in C, if $\{r'_n\}_{n\geq 1} \in [\{r_n\}]$ and $\{s'_n\}_{n\geq 1} \in [\{s_n\}]$ then $[\{r_n + s_n\}] = [\{s'_n + r'_n\}]$. Thus for $[\{r_n\}], [\{s_n\}]$ in $\mathbb{R}$, addition given by

$$[\{r_n\}] \oplus [\{s_n\}] := [\{r_n + s_n\}]$$

is a well-defined binary operation on $\mathbb{R}$.

(iii) $[\{r_n\}] \oplus [\{s_n\}] = [\{s_n\}] \oplus [\{r_n\}]$.

(iv) $([\{r_n\}] \oplus [\{s_n\}]) \oplus [\{t_n\}] = [\{r_n\}] \oplus ([\{s_n\}] \oplus [\{t_n\}])$.

(v) Let $[\{0\}]$ denote the equivalence class containing the constant sequence $\{s_n\}_{n\geq 1}$, where $s_n = 0 \; \forall \, n \geq 1$. Then

$$[\{r_n\}] \oplus [\{0\}] = [\{0\}] \oplus [\{r_n\}] = [\{r_n\}].$$

(vi) For every sequence $\{r_n\}_{n\geq 1}$ in C, consider the sequence $\{s_n\}_{n\geq 1}$ in C where $s_n := -r_n \; \forall \, n \geq 1$. Then

$$[\{r_n\}] \oplus [\{s_n\}] = [\{s_n\}] + [\{r_n\}] = [\{0\}].$$

Properties (iii), (iv), (v) and (vi) tell us that under $\oplus$, the binary operation of addition given by (ii), $\mathbb{R}$ is a commutative group.

(vii) For $r, s \in \mathbb{Q}$,
$$[r] \oplus [s] = [r+s].$$

Proof. (i) Let $\{r_n\}_{n\geq 1}$ and $\{s_n\}_{n\geq 1}$ be Cauchy sequences. Then, given any rational $\epsilon > 0$, $\exists$ positive integers n_1 and n_2 such that

$$|r_n - r_m| < \epsilon/2 \quad \forall \, n \geq n_1 \quad \text{and} \quad |s_n - s_m| < \epsilon/2 \quad \forall \, n \geq n_2.$$

If we take $n_0 := \max\{n_1, n_2\}$, then for every $n \geq n_0$,

$$|(r_n + s_n) - (r_m + s_m)| \leq |r_n - r_m| + |s_n - s_m|$$
$$< \epsilon/2 + \epsilon/2 = \epsilon.$$

Hence $\{(r_n + s_n)\}_{n \geq 1}$ is also a Cauchy sequence of rationals.

(ii) If $\{r'_n\}_{n \geq 1} \in [r_n]$ and $\{s'_n\}_{n \geq 1} \in [s_n]$, then $\{r'_n\}_{n \geq 1} \sim \{r_n\}_{n \geq 1}$ and $\{s'_n\}_{n \geq 1} \sim \{s_n\}_{n \geq 1}$. Thus given $\epsilon > 0$, $\exists\ n_1$ and n_2 positive integers such that

$$|r'_n - r_n] < \epsilon/2 \quad \forall\ n \geq n_1 \quad \text{and} \quad |s_n - s'_n| < \epsilon/2 \quad \forall\ n \geq n_2.$$

Thus for $n \geq n_0 := \max\{n_1, n_2\}$, we have

$$|(r_n + s_n) - (r'_n + s'_n)| \leq |r_n - r'_n| + |s_n - s'_n|$$
$$< \epsilon/2 + \epsilon/2 = \epsilon.$$

Hence $\{(r'_n + s'_n)\}_{n \geq 1} \sim \{(r_n + s_n)\}_{n \geq 1}$, i.e., $[\{r'_n + s'_n\}] = [\{r_n + s_n\}]$. Thus given $[\{r_n\}], [\{s_n\}] \in \mathbb{R}$,

$$[\{r_n\}] + [\{s_n\}] := [\{r_n + s_n\}]$$

is well-defined. This proves (ii). Proof of (iii), (iv), (v), (vi) and (vii) are simple and are left as exercises. ∎

5.3.5 Theorem (Multiplication on $\mathbb{R}$).

(i) Let $\{r_n\}_{n \geq 1}$ and $\{s_n\}_{n \geq 1}$ be elements of C. Then $\{(r_n s_n)\}_{n \geq 1}$ is also an element of C.

(ii) Let $\{r'_n\}_{n \geq 1} \in [\{r_n\}]$ and $\{s'_n\}_{n \geq 1} \in [\{s_n\}]$. Then $\{r'_n s'_n\}_{n \geq 1} \sim \{r_n s_n\}_{n \geq 1}$ and hence for $[\{r_n\}], [\{s_n\}] \in \mathbb{R}$, the multiplication given by

$$[\{r_n\}] \odot [\{s_n\}] := [\{r_n s_n\}]$$

is a well-defined binary operation on $\mathbb{R}$.

(iii) $[\{r_n\}] \odot [\{s_n\}] = [\{s_n\}][\{r_n\}]$.

5.3 Construction by Cantor's method

(iv) $[\{r_n\}] \odot ([\{s_n\}] \odot [\{t_n\}]) = ([\{r_n\}] \odot [\{s_n\}]) \odot [\{t_n\}]$.

(v) $[\{r_n\}] \odot [1] = [1] \odot [\{r_n\}] = [\{r_n\}]$.

(vi) If $[\{r_n\}] \neq [0]$, *then* $\exists$ *a unique* $[\{s_n\}] \in \mathbb{R}$ *such that* $[\{s_n\}] \neq [0]$ *and*
$$[\{r_n\}] \odot [\{s_n\}] = [\{s_n\}] \odot [\{r_n\}] = [1].$$

This unique element is denoted by $[r_n]^{-1}$ *and is called the* **multiplicative inverse** *of* $[r_n]$.

(vii) $[\{r_n\}] \odot [0] = [0]$.

(viii) $[r] \odot [s] = [rs]\ \forall\ r, s \in \mathbb{Q}$.

Proof. In the proofs of all the statements one uses the fact that every Cauchy sequence $\{r_n\}_{n\geq 1}$ of rational numbers is bounded, i.e., $\exists\ M \in \mathbb{Q}$ such that $|r_n| \leq M\ \forall\ n \geq 1$. The proof of this statement is already given in theorem 4.8.7.

(i) Since $\{r_n\}_{n\geq 1}$ and $\{s_n\}_{n\geq 1}$ are Cauchy, they are bounded, i.e., $\exists\ M_1, M_2 \in \mathbb{Q}$ such that $|r_n| \leq M_1$ and $|s_n| \leq M_2\ \forall\ n \geq 1$. Let $M := \max\{M_1, M_2\}$. Then $|r_n| \leq M$ and $|s_n| \leq M\ \forall\ n \geq 1$. Let $\epsilon > 0$ be an arbitrary rational. Choose positive integers n_1 and n_2, such that

$$|r_n - r_m| < \frac{\epsilon}{2M}\quad \forall\ n, m \geq n_1\ \text{ and }\ |s_n - s_m| < \frac{\epsilon}{2M}\quad \forall\ n, m \geq n_2.$$

Then $\forall\ n \geq n_0 := \max\{n_1, n_2\}$,

$$\begin{aligned}
|r_n s_n - r_m s_m| &= |r_n s_n - r_n s_m + r_n s_m - r_m s_m| \\
&\leq |r_n s_n - r_n s_m| + |r_n s_m - r_m s_m| \\
&\leq |r_n||s_n - s_m| + |r_n - r_m||s_m| \\
&\leq \frac{M\epsilon}{2M} + \frac{M\epsilon}{2M} = \epsilon.
\end{aligned}$$

Hence the sequence $\{r_n s_n\}_{n\geq 1}$ is Cauchy. This proves (i).

(ii) Let $\{r'_n\}_{n\geq 1} \in [\{r_n\}]$ and $\{s'_n\}_{n\geq 1} \in [\{s_n\}]$. Then $\{r'_n\}_{n\geq 1} \sim \{r_n\}_{n\geq 1}$ and $\{s'_n\}_{n\geq 1} \sim \{s_n\}_{n\geq 1}$. Since $\{r_n\}_{n\geq 1}, \{r'_n\}_{n\geq 1}, \{s_n\}_{n\geq 1}$ and $\{s'_n\}_{n\geq 1}$ are all Cauchy, they are bounded. Thus $\exists$ positive rationals M_1, M_2, M_3, M_4 such that $|r_n| \leq M_1, |r'_n| < M_2, |s_n| \leq M_3$, and $|s'_n| \leq M_4$ $\forall$ $n \geq 1$. Let $M := \max\{M_1, M_2, M_3, M_4\}$. Then $\forall$ $n \geq 1$, each of $|r_n|, |r'_n|, |s_n|, |s'_n|$ is less than or equal to M. Now let $\epsilon > 0$ be any rational. Then we can choose positive integers n_1 and n_2 such that

$$|r_n - r'_n| < \frac{\epsilon}{2M} \quad \forall\, n \geq n_1 \quad \text{and} \quad |s_n - s'_n| < \frac{\epsilon}{2M} \quad \forall\, n \geq n_2.$$

Then $\forall\, n \geq n_0 := \max\{n_1, n_2\}$, we have

$$\begin{aligned}
|r_n s_n - r'_n s'_n| &= |r_n s_n - r'_n s_n + r'_n s_n - r'_n s'_n| \\
&\leq |r_n s_n - r'_n s_n| + |r'_n s_n - r'_n s'_n| \\
&= |r_n - r'_n||s_n| + |r'_n||s_n - s'_n| \\
&\leq \frac{M\epsilon}{2M} + \frac{M\epsilon}{2M} = \epsilon.
\end{aligned}$$

Thus $\{r'_n s'_n\}_{n\geq 1} \sim \{r_n s_n\}_{n\geq 1}$ and hence $[\{r_n\}] \odot [\{s_n\}] := [\{r_n s_n\}]$ is well-defined. This proves (ii).

Proofs of (iii), (iv), (v), (vii) and (viii) are easy and are left as exercises. We prove (vi). Let $[\{r_n\}] \in \mathbb{R}$ and $[\{r_n\}] \neq [0]$. Then by proposition 5.3.2 (i) the sequence $\{r_n\}_{n\geq 1}$ does not converge to $0 \in \mathbb{Q}$. Using theorem 5.2.20 (iii), $\exists$ a rational $K > 0$ and a positive integer n_1 such that $\forall\, n \geq n_1$,

$$|r_n| > K > 0.$$

Consider the sequence $\{s_n\}_{n\geq 1}$ defined by

$$s_n := \begin{cases} 0 & \text{if } 1 \leq n < n_1, \\ 1/r_n & \text{if } n \geq n_1. \end{cases}$$

We first show that $\{s_n\}_{n\geq 1}$ is a Cauchy sequence. Since $\{r_n\}_{n\geq 1}$ is Cauchy, given $\epsilon > 0$ any arbitrary rational we can choose a positive integer n_2 such that $\forall\, n \geq n_2$,

$$|r_n - r_m| < \epsilon K^2 \quad \forall\, n \geq n_2.$$

5.3 Construction by Cantor's method 213

Then $\forall n \geq n_3 := \max\{n_1, n_2\}$, we have

$$\begin{aligned} |s_n - s_m| &= |1/r_n - 1/r_m| \\ &= \frac{|r_m - r_n|}{|r_n r_m|} \\ &< \frac{\epsilon K^2}{K^2} = \epsilon. \end{aligned}$$

Hence $\{s_n\}_{n \geq 1}$ is a Cauchy sequence. Further $\forall n \geq n_3, \dfrac{r_n}{s_n} = 1$ and hence

$$[\{r_n\}] \odot [\{s_n\}] = [1].$$

This also proves (vi). ∎

5.3.6 Theorem. $(\mathbb{R}, \oplus, \odot)$ *is a field with the binary operation of addition $\oplus$ given by theorem 5.3.4 and of multiplication $\odot$ given by theorem 5.3.5.*

Proof. It follows from theorem 5.3.4 that $(\mathbb{R}, \oplus)$ is a commutative group and theorem 5.3.5 tells us that $(\mathbb{R} \setminus [0], \odot)$ is also a commutative group. The only property left to be checked is that multiplication distributes over addition, i.e., for $[r_n], [s_n]$ and $[t_n]$ in $\mathbb{R}$,

$$\begin{aligned} [\{r_n\}] &\odot ([\{t_n\}] \oplus [\{s_n\}]) \\ &= ([\{r_n\}] \odot [\{t_n\}]) \oplus ([\{r_n\}] \odot [\{s_n\}]) \end{aligned} \quad (5.1)$$

and

$$\begin{aligned} ([\{t_n\}] &\oplus [\{s_n\}]) \odot [\{r_n\}] \\ &= ([\{t_n\}] \odot [\{r_n\}]) \oplus ([\{s_n\}] \odot [\{r_n\}]). \end{aligned} \quad (5.2)$$

To prove (5.1), we have to show that $\{a_n\}_{n \geq 1} \sim \{b_n\}_{n \geq 1}$ for every $\{a_n\}_{n \geq 1} \in [\{r_n\}] \odot ([\{t_n\}] \oplus [\{s_n\}])$ and every $\{b_n\}_{n \geq 1} \in ([\{r_n\}] \odot [\{t_n\}]) \oplus ([\{r_n\}] \odot [\{s_n\}])$. By definition $\{a_n\}_{n \geq 1}$ and $\{b_n\}_{n \geq 1}$ are given by $a_n = r'_n(t'_n + s'_n)$ and $b_n = r''_n t''_n + r'''_n s''_n$ for all $n \geq 1$, for some $\{r'_n\}_{n\geq 1}, \{r''_n\}_{n\geq 1}$ in $[\{r_n\}], \{t'_n\}_{n\geq 1}, \{t''_n\}_{n\geq 1}$ in $[\{t_n\}]$ and $\{s'_n\}_{n\geq 1}, \{s''_n\}_{n\geq 1}$ in $[\{s_n\}]$. Since $\{r'_n\}_{n\geq 1} \sim \{r_n\}_{n\geq 1} \sim \{r''_n\}_{n\geq 1}$, $\{s'_n\}_{n\geq 1} \sim \{s_n\}_{n\geq 1} \sim \{s''_n\}_{n\geq 1}$ and $\{t'_n\}_{n\geq 1} \sim \{t_n\}_{n\geq 1} \sim \{t''_n\}_{n\geq 1}$, using theorem 5.3.4 (ii) and 5.3.5 (ii), we have

$$\begin{aligned}
\{a_n\}_{n\geq 1} &\sim \{r'_n(t'_n + s'_n)\}_{n\geq 1} \\
&\sim \{r'_n t'_n + r'_n s'_n\}_{n\geq 1} \\
&\sim \{r_n t_n + r_n s_n\}_{n\geq 1} \\
&\sim \{r''_n t''_n + r''_n s''_n\}_{n\geq 1} \\
&\sim \{r''_n(t''_n + s''_n)\}_{n\geq 1} \\
&\sim \{b_n\}_{n\geq 1}.
\end{aligned}$$

This proves (5.1). Proof of (5.2) is similar. ∎

5.3.7 Definition. Let $\{r_n\}_{n\geq 1}$ be any Cauchy sequence of rationals. We say $\{r_n\}_{n\geq 1}$ is a **positive sequence** if $\exists$ some rational $s > 0$ and a positive integer n_0 such that $r_n > s \ \forall \ n \geq n_0$. We say $\{r_n\}_{n\geq 1}$ is a **negative sequence** if $\{-r_n\}_{n\geq 1}$ is a positive sequence.

5.3.8 Theorem.

(i) Let $\{r_n\}_{n\geq 1}, \{s_n\}_{n\geq 1} \in \mathcal{C}$ be such that $\{r_n\}_{n\geq 1} \sim \{s_n\}_{n\geq 1}$ and $\{r_n\}_{n\geq 1}$ is positive. Then $\{s_n\}_{n\geq 1}$ is also positive.

(ii) Given any sequence $\{r_n\}_{n\geq 1} \in \mathcal{C}$, either $\{r_n\}_{n\geq 1}$ is positive, or $\{r_n\}_{n\geq 1}$ is negative or $\{r_n\}_{n\geq 1} \sim \{0\}$. Further, only one of these alternatives holds.

(iii) If $\{r_n\}_{n\geq 1}, \{s_n\}_{n\geq 1} \in \mathcal{C}$ are both positive, so are the sequences $\{r_n + s_n\}_{n\geq 1}$ and $\{r_n s_n\}_{n\geq 1}$.

Proof. (i) Since $\{r_n\}_{n\geq 1}$ is positive, we can find a rational $s > 0$ and a positive integer n_0 such that $r_n > s > 0 \ \forall \ n > n_1$. Since $\{r_n\}_{n\geq 1} \sim \{s_n\}_{n\geq 1}$, given $\epsilon = \dfrac{s}{2}$ we can choose a positive integer

5.3 Construction by Cantor's method

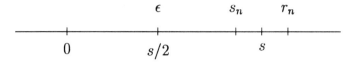

Figure 5.5

n_2 such that $|r_n - s_n| < s/2 \; \forall \; n \geq n_2$. Let $n_0 := \max\{n_1, n_2\}$. Then $\forall \; n \geq n_0 \; s_n > r_n - s/2 > s/2$. Hence $\{s_n\}_{n \geq 1}$ is also positive. This proves (i).

To prove (ii), we note that if $\{r_n\}_{n \geq 1} \in \mathcal{C}$ and is not equivalent to $\{0\}$, then $\exists$ a positive rational $\eta > 0$ such that $\forall$ positive integer n, $\exists$ a positive integer $n' > n$ such that

$$|r_{n'}| > 2\eta.$$

Also, since $\{r_n\}_{n \geq 1}$ is Cauchy, we can choose positive integer n_0 such that

$$|r_n - r_m| < \eta \quad \forall \; n, m \geq n_0.$$

In particular, $\forall \; n \geq n_0$

$$|r_n - r_{n_0}| < \eta, \quad \text{i.e.,} \quad r_{n_0} + \eta > r_n > r_{n_0} - \eta.$$

Choose $n'_0 > n_0$ as given above. Then

$$|r_{n'_0}| > 2\eta.$$

Now two possibilities arise. Either $r_{n'_0} < 0$ or $r_{n'_0} > 0$. In case $r_{n'_0} > 0$, we have $\forall \; n \geq n_0$

$$\begin{aligned} r_n &> r_{n'_0} - \eta \\ &> 2\eta - \eta = \eta. \end{aligned}$$

Hence $\{r_n\}_{n \geq 1}$ is a positive sequence. In case $r_{n'_0} < 0$, we will get $\forall \; n \geq n_0$,

$$\begin{aligned} -r_n &> -r_{n'_0} - \eta \\ &> 2\eta - \eta = \eta. \end{aligned}$$

Hence $\{r_n\}_{n\geq 1}$ will be a negative sequence in this case. This proves that one of the three alternatives holds. Clearly, only one of them can hold. This proves (ii).

Proof of (iii) is easy (keeping in mind that every Cauchy sequence of rationals is bounded), and is left as an exercise. ∎

5.3.9 Definition. A real number $\alpha \in \mathbb{R}$ is said to be **positive** if there exists $\{r_n\}_{n\geq 1} \in \alpha$ such that $\{r_n\}_{n\geq 1}$ is positive. We write this as $\alpha > [0]$. For real numbers $\alpha, \beta \in \mathbb{R}$, we say α is **larger** than β if $(\alpha - \beta) > [0]$ and write this as $\alpha > \beta$. We write $\alpha \geq \beta$ or $\beta \leq \alpha$ if either $\alpha > \beta$ or $\alpha = \beta$.

Note that if $\alpha > [0]$ then every $\{r_n\}_{n\geq 1} \in \alpha$ is a positive sequence because of theorem 5.3.8 (i).

5.3.10 Proposition. *For $\alpha, \beta \in \mathbb{R}, \alpha > \beta$ iff for every sequence $\{r_n\}_{n\geq 1} \in \alpha$ and $\{s_n\}_{n\geq 1} \in \beta$, $\{r_n - s_n\}_{n\geq 1}$ is positive.*

Proof. For $\alpha, \beta \in \mathbb{R}$, let $\alpha > \beta$. Then $(\alpha - \beta) > [0]$, i.e., $\exists$ a sequence $\{t_n\}_{n\geq 1} \in (\alpha - \beta)$ such that $\{t_n\}_{n\geq 1}$ is positive. Thus for $\{r_n\}_{n\geq 1} \in \alpha$ and $\{s_n\}_{n\geq 1} \in \beta$, since $\{r_n - s_n\}_{n\geq 1} \in (\alpha - \beta)$, and $\{t_n\}_{n\geq 1} \sim \{r_n - s_n\}_{n\geq 1}$, by theorem 5.3.8(i), $\{r_n - s_n\}_{n\geq 1}$ is positive. Conversely, for every sequence $\{r_n\}_{n\geq 1} \in \alpha$ and $\{s_n\}_{n\geq 1} \in \beta$, $\{r_n - s_n\}_{n\geq 1}$ positive implies that $(\alpha - \beta) > [0]$, i.e., $\alpha > \beta$. ∎

5.3.11 Theorem. *For $\alpha, \beta \in \mathbb{R}$, the following hold:*

(i)   *$\alpha < \alpha$ is never true, i.e., $<$ is irreflexive.*

(ii) *If $\alpha < \beta$ and $\beta < \alpha$, then $\alpha = \beta$, i.e., $<$ is anti-symmetric.*

(iii) *If $\alpha < \beta$ and $\beta < \gamma$, then $\alpha < \gamma$, i.e., $<$ is transitive.*

(iv) *One and only one of the alternatives holds: $\alpha = [0], \alpha > [0]$, or $[0] > \alpha$.*

(v)   *If $\alpha \geq [0]$ and $\beta \geq [0]$, then $(\alpha \oplus \beta) \geq [0]$ and $(\alpha \odot \beta) \geq [0]$.*

5.3 Construction by Cantor's method

(vi) For every $r, s \in \mathbb{Q}, [r] > [s]$ iff $r > s$.

Proof. Proof of (i) is obvious. To prove (ii), let $\alpha < \beta$ and $\beta < \alpha$. Let $\{r_n\}_{n \geq 1} \in \alpha$ and $\{s_n\}_{n \geq 1} \in \beta$. Then by proposition 5.3.10, both $\{r_n - s_n\}_{n \geq 1}$ and $\{s_n - r_n\}_{n \geq 1}$ are positive sequences. Hence there exists a positive integer n_0 such that $r_n = s_n \ \forall \ n \geq n_0$. Thus $\{r_n\}_{n \geq 1} \sim \{s_n\}_{n \geq 1}$ and hence $\alpha = \beta$. This proves (ii). Next, if $\alpha < \beta$ and $\beta < \gamma$, then $(\beta - \alpha) > [0]$ and $(\gamma - \beta) > [0]$. Thus the required claim will follow from (v). Statement (iv) follows from theorem 5.3.8 (ii). Statement (v), is obvious if either $\alpha = [0]$ are $\beta = [0]$, for in that case $\alpha \oplus \beta = \alpha$ or β and $\alpha \odot \beta = [0]$. Suppose $\alpha > [0]$ and $\beta > [0]$. Then $\exists$ positive sequences $\{r_n\}_{n \geq 1} \in \alpha$ and $\{s_n\}_{n \geq 1} \in \beta$. Now it follows from theorem 5.3.8 (iii) that both $\{r_n + s_n\}_{n \geq 1}$ and $\{r_n s_n\}_{n \geq 1}$ are positive sequences. Since $\{r_n + s_n\}_{n \geq 1} \in (\alpha \oplus \beta)$ and $\{r_n s_n\}_{n \geq 1} \in \alpha \odot \beta$, we have $(\alpha \oplus \beta) > 0$ and $\alpha \odot \beta > 0$. This proves (v). Proof of (vi) follows from proposition 5.3.10. ∎

5.3.12 Theorem.

(i) $(\mathbb{R}, \oplus, \odot, <)$ *is an ordered field with addition $\oplus$ as defined by theorem 5.3.4, multiplication $\odot$ as defined by theorem 5.3.5 and order $<$ as defined in 5.3.10.*

(ii) $\{[r] \mid r \in \mathbb{Q}\}$ *is a subfield of $\mathbb{R}$ isomorphic to $\mathbb{Q}$, i.e., $\mathbb{R}$ contains a copy of $\mathbb{Q}$.*

Proof. Statement (i) is already proved in theorems 5.3.4, 5.3.5 and 5.3.10. To prove (ii) we note that $r \longmapsto [r]$ is a map from $\mathbb{Q}$ into $\mathbb{R}$, which is one-one and respects the order and the field properties: i.e., $\forall \ r, s \in \mathbb{Q}$

$$[r+s] = [r] \oplus [s], \ [rs] = [r] \odot [s] \quad \text{and} \quad [r] \geq [s] \text{ iff } r > s.$$

Clearly $\{[r] \mid r \in \mathbb{Q}\}$ is a subfield of $\mathbb{R}$. Thus we can treat $\{[r] \mid r \in \mathbb{Q}\}$ as a copy of $\mathbb{Q}$ inside $\mathbb{R}$. ∎

In view of the above theorem, for $r \in \mathbb{Q}$ we shall denote

$[r] \in \mathbb{R}$ by r itself. Further for $\alpha, \beta \in \mathbb{R}$ we shall denote $\alpha \oplus \beta$ by $\alpha + \beta$ and $\alpha \odot \beta$ by $\alpha\beta$, respectively.

5.3.13 Theorem (Denseness of rationals). *Let α and β be real numbers such that $\alpha > \beta$. Then $\exists\, r \in \mathbb{Q}$ such that $\alpha > r > \beta$.*

Proof. Let $\{r_n\}_{n\geq 1} \in \alpha$ and $\{s_n\}_{n\geq 1} \in \beta$. Since $\alpha > \beta$, $\exists$ a rational $\eta > 0$ and a positive integer n_1 such that $(r_n - s_n) > \eta\ \forall\, n \geq n_1$. Also $\{r_n\}_{n\geq 1}$ and $\{s_n\}_{n\geq 1}$ being Cauchy sequences, we can choose positive integers n_2 and n_3 such that

$$|r_n - r_m| < \eta/3\ \forall\, n, m \geq n_2 \quad \text{and} \quad |s_n - s_m| < \eta/3\ \forall\, n, m \geq n_3.$$

Let $n_0 := \max\{n_1, n_2, n_3\}$. Then $\forall\, n \geq n_0$

$$s_{n_0} - \eta/3 < s_n < s_{n_0} + \eta/3 \quad \text{and} \quad (r_{n_0} - s_{n_0}) > \eta.$$

Thus $\forall\, n \geq n_0$,

$$\frac{r_{n_0} + s_{n_0}}{2} - s_n = \frac{1}{2}[(r_{n_0} - s_n) + (s_{n_0} - s_n)]$$
$$> \frac{1}{2}[(r_{n_0} - s_{n_0} - \eta/3) - \eta/3]$$
$$> \frac{1}{2}[\eta - 2\eta/3] = \eta/6.$$

Hence for $r := \dfrac{r_{n_0} + s_{n_0}}{2} \in \mathbb{Q}$, $[r] - [\{s_n\}] = [r] - \beta > 0$. Similarly $\forall\, n \geq n_0$,

$$\left(r_n - \frac{r_{n_0} + s_{n_0}}{2}\right) > \eta/6,$$

i.e., $[r] < \alpha$. Hence $\beta < [r] = r < \alpha$. ∎

5.3.14 Exercise. Let $\alpha \in \mathbb{R}$. Show that there exist rationals $r_1, r_2 \in \mathbb{Q}$ such that $r_1 < \alpha < r_2$.

Since $\mathbb{R}$ is an ordered field, we can define the notion of absolute value for elements of $\mathbb{R}$, as we did for the ordered field of rationals in section 4.7.

5.3 Construction by Cantor's method

5.3.15 Definition. For $\alpha \in \mathbb{R}$, define the **absolute value** of α, denoted by $|\alpha|$, as follows:

$$|\alpha| := \begin{cases} \alpha & \text{if } \alpha \geq [0], \\ -\alpha & \text{if } \alpha < [0]. \end{cases}$$

Note that if $\{r_n\}_{n\geq 1} \in \alpha$, then

$$|\alpha| := \begin{cases} [\{r_n\}] & \text{if } \{r_n\}_{n\geq 1} \geq 0, \\ -[\{r_n\}] := [\{-r_n\}] & \text{if } \{r_n\}_{n\geq 1} < 0. \end{cases}$$

5.3.16. Proposition. *For $\alpha, \beta \in \mathbb{R}$ and $r \in \mathbb{Q}$ the following hold:*

(i) $|\alpha + \beta| \leq |\alpha| + |\beta|$.

(ii) $|\alpha \beta| = |\alpha|\,|\beta|$.

(iii) $|[r]| = [|r|]$.

Proof. Proofs of (i) and (ii) are the same as in the case of rationals (see theorem 4.7.2). Proof of (iii) is obvious in view of the remark after theorem 5.3.12. ∎

Using this notion of absolute value we can consider convergence and Cauchyness of sequences in $\mathbb{R}$, as was done for the rationals in section 4.8.

5.3.17. Definition.

(i) A sequence $\{\alpha_n\}_{n\geq 1}$ of elements of $\mathbb{R}$ is called a **Cauchy sequence** if $\forall$ real number $\epsilon > 0$, $\exists$ a positive integer n_0 such that

$$|\alpha_n - \alpha_m| < \epsilon \quad \forall\, n, m \geq n_0.$$

(ii) A sequence $\{\alpha_n\}_{n\geq 1}$ in $\mathbb{R}$ is said to be **convergent** to a real number $\alpha \in \mathbb{R}$ if $\forall$ real numbers $\epsilon > 0$, $\exists$ a positive integer n_0 such that
$$|\alpha_n - \alpha| < \epsilon \quad \forall\, n \geq n_0.$$

The relation between Cauchy sequences in $\mathbb{Q}$ and Cauchy sequences in $\mathbb{R}$ is given by the next theorem.

5.3.18 Theorem.

(i) Let $r_n, r \in \mathbb{Q}, n = 1, 2, \ldots$. Then the following hold:

(a) $\{[r_n]\}_{n\geq 1}$ is a Cauchy sequence in $\mathbb{R}$ iff $\{r_n\}_{n\geq 1}$ is a Cauchy sequence in $\mathbb{Q}$.

(b) The sequence $\{[r_n]\}_{n\geq 1}$ converges to $[r]$ in $\mathbb{R}$ iff $\{r_n\}_{n\geq 1}$ converges to $r \in \mathbb{Q}$.

(c) $\{r_n\}_{n\geq 1} \in [r]$ iff $\{[r_n]\}_{n\geq 1}$ converges to $[r]$.

(ii) Let $\alpha \in \mathbb{R}$ and $\{r_n\}_{n\geq 1} \in \alpha$. For $\alpha_n := [r_n]$, $\{\alpha_n\}_{n\geq 1}$ is a Cauchy sequence in $\mathbb{R}$ and $\{\alpha_n\}_{n\geq 1}$ converges to α in $\mathbb{R}$.

Proof. (i) (a) Let the sequence $\{[r_n]\}_{n\geq 1}$ be Cauchy in $\mathbb{R}$. Then given any $\epsilon \in \mathbb{Q} \subseteq \mathbb{R}$ with $\epsilon > 0$, we can find a positive integer n_0 such that
$$|[r_n] - [r_m]| < \epsilon \quad \forall\, n \geq n_0.$$
Since $[r_n] - [r_m] = [r_n - r_m]$, it follows from proposition 5.3.16(iii) that
$$|r_n - r_m| < \epsilon \quad \forall\, n \geq n_0.$$
Hence $\{r_n\}_{n\geq 1}$ is Cauchy in $\mathbb{Q}$. Conversely, let $\{r_n\}_{n\geq 1}$ be Cauchy and $\epsilon > 0$ be any real number. Using theorem 5.3.13 choose a rational $\epsilon' > 0$, such that $\epsilon > [\epsilon'] > 0$. Since $\{r_n\}_{n\geq 1}$ is Cauchy, we can find a positive integer n_0 such that
$$|r_n - r_m| < \epsilon' \quad \forall\, n \geq n_0.$$

5.3 Construction by Cantor's method

Once again, since $|r_n - r_m| = |[r_n - r_m]| = |[r_n] - [r_m]|$, we have
$$\|[r_n] - [r_m]\| < [\epsilon'] < \epsilon \quad \forall n \geq n_0.$$
Hence $\{[r_n]\}_{n \geq 1}$ is Cauchy in $\mathbb{R}$. This proves (a). Proof of (b) is similar and is left as an exercise. To prove (c) note that if $\{r_n\}_{n \geq 1} \in [r]$, then $\{r_n\}_{n \geq 1}$ is equivalent to the constant sequence $\{r\}$ and hence converges to r by proposition 5.3.2. Conversely, if $\{r_n\}_{n \geq 1}$ converges to $r \in \mathbb{Q}$, then obviously $\{r_n\}_{n \geq 1}$ is equivalent to the constant sequence $\{r\}$ and hence $\{r_n\}_{n \geq 1} \in [r]$.

To prove (ii), let $\{r_n\}_{n \geq 1} \in \alpha$. Since $\{r_n\}_{n \geq 1}$ is a Cauchy sequence in $\mathbb{Q}$, $\{[r_n]\}_{n \geq 1}$ is also a Cauchy sequence in $\mathbb{R}$ by (i) (a) above. Let $\epsilon > 0$ be any real number. Using theorem 5.3.11 choose a rational $\epsilon' > 0$ such that $0 < [\epsilon'] < \epsilon$. Since $\{r_n\}_{n \geq 1}$ is Cauchy in $\mathbb{Q}$, choose a positive integer n_0 such that
$$|r_n - r_m| < \epsilon' \quad \forall n, m \geq n_0. \tag{5.3}$$
Let $\alpha_n := [r_n]$. Then $\{r_n - r_m\}_{m \geq 1} \in (\alpha_n - \alpha)$ and $r_n < r_m + \epsilon' \ \forall n, m \geq n_0$, by (5.3). Thus $\forall n \geq n_0, \alpha_n < [\{r_m + \epsilon'\}] = [\{r_m\}] + [\epsilon'] = \alpha + [\epsilon'] < \alpha + \epsilon$. Similarly $\forall n \geq n_0, \alpha - \epsilon < \alpha_n$. Hence $\{\alpha_n\}_{n \geq 1}$ converges to α. ∎

5.3.19 Theorem (Cauchy completeness). *Let $\{\alpha_n\}_{n \geq 1}$ be a Cauchy sequence of real numbers. Then there exists $\alpha \in \mathbb{R}$ such that $\{\alpha_n\}_{n \geq 1}$ converges to α.*

Proof. Since $\{\alpha_n\}_{n \geq 1}$ is Cauchy, given any real $\epsilon > 0$ we can find a positive integer n_0 such that
$$|\alpha_n - \alpha_m| < \epsilon/3.$$
Also by theorem 5.3.13, $\forall n$ we can find a rational r_n such that
$$\alpha_n - \epsilon/3 < [r_n] < \alpha_n + \epsilon/3, \text{ i.e.,} \qquad |\alpha_n - [r_n]| < \epsilon/3.$$
Thus $\forall n, m \geq n_0$,
$$\begin{aligned}\|[r_n] - [r_m]\| &\leq \|[r_n] - \alpha_n| + |\alpha_n - \alpha_m| + \|[r_m] - \alpha_m| \\ &< \epsilon/3 + \epsilon/3 + \epsilon/3 = \epsilon.\end{aligned}$$

Hence $\{[r_n]\}_{n\geq 1}$ is a Cauchy sequence and it follows from theorem 5.3.1(i)(a) that $\{r_n\}_{n\geq 1}$ is also a Cauchy sequence in $\mathbb{Q}$. Let $\alpha := [\{r_n\}]$. Then again by theorem 5.3.18, the sequence $\{[r_n]\}_{n\geq 1}$ converges to α. Thus we can find a positive integer n_1 such that

$$|[r_n] - \alpha| < \epsilon/3 \quad \forall\, n \geq n_1.$$

Then $\forall\, n \geq n_1$

$$\begin{aligned}|\alpha_n - \alpha| &\leq |\alpha_n - [r_n]| + |[r_n] - \alpha| \\ &< \epsilon/3 + \epsilon/3 < \epsilon.\end{aligned}$$

Hence $\{\alpha_n\}_{n\geq 1}$ converges to α. ∎

$(\mathbb{R}, \oplus, \odot, <)$ being an ordered field, we can analyze the existence of upper (lower) bounds for subsets of $\mathbb{R}$ with respect to the order on $\mathbb{R}$. We shall prove that $\mathbb{R}$ is also order complete. For that we need the following property of $\mathbb{R}$. For $n \in \mathbb{N}$ and $\alpha \in \mathbb{R}$, we write $n\alpha$ for $[n]\alpha$.

5.3.20 Theorem (Archimedean property). *For every $\alpha, \beta \in \mathbb{R}$ with $\alpha > 0$, there exists a positive integer $n \in \mathbb{N}$ such that $n\alpha > \beta$.*

Proof. If $\alpha > \beta$, then there is nothing to prove. Suppose $\beta \geq \alpha > 0$. Using theorem 5.3.13 and exercise 5.3.14, choose $r, s \in \mathbb{Q}$ such that $[s] \geq \beta \geq \alpha > [r] > 0$. Now by the Archimedean property of $\mathbb{Q}$, $\exists$ a positive integer n such that $nr > s$. Hence

$$n\alpha > n[r] = [nr] > [s] > \beta. \quad \blacksquare$$

5.3.21 Theorem (Order completeness). *Every nonempty subset of $\mathbb{R}$ that is bounded above has least upper bound.*

Proof. Let $S \neq \emptyset$ be a subset of $\mathbb{R}$ which is bounded above. Let $\alpha \in S$ and β an upper bound for S. By the Archimedean property,

5.3 Construction by Cantor's method

there exists a positive integer $-m$, such that $-m > -\alpha$ and hence $m < \alpha$. Thus m is not an upper bound for S. We put $\alpha_0 := m$, $\beta_0 := \beta$ and $N := \beta_0 - \alpha_0$. We construct real numbers α_n and β_n inductively as follows:

$$\alpha_{n+1} := \begin{cases} \frac{1}{2}(\alpha_n + \beta_n) & \text{if } \frac{1}{2}(\alpha_n + \beta_n) \text{ is not an upper bound for } S, \\ \alpha_n & \text{if } \frac{1}{2}(\alpha_n + \beta_n) \text{ is an upper bound for } S. \end{cases}$$

$$\beta_{n+1} := \begin{cases} \frac{1}{2}(\alpha_n + \beta_n) & \text{if } \frac{1}{2}(\alpha_n + \beta_n) \text{ is an upper bound for } S, \\ \beta_n & \text{if } \frac{1}{2}(\alpha_n + \beta_n) \text{ is not an upper bound for } S. \end{cases}$$

Then $\{\alpha_{n+1}\}_{n \geq 1}$ and $\{\beta_{n+1}\}_{n \geq 1}$ are sequences of real numbers such that each β_n is an upper bound for S and no α_n is an upper bound for S. We shall show that $\{\beta_n\}_{n \geq 1}$ is a Cauchy sequence and hence converges to some $\gamma \in \mathbb{R}$, by theorem 5.3.19, with $\gamma = lub(S)$. It is easy to see that $\forall n$, $\beta_{n+1} - \alpha_{n+1} = \dfrac{(\beta_n - \alpha_n)}{2}$ and hence $\beta_n - \alpha_n = \dfrac{N}{2^n}$. Thus $\alpha_{n+1} < \beta_{n+1}$ $\forall n$. Since for every n either $\beta_n = \beta_{n+1}$ or

$$\begin{aligned} \beta_n - \beta_{n+1} &= \beta_n - \frac{1}{2}(\beta_n + \alpha_n) \\ &= \frac{1}{2}(\beta_n - \alpha_n) \\ &= \frac{N}{2^{n+1}}, \end{aligned}$$

thus $\beta_{n+1} \leq \beta_n$ $\forall n$. Similarly $\alpha_{n+1} \geq \alpha_n$ $\forall n$. If m and n are positive integer such that $m < n$, then

$$\beta_m - \beta_n < \beta_m - \alpha_n < \beta_m - \alpha_m = \frac{N}{2^m}.$$

Hence

$$|\beta_m - \beta_n| < \frac{N}{2^m} \quad \forall\, m < n.$$

Let $\epsilon > 0$ be any real number. We can choose m_0 sufficiently large such that $\dfrac{N}{2^{m_0}} < \epsilon$, which is possible by theorem 5.3.20. Then $\forall\, n, m \geq m_0$
$$|\beta_m - \beta_n| < \epsilon.$$
This proves that $\{\beta_n\}_{n\geq 1}$ is a Cauchy sequence. Let $\{\beta_n\}_{n\geq 1}$ converge to $\gamma \in \mathbb{R}$. We show that $\gamma = lub(S)$. First we claim that γ is an upper bound for S. Suppose not, i.e., $\exists\, \alpha \in S$ such that $\alpha > \gamma$. Choose a real number η such that $(\alpha - \gamma) > \eta > 0$. Since $\forall\, n, \beta_n$ is an upper bound for S, $\beta_n \geq \alpha$ and hence $(\beta_n - \gamma) > (\alpha - \gamma) > \eta$, i.e., $\beta_n > \gamma + \eta$. Since $\{\beta_n\}_{n\geq 1}$ converges to γ, we can choose a positive integer n such that $(\beta_n - \gamma) < \dfrac{\eta}{2}$ for every $n \geq n_0$. But then
$$\frac{\eta}{2} > (\beta_{n_0} - \gamma) > \eta,$$
not possible. Hence γ is an upper bound for S. Next, suppose $\tilde{\beta}$ is any upper bound for S and $\tilde{\beta} < \gamma$, i.e., $(\gamma - \tilde{\beta}) > 0$. Since $\beta_n - \alpha_n = \dfrac{N}{2^n}$, it is clear that $\{\beta_n - \alpha_n\}_{n\geq 1}$ converges to zero. Thus there exists positive integer n_1 such that $\forall\, n \geq n_1$.
$$\beta_n - \alpha_n < (\gamma - \tilde{\beta}), \quad \text{i.e.,} \quad (\beta_n - \gamma) < (\alpha_n - \tilde{\beta}).$$
But, $\{\beta_n\}_{n\geq 1}$ is a decreasing sequence and hence $(\beta_n - \gamma) \geq 0\ \forall\, n$. Thus for all $n \geq n_1$ $(\alpha_n - \tilde{\beta}) > 0$, i.e., $\alpha_n > \tilde{\beta}$. Since α_n is not an upper bound for S, $\tilde{\beta}$ cannot be an upper bound for S, a contradiction. Hence $\gamma = lub(S)$. ∎

5.4 Uniqueness of the real number system

We discussed in sections 5.2 and 5.3 two methods of constructing Archimedean ordered fields which include a copy of $\mathbb{Q}$ and are both Cauchy and order complete. Let us denote by $\mathbb{R}_D$ the ordered

5.4 Uniqueness of the real number system

field constructed by the Dedekind method and let $\mathbb{R}_C$ denote the ordered field constructed by the Cantor method. We shall show that both these ordered fields are essentially the same, i.e., they are isomorphic to each other, and we call either of them the field of **real numbers**, denoted by $\mathbb{R}$. We first make the following simple observations regarding *glb* and *lub* of sets.

5.4.1 Theorem. *Let S be a nonempty subset of $\mathbb{R} (= \mathbb{R}_C$ or $\mathbb{R}_D)$ and $\alpha \in \mathbb{R}$ be an upper bound of S. Then the following statements are equivalent:*

(i) $\alpha = lub(S)$.

(ii) *For every real number $\epsilon > 0$, there exists a real number $\beta \in S$ such that $\alpha - \epsilon < \beta$.*

(iii) *If $\gamma \in \mathbb{R}$ is such that $\gamma < \alpha$, then there exists $\beta \in S$ such that $\gamma < \beta$.*

Proof. We shall prove the implications: (i) $\Rightarrow$ (ii) $\Rightarrow$ (iii) $\Rightarrow$ (i). Assume that $\alpha = lub(S)$ and $\epsilon > 0$ is given. If $\forall \beta \in S, \alpha - \epsilon \geq \beta$ then clearly $(\alpha - \epsilon)$ is an upper bound for S and $(\alpha - \epsilon) < \alpha$, contradicting the fact that $\alpha = lub(S)$. Hence $\exists \beta \in S$ such that $\alpha - \epsilon < \beta$. Thus (i) $\Rightarrow$ (ii). Next, let (ii) hold and $\gamma \in \mathbb{R}$ be such that $\gamma < \alpha$. Let $\epsilon := \alpha - \gamma$. Using (ii) choose $\beta \in S$ such that $(\alpha - \epsilon) < \beta$. Then $\gamma = (\alpha - \epsilon) < \beta$. Thus (iii) holds. Finally, if (iii) holds and γ is any upper bound of S, then $\alpha \leq \gamma$ for if not, i.e., $\gamma < \alpha$ then by (iii), we can find $\beta \in S$ such that $\gamma < \beta$ contradicting the fact that γ is an upper bound of S. Hence (iii) implies (i). ∎

5.4.2 Exercise. State and prove the result similar to that of theorem 5.4.1 for $glb(S)$.

5.4.3 Exercise. Let A and B be nonempty subsets of $\mathbb{R} (= \mathbb{R}_C$ or $\mathbb{R}_D)$ and $\alpha \in \mathbb{R}$. Let $-A := \{-\alpha | \alpha \in A\}, \alpha + A := \{\alpha + \beta | \beta \in A\}$,

$AB := \{\alpha\beta | \alpha \in A, \beta \in B\}$ and $A + B := \{\alpha + \beta | \alpha \in A, \beta \in B\}$. Prove the following:

(i) A is bounded above if $-A$ is bounded below and in that case $-lub(A) = glb(-A)$.

(ii) $lub(\alpha + A) = \alpha + lub(A)$ if A is bounded above.

(iii) $lub(A + B) = lub(A) + lub(B)$ if A and B both are bounded above.

(iv) $lub(AB) = lub(A)lub(B)$ if A and B are both bounded subsets of positive reals.

(v) $lub(A) \geq lub(B)$ if $B \subseteq A$ and A is bounded above.

5.4.4 Theorem. *For $\alpha \in \mathbb{R}(= \mathbb{R}_C$ or $\mathbb{R}_D)$, let*

$$L_\alpha := \{r \in \mathbb{Q} \mid r < \alpha\} \quad \text{and} \quad U_\alpha := \{r \in \mathbb{Q} \mid r \geq \alpha\}.$$

Then $lub(L_\alpha) = \alpha = glb(U_\alpha)$.

Proof. We first note that L_α is not empty because $\forall\, \alpha \in \mathbb{R}$, using the denseness of rationals, we can find a rational r such that $(\alpha - 1) < r < \alpha$. Also, L_α is bounded above by α and hence by the order completeness $lub(L_\alpha)$ exists. Clearly $lub(L_\alpha) \leq \alpha$. If $lub(L_\alpha) < \alpha$, again by the denseness of rationals $r \in \mathbb{Q}$ such that $lub(L_\alpha) < r < \alpha$, not possible, for then $r \in L_\alpha$. Hence $lub(L_\alpha) = \alpha$. The proof that $glb(U_\alpha) = \alpha$ is similar. ∎

5.4.5 Theorem. *There exists a map $\phi : \mathbb{R}_C \to \mathbb{R}_D$ with the following properties:*

(i) ϕ is one-one and onto.

(ii) $\phi(\alpha + \beta) = \phi(\alpha) + \phi(\beta) \quad \forall\, \alpha, \beta \in \mathbb{R}_C$.

(iii) $\phi(\alpha\beta) = \phi(\alpha)\phi(\beta) \quad \forall\, \alpha, \beta \in \mathbb{R}_C$.

5.4 Uniqueness of the real number system

(iv) $\phi(\alpha) > 0$ iff $\alpha > 0$.

In other words, the ordered fields $\mathbb{R}_C$ and $\mathbb{R}_D$ are isomorphic.

Proof. We first define ϕ on $\mathbb{Q} \subset \mathbb{R}_C$. Let $r \in \mathbb{Q}$. Then as a member of $\mathbb{R}_C$, it is $[r]$. Let $\alpha_r := \{s \in \mathbb{Q} | s < r\}$ the cut induced by the rational r, as in example 5.2.4(i). We define $\phi : \mathbb{Q} \to \mathbb{R}_D$ as follows:

$$\phi([r]) := \alpha_r.$$

We first check that ϕ has properties (i),(ii), (iii) and (iv), as a map from $\mathbb{Q}$ in $\mathbb{R}_C$ to $\mathbb{Q}$ in $\mathbb{R}_D$. If $\alpha_r = \alpha_s$, then we know $r = s$ and hence $[r] = [s]$. Thus ϕ is one-one on $\mathbb{Q}$. Also given $\alpha_r \in \mathbb{R}_D$ for $r \in \mathbb{Q}$, we can consider $[r] \in \mathbb{R}_C$ and get $\phi([r]) = \alpha_r$. Hence ϕ is also onto. Since $\alpha_r + \alpha_s = \alpha_{r+s}$ and $\alpha_r \alpha_s = \alpha_{rs}$ $\forall$ $r, s \in \mathbb{Q}$, ϕ also has properties (ii) and (iii). Property (iv) is obvious. Thus ϕ is an isomorphism of the field of rationals in $\mathbb{R}_C$ onto the field of rationals in $\mathbb{R}_D$. We now extend it to all of $\mathbb{R}_C$.

Let $\alpha \in \mathbb{R}_C$. Consider $L_\alpha := \{r \in \mathbb{Q} \subset \mathbb{R}_C | [r] < [\alpha]\}$ and $U_\alpha := \{r \in \mathbb{Q} \subset \mathbb{R}_C | [r] \geq \alpha\}$. By theorem 5.4.1 we know that $lub(L_\alpha) = \alpha = glb(U_\alpha)$. Since $r < s$ $\forall$ $r \in L_\alpha$ and $s \in U_\alpha$, we have $\phi(r) < \phi(s)$. Thus the set $\phi(L_\alpha)$ is bounded above and $\phi(U_\alpha)$ is bounded below. Hence $lub(\phi(L_\alpha))$ and $glb(\phi(U_\alpha))$ exist in $\mathbb{R}_D$ and $lub(\phi(L_\alpha)) \leq glb(\phi(U_\alpha))$. Suppose $lub(\phi(L_\alpha)) < glb(\phi(U_\alpha))$. Then using the denseness of rationals in $\mathbb{R}_D$, we can find rationals r_1 and r_2 such that

$$lub(\phi(L_\alpha)) < \alpha_{r_1} < \alpha_{r_2} < glb(\phi(U_\alpha)).$$

Thus $\forall$ $r \in L_\alpha, \phi(r) < \alpha_{r_1}$, i.e., $\alpha_r < \alpha_{r_1}$ and hence $r < r_1$. Thus $lub(L_\alpha) \leq r_1$. Similarly $glb(U_\alpha) \geq r_2$. But then

$$lub(L_\alpha) \leq r_1 < r_2 \leq glb(U_\alpha),$$

not true. Hence $lub(\phi(L_\alpha)) = glb(\phi(U_\alpha))$. We define for $\alpha \in \mathbb{R}_C$

$$\phi(\alpha) := lub(\phi(L_\alpha)) = glb(\phi(U_\alpha)).$$

Note that if $\alpha = r \in \mathbb{Q}$, then clearly $lub(\phi(L_\alpha)) = \alpha_r$ and hence ϕ is consistently defined, i.e., it is an extension of ϕ as defined above on $\mathbb{Q}$. We check that ϕ has the required properties. We first show that $\alpha < \beta$ implies $\phi(\alpha) < \phi(\beta)$. Since $\alpha < \beta$, by the denseness of rationals we can choose $r_1, r_2 \in \mathbb{Q}$ such that $\alpha < [r_1] < [r_2] < \beta$. Then $r_1 \in U_\alpha$ and $\phi(r_1) \in \phi(U_\alpha)$. Thus $\phi(\alpha) = glb(\phi(U_\alpha)) \leq \phi(r_1)$. Similarly $r_2 \in L_\beta$ and $\phi(\beta) > \phi(r_2)$. Thus

$$\phi(\alpha) \leq \phi(r_1) < \phi(r_2) \leq \phi(\beta).$$

Hence ϕ preserves order. This implies that ϕ is also one-one, because if $\alpha, \beta \in \mathbb{R}_C$ and $\alpha \neq \beta$ then either $\alpha > \beta$ or $\alpha < \beta$ and hence either $\phi(\alpha) < \phi(\beta)$ or $\phi(\alpha) > \phi(\beta)$. To show that ϕ is onto, let $\beta \in \mathbb{R}_D$, i.e., β is a cut. Let

$$A := \{\alpha \in \mathbb{R}_C | \phi(\alpha) \geq \beta\}.$$

The set A is nonempty because by denseness of rationals we can find rationals r and s such that $\alpha_r < \beta < \alpha_s$, i.e., $[s] \in A$ and $[r]$ is a lower bound for A. Thus A is also bounded below. By the completeness of $\mathbb{R}_C$, $\alpha' := glb(A)$ exists. We claim that $\phi(\alpha') = \beta$. Suppose if possible $\phi(\alpha') < \beta$. Then, again by denseness of rationals we can find an $r \in \mathbb{Q}$ such that $\phi(\alpha') < \alpha_r < \beta$, i.e., $\phi(\alpha') < \phi(r) < \beta$. But then $[r]$ is a lower bound for A and $\alpha' < r$, not possible. Thus $\phi(\alpha') < \beta$ is not possible. Let us suppose next that $\phi(\alpha') > \beta$. Then again we can find a rational s such that $\phi(\alpha') > \alpha_s > \beta$. But then $\alpha' > [s]$ and $\phi(s) = \alpha_s > \beta$, i.e., $[s] \in A$. Thus $\alpha' > [s] \in A$, not possible. Thus $\phi(\alpha') > \beta$ is also not possible. Hence $\phi(\alpha') = \beta$. This proves that ϕ is onto. To prove (ii), let $\alpha, \beta \in \mathbb{R}_C$. By exercise 5.4.3 we have

$$\phi(\alpha) + \phi(\beta) = lub(L_\alpha) + lub(L_\beta) = lub(L_\alpha + L_\beta).$$

Also

$$\begin{aligned} L_\alpha + L_\beta &= \{r+s | [r] < \alpha, [s] < \beta\} \\ &\subseteq \{r+s | [r+s] < (\alpha+\beta)\} \\ &= L_{\alpha+\beta}. \end{aligned}$$

5.5 Abstractions

Thus $lub(L_{\alpha+\beta}) \leq lub(L_\alpha + L_\beta)$. Hence $\phi(\alpha) + \phi(\beta) \geq lub(L_{\alpha+\beta}) = \phi(\alpha + \beta)$. Similarly $\phi(\alpha) + \phi(\beta) \leq glb(L_{\alpha+\beta}) = \phi(\alpha + \beta)$. Thus $\phi(\alpha+\beta) = \phi(\alpha) + \phi(\beta)$. Finally to prove (iv), let $\alpha, \beta \in \mathbb{R}_C$ be such that $\alpha > 0, \beta > 0$. Consider rationals r, s such that $\alpha > [s] > 0$ and $\beta > [r] > 0$. Then $0 < [rs] < \alpha\beta$ and $\phi(r)\phi(s) = \phi(rs) < \phi(\alpha\beta)$. As $r > 0, s > 0$ we have $\phi(r) > 0, \phi(s) > 0$ and hence

$$\phi(r) < \phi(\alpha\beta)/\phi(s).$$

Since this holds $\forall r$ such that $\alpha > [s] > 0$, we have $\phi(\alpha) = lub\{\phi(r)|\alpha > [r]\} \leq \phi(\alpha\beta)/\phi(s)$, i.e., $\phi(\alpha)\phi(r) \leq \phi(\alpha\beta)$. Since this holds $\forall r \in \mathbb{Q}$ with $0 < [r] < \beta$, we have

$$\phi(\alpha)\phi(\beta) \leq \phi(\alpha\beta).$$

Similarly, $\phi(\alpha)\phi(\beta) \geq \phi(\alpha\beta)$. Hence $\phi(\alpha\beta) = \phi(\alpha)\phi(\beta)$ if $\alpha, \beta \in \mathbb{R}_C, \alpha > 0, \beta > 0$. If either $\alpha = 0$ or $\beta = 0$, it follows from (iii) that $\phi(0) = 0$ and hence $\phi(\alpha\beta) = \phi(0) = \phi(\alpha)\phi(\beta)$. Finally, if $\alpha < 0$ and $\beta > 0$, noting that $\phi(-\alpha) = -\phi(\alpha) \; \forall \; \alpha \in \mathbb{R}_C$ (by (ii)), we have $\phi(\alpha\beta) = \phi(-\alpha\beta) = \phi(-\alpha)\phi(\beta) = -\phi(\alpha)\phi(\beta)$, i.e., $\phi(\alpha\beta) = \phi(\alpha)\phi(\beta)$. For $\alpha > 0$ and $\beta > 0$, the proof is similar. ∎

5.5 Abstractions

In sections 5.3 and 5.4 we explicitly constructed the real numbers from a given field of rationals by two different methods. Each method gave us an ordered field having the *lub*-property and being Cauchy complete. In fact, the ordered field $\mathbb{R}_D$ constructed in section 5.3 was first shown to be order complete (i.e., $\mathbb{R}_D$ has the *lub*-property) and as a consequence we deduced that $\mathbb{R}_D$ has the Archimedean property and is also Cauchy complete. The ordered field $\mathbb{R}_C$ constructed in section 5.4 was first shown to be Cauchy complete and shown to have the Archimedean property. As a consequence it was shown that $\mathbb{R}_C$ is also order complete. A careful analysis of the proofs of these facts will reveal that the arguments presented in these proofs work on general ordered fields. Recall, as pointed in note 4.10.13, given an ordered field $(F, +, \cdot, <)$, we can analyze its order completeness and Cauchy completeness (via the absolute value function given by its order). Hence the following definition makes sense.

5.5.1 Definition.

(i) An ordered field $(F, +, \cdot, <)$ is said to be **order complete** if every nonempty bounded subset of F which is bounded above has lub.

(ii) An ordered field $(F, +, \cdot, <)$ is said to be **Cauchy complete** if every Cauchy sequence in F converges to a point in F.

5.5.2 Exercise.

(i) Show that an ordered field $(F, +, \cdot, <)$ is order complete iff every nonempty subset of F which is bounded below has glb .

(ii) Show that $lub(A)$ and $glb(A)$ are unique whenever they exist for a nonempty subset A of an ordered field $(F, +, \cdot, <)$.

Theorems 5.2.17, 5.2.22 and 5.3.18 motivate the following theorem.

5.5.3 Theorem. *Let $(F, +, \cdot, <)$ be an ordered field. Then the following statements are equivalent:*

(i) F is order complete.

(ii) Every bounded monotone sequence in F is convergent.

(iii) F has the Archimedean property and the Bolzano-Weierstrass property (i.e., every bounded sequence in F has a convergent subsequence).

(iv) F has the Archimedean property and is Cauchy complete.

5.5 Abstractions

Proof. The proof of the implication (i) $\Rightarrow$ (ii) is as in theorem 5.2.22 (iii) and (iv). To prove that (ii) $\Rightarrow$ (iii), we note that (ii) implies F has Bolzano-Weierstrass property, as proved in theorem 5.2.22 (vi). We show that F also has the Archimedean property which is equivalent, by theorem 4.9.13, to showing that $\mathbb{N}_F$, the set of natural numbers in F, is not bounded above. Assume that the set $\mathbb{N}_F$ is bounded above. Then the sequence $\{n 1_F\}_{n \geq 1}$ in $\mathbb{N}_F \subseteq F$ is monotonically increasing and is bounded above, and hence is convergent by (ii). Since every convergent sequence is also Cauchy, given $\epsilon = (21_F)^{-1}$, we can choose $n_0 \in \mathbb{N}$ such that $\forall\, n, m \geq n_0$, $|n 1_F - m 1_F| < (21_F)^{-1}$. In particular, $1_F = |(n_0+1)1_F - n 1_F| < (21_F)^{-1}$, a contradiction. Hence F has the Archimedean property. Thus (ii) $\Rightarrow$ (iii). That (iii) $\Rightarrow$ (iv) is as proved in theorem 5.2.22 (vii). Finally, (iv) $\Rightarrow$ (i) can be proved as in theorem 5.3.21. ∎

5.5.4 Corollary. *The field of rationals is not Cauchy complete.*

Proof. Follows from theorems 5.5.3 and 4.4.6.

5.5.5 Definition. Let $(F, +, \cdot, <)$ be an ordered field. We say F is a **complete ordered field** if it has one of the properties (i) to (iv) of theorem 5.5.3.

Our next theorem says that any two ordered fields are isomorphic, as were $\mathbb{R}_D$ and $\mathbb{R}_C$ in theorem 5.4.5.

5.5.6 Theorem. *Any two complete ordered fields $(F, +, \cdot, <)$ and $(F, \oplus, \odot, \prec)$ are isomorphic.*

Proof. Let $\mathbb{Q}_1$ and $\mathbb{Q}_2$ denote the field of rationals of F_1 and F_2 (see theorem 4.9.7). By theorem 4.9.9, $\mathbb{Q}_1$ and $\mathbb{Q}_2$ are isomorphic as ordered fields, i.e., there exists a one-to-one map $\phi : \mathbb{Q}_1 \to \mathbb{Q}_2$ such that $\forall\, x, y \in \mathbb{Q}_1$ the following hold:

(i) $\phi(x+y) = \phi(x) \oplus \phi(y)$,

(ii) $\phi(xy) = \phi(x) \odot \phi(y)$,

(iii) $\phi(x) \preceq \phi(y)$ iff $x \leq y$.

Now ϕ can be extended to an isomorphism from F_1 to F_2 as in theorem 5.4.5. ∎

5.5.7 Note. Constructions of sections 5.2 and 5.3 tell us that complete ordered fields exist and any two complete ordered fields are isomorphic. Thus the set $\mathbb{R}$ of real numbers can be defined to be any complete ordered field.

Chapter 6

PROPERTIES OF REAL NUMBERS

6.1 Historical comments

As stated in section 4.1, by the end of the 18th century mathematicians were worried about the tremendous obscurity in various branches of analysis. Efforts were made to give rigorous proofs of properties of functions on real numbers. The concept of function at that time was understood as a relation between variables in terms of algebraic equations. The need to define a function precisely was felt during the second half of the eighteenth century when the works of Jean le d'Alembert (1717–83), Leonhard Euler (1707–83) and Daniell Bernoulli (1700–82) on the vibrating string problem started a debate about the concept of a function. By the end of the 18th century, the accepted concept of a function was as proposed by Euler in 1755: "If some quantities depend upon the others in such a way as to undergo variations when the latter are varied, then the former are called functions of the latter". However, it was still believed that the graph of every function could be traced from a 'free motion of the hand', i.e., functions were believed to be 'piecewise smooth'. Interest in the notion of function was revived with the work of Joseph Fourier (1768–1830) on the 'Analytic Theory of Heat' in 1807. It was only in 1837 that Gustav Lejeune-Dirichlet (1805–59) gave the definition of a function, employed most often now. He also gave the example of the function $f : [0,1] \longrightarrow \mathbb{R}$ given by $f(x) := 1$ if x is a rational, and $f(x) := 0$ if x is an irrational. This is an example of a function which is neither defined by a formula nor could its graph be drawn. This example not only crystallized the concept of a function, but also made mathematicians treat continuous and discontinuous functions with equal vigor, and later this motivated Bernhard Riemann (1826–66) to define the notion of integral.

A proper definition of continuity for functions on an interval was given by Bernard Bolzano (1781–1848) in 1817: "$f(x)$ is continuous in x in an interval if at any point x in the interval the value $f(x+y) - f(x)$ can be made as small as one wishes by taking y sufficiently small". A similar definition was given by Cauchy in 1821. The 'small' and 'sufficiently small' was made precise by Weierstrass later around the middle of 19th century. Efforts were also on to give rigorous proofs of intuitively obvious theorems about continuous functions. For example in 1817 Bolzano sought to prove that if a continuous function $f(x)$ is positive for $x = a$ and negative for $x = b$, then $f(x) = 0$ for some value of x between a and b. This came to be known as the 'intermediate value theorem'. In the proof of this theorem Bolzano assumed the 'Cauchy completeness' of $\mathbb{R}$, which was not known at that time. Similar arguments were used by Weierstrass in the 1860s in his Berlin lectures to prove that every bounded infinite set has a limit point. Cantor defined the notions of limit point of a set, the closed set, etc., in 1872 in order to prove the uniqueness of trigonometric series representations of functions. The fact that every continuous function on a closed interval has a minimum was assumed by Cauchy in his works. It was proved by Weierstrass in his Berlin lectures. The result that every countable covering of a closed bounded interval by open intervals has a finite sub-covering was proved by (Heinrich) Eduard Heine (1821–81) in 1872. He proved this in order to prove that every continuous function on a closed bounded interval is uniformly continuous. Emile Borel (1871–1956) stated this as an independent theorem and it came to be known as 'Heine-Borel theorem'. The extension of Heine-Borel theorem to the case when the given covering is allowed to be that of open sets was given by Pierre Cousin (1837–1933) in 1895 and Henri Lebesgue (1875–1941) in 1898. The notion of connected subsets of the real numbers was discovered by Camilla Jordan (1838–1927) in 1892.

Though throughout the works of Bolzano, Cauchy, Weierstrass and others, rigor was installed in many of the results, the existence of real number system was presupposed. It was only late in 19th century that real numbers were constructed from the rationals, as described in chapter 5.

Bernard Bolzano (1781–1848)

Bolzano was born in Prague and studied theology and mathematics. He became a priest in 1804. In 1805 he was appointed to the chair of philosophy and religion but was dismissed in 1819 because of his non-conformist ideas. He initiated the rigorous study of properties of functions and conceived many original ideas like that of limit, continuity and derivative of functions. Though very much in advance of his age, his ideas were only developed later by others.

Emile Borel (1871–1956)

Borel was born at Saint-Affrique in the Aveyron. He studied at the Ecole Normale Supérieure. He taught at the University of Lille, at the Ecole Normale Supérieure, and then at the Sorborne from 1909. He was a member of the Chamber of Deputies from 1924–36, and was the Minister of Navy in 1925. He founded the Center National

de la Recherche Scientifique and contributed to the planning of the Institute Henri Poincaré, acting as its director from 1928 until his death. He was one of the first to apply Cantor's ideas to analysis and to probability theory.

Augustin-Louis Cauchy (1789–1857)

Cauchy was born in Paris in 1789 and in 1805 he entered the Ecole Polytechnique to study engineering. Due to his poor health, Lagrange and Laplace advised him to devote himself to mathematics. In 1810 upon completing his training in civil engineering from the Ecole des Ponts et Chaussées, he was given commission in Napoleon's army as a military engineer. He left Paris for Cherbourg on his first assignment. Despite a busy schedule as an engineer, he found time to assist local authorities in conducting school examinations and doing research. In 1811 he submitted his first work on the theory of polyhedra to the Académie des Sciences. The second part of his work was submitted in 1812. In 1813 he returned to Paris and became a Professor at the Ecole Polytechnique. Later he taught at the Faculté des Sciences and at the Collége de France. In 1830 after Charles was unseated, Cauchy who had sworn a solemn oath of allegiance to Charles, resigned his professorship and exiled himself to Turin. There he taught Latin and Italian for

some years. In 1833 he tutored the grandson of Charles X at Prague. In 1838 he returned to Paris where he served as professor in several religious institutions. In 1848 after the revolution, government did away with oaths of allegiance and Cauchy took over the chair of mathematical astronomy at Faculty de Science at Sorbonne. He produced over 500 papers in diverse branches of mathematics in the last 19 years of his life and died in 1857. He was the pioneer of rigor in mathematical analysis, created abstract theory of groups and founded the theory of elasticity. He advanced the theory of determinant and contributed basic theorems in ordinary and partial differential equations and complex function theory.

6.2 Axiomatic definition of reals and its consequences

In view of the discussions in section 4.5 (see also note 5.5.7), we can define $\mathbb{R}$, the real numbers to be elements of any complete ordered field. Thus $\mathbb{R}$ is a set on which two binary operations addition, $+$, and multiplication, $\cdot$, are defined such that the following properties hold: for $x, y \in \mathbb{R}$, let us denote addition of x and y by $x + y$ and multiplication of x with y by xy, then

(i) $(\mathbb{R}, +, \cdot)$ is a field.

Note that if 0 is the additive identity and 1 its multiplicative identity of $\mathbb{R}$, then $1 \neq 0$. Also on $\mathbb{R}$, there exists a relation $<$ such that the following hold:

(ii) For every $x, y \in \mathbb{R}$, exactly one of the following is true: $x < y, x = y$ or $x > y$. It is known as law of trichotomy.

(iii) For $x, y, z \in \mathbb{R}$, $x < x$ is never true; $x < y$ and $y < x$ imply $x = y$; and $x < y$ and $y < z$ imply $x < z$.

(iv) If $x > 0, y > 0$, then $(x + y) > 0$ and $xy > 0$.

(v) Every nonempty subset of $\mathbb{R}$ which is bounded above has lub. This is called the lub-**property** or the **completeness axiom** of $\mathbb{R}$.

For $x \in \mathbb{R}$, we shall write $-x$ for the unique inverse of x with respect to $+$ and for $x \neq 0$, x^{-1} will denote the unique inverse of x with respect to multiplication. We also write x/y for xy^{-1}.

$\mathbb{R}$ being an ordered field, in particular it is an ordered integral domain, as stated in section 3.6. Thus $1 > 0$. In view of theorem 3.6.17, if we write $\tilde{n}$ for the n-fold sum of 1 with itself, then $\mathbb{N}_\mathbb{R} := \{\tilde{n} \in \mathbb{R} | n \in \mathbb{N}\}$ is an infinite subset of $\mathbb{R}$ and is a copy of $\mathbb{N}$, the natural numbers. Similarly $\mathbb{Z}_\mathbb{R} := \mathbb{N}_\mathbb{R} \cup \{0\} \cup \{-\tilde{n} | n \in \mathbb{N}_\mathbb{R}\}$ is a copy of the integers inside $\mathbb{R}$ and $\mathbb{Q}_\mathbb{R} := \{xy^{-1} | x, y \in \mathbb{Z}_\mathbb{R}, y \neq 0\}$ is a copy of the field of rationals $\mathbb{Q}$ inside $\mathbb{R}$. We shall write $\mathbb{N}, \mathbb{Z}$ and $\mathbb{Q}$ for $\mathbb{N}_\mathbb{R}, \mathbb{Z}_\mathbb{R}$ and $\mathbb{Q}_\mathbb{R}$, respectively. It is easy to show that the lub-property of $\mathbb{R}$ is equivalent to the glb-property, i.e., every nonempty subset A of $\mathbb{R}$ which is bounded below has a glb and that $lub(A)$ and $glb(A)$ are unique whenever they exist. In the next theorem we point out some of the properties of $\mathbb{N}, \mathbb{Z}, \mathbb{Q}$ and $\mathbb{R}$ (which we had proved earlier), as consequence of the lub-property of the real numbers.

6.2.1 Theorem.

(i) For every $x, y \in \mathbb{R}, x > 0$, $\exists$ a positive integer n such that $nx > y$. This is called the **Archimedean property** of the real numbers.

(ii) The set $\mathbb{N}$ is the smallest inductive subset of $\mathbb{R}$. It is an infinite set and every nonempty subset of $\mathbb{N}$ has a smallest element.

(iii) Every nonempty subset S of $\mathbb{Z}$ which is bounded above has a largest element, i.e., $lub(S) \in S$.

(iv) Every nonempty subset S of integers which is bounded below has a smallest number, (i.e., $glb(S) \in S$).

(v) For every $x \in \mathbb{R}$ there exists a unique integer n such that $n \leq x < n+1$. This n is called the **integral part** of x and is denoted by $[x]$.

(vi) Let $x, y \in \mathbb{R}$ be such that $y > x+1$. Then there exists a positive integer n such that $x < n < y$.

(vii) For every $x, y \in \mathbb{R}$ if $x > y$, then there exists a rational r such that $x > r > y$. This is called the **denseness of rationals** in $\mathbb{R}$.

Proof. (i) Proof is the same as that of theorem 5.2.17.

(ii) That $\mathbb{N}$ is the smallest inductive (and hence infinite) subset of $\mathbb{R}$ follows from theorem 3.6.22 and exercise 3.6.23. That every nonempty subset of $\mathbb{N}$ has a smallest element is proved as in theorem 2.4.15.

(iii) Since $S \subseteq \mathbb{Z}$ is such that $S \neq \emptyset$ and is bounded above, $\exists\ n \in S$ and $x \in \mathbb{R}$ such that $n < x$. Using (i), we can find a positive integer $m \geq x$. If $m = x$ and $m \in S$, then clearly $m = lub(S) \in S$. If not, then $m > x > n$. Consider $m - 1$. Either $(m - 1) \in S$, in which case $lub(S) = (m - 1) \in S$. If not, consider

$m-2$, and so on. If k is the first natural number such that $(m-k) \in S$, then $lub(S) = m - k \in S$.

(iv) Consider $-S := \{-s \mid s \in S\}$ and apply (iii).

(v) Let $x \in \mathbb{R}$. If x is itself an integer, then $n := x$ will prove the result. If $x \notin \mathbb{Z}$, let $S := \{n \in \mathbb{Z} \mid n \leq x\}$. Then S is a nonempty set because given $-x$, by (i) we can find a positive integer n such that $n > -x$ and hence $-n < x$. Also S is bounded above. Hence by (iii), $n_0 := lub(S)$ exists and $n_0 \in S$. Thus $n_0 \leq x$. Since x is not an integer, $(n_0 + 1) \neq x$. In fact $x < (n_0 + 1)$ for if not, then $(n_0 + 1) \leq x$ implying $(n_0 + 1) \in S$, contradicting the fact that $n_0 = lub(S)$. Hence $n_0 \leq x < (n_0 + 1)$.

(vi) Let $x, y \in \mathbb{R}$ be such that $y > (x+1)$. Let $S := \{n \in \mathbb{Z} \mid n < y\}$. Then $S \neq \emptyset$, for we can find by (i) a positive integer n such that $n > -y$, i.e., $-n < y$. Also by definition S is bounded above by y. Thus by (ii), $n_0 := lub(S)$ exists and $n_0 \in S$. Thus $n_0 < y$. Also $(n_0+1) \geq y$, for if not then $(n_0+1) < y$, contradicting the fact that $n_0 = lub(S)$. Hence

$$x < y - 1 \leq n_0 < y,$$

proving the required claim.

(vii) Let $x > y$ be reals. Then using (i) we can find a positive integer n such that $n(y-x) > 1$. Thus $ny - nx > 1$. Now by (vi), there exists an integer m such that $ny > m > nx$. Hence $y > \dfrac{m}{n} > x$. Let $r := \dfrac{m}{n}$. ∎

6.2.2 Exercise. Let x be any real number and let $L_x := \{r \in \mathbb{Q} \mid r < x\}$. Using the Archimedean property of $\mathbb{R}$, show that L_x is a nonempty set which is bounded above with $lub(L_x) = r$. Using this to deduce that for $a, b \in \mathbb{R}$ with $a < b$, there exists a rational r such that $a < r < b$.

We had seen in chapter 4 that there does not exist any rational r such that $r^2 = 2$ or $r^3 = 2$. In fact, whenever p is not the square of a positive integer, there does not exist any rational r

6.2 Axiomatic definition of reals

such that $r^2 = p$. We show in the next theorem how this becomes possible in the set of real numbers. For $\beta \in \mathbb{R}$ and $n \in \mathbb{N}$, we denote by β^n the n-fold product of β with itself. More precisely, we define for $n \geq 1$, $\beta^{n+1} := \beta^n \beta$.

6.2.3 Lemma. *Let $\alpha \in \mathbb{R}$ and $n \in \mathbb{N}$. Then the following hold:*

(i) *For $0 \leq \alpha \leq 1$, $\alpha^n \leq \alpha \leq 1$.*

(ii) *For $\alpha \geq 1$, $1 \leq \alpha \leq \alpha^n$.*

(iii) *For $0 < \alpha < 1$, $(1+\alpha)^n < 1 + 3^n \alpha$.*

(iv) *For $\alpha > -1$, $(1+\alpha)^n \geq 1 + n\alpha$.*

Proof. All these inequalities can be proved by induction on n.

(i) If $0 \leq \alpha \leq 1$, then clearly $\alpha^n \leq 1$ for $n = 1$. Suppose $\alpha^{n-1} \leq \alpha < 1$ holds. Then

$$\begin{aligned} \alpha^n = \alpha^{n-1}\alpha &\leq \alpha^{n-1} \quad \text{(as } 0 \leq \alpha \leq 1\text{)} \\ &\leq \alpha \leq 1 \quad \text{(by induction hypothesis)}. \end{aligned}$$

Hence $\alpha^n \leq \alpha \leq 1 \; \forall \, n$. This proves (i). Proof of (ii) is similar to that of (i).

(iii) If $n = 1$, then for $0 < \alpha < 1$, clearly $1 + \alpha < 1 + 3\alpha$. Thus the required inequality holds for $n = 1$. Suppose it also holds for $n - 1$ with $n \geq 2$, i.e., $(1+\alpha)^{n-1} < 1 + 3^{n-1}\alpha$. Then using the fact that $\alpha > 0$ and by induction hypothesis we get

$$\begin{aligned} (1+\alpha)^n &= (1+\alpha)^{n-1}(1+\alpha) \\ &\leq (1+3^{n-1}\alpha)(1+\alpha) \\ &= 1 + 3^{n-1}\alpha + \alpha + 3^{n-1}\alpha^2 \\ &= 1 + (3^{n-1} + 1 + 3^{n-1}\alpha)\alpha \\ &\leq 1 + (3^{n-1} + 3^{n-1} + 3^{n-1})\alpha \\ &= 1 + 3^n \alpha. \end{aligned}$$

Hence the required inequality holds $\forall \, n$ by induction.

(iv) Clearly $(1+\alpha)^n \geq 1+n\alpha$ holds for every $n = 1$. Assume that it holds for $n-1, n \geq 2$. Then using induction hypothesis and that $(1+\alpha) > 0$, we get

$$\begin{aligned}
(1+\alpha)^n &= (1+\alpha)^{n-1}(1+\alpha) \\
&\geq [1+(n-1)\alpha](1+\alpha) \\
&= 1+(n-1)\alpha + \alpha + (n-1)\alpha^2 \\
&= 1+n\alpha + (n-1)\alpha^2 \\
&\geq 1+n\alpha.
\end{aligned}$$

Note that in the last inequality we have used the fact that $\alpha^2 \geq 0$ $\forall \alpha \in \mathbb{R}$. ∎

6.2.4 Theorem (Existence of nth-root). *Let $n \in \mathbb{N}$ and $\alpha \in \mathbb{R}$ be such that $\alpha \geq 0$. Then $\exists$ a unique real number $\beta \in \mathbb{R}$ such that $\beta \geq 0$ and $\beta^n = \alpha$.*

Proof. We first show the existence of $\beta \in \mathbb{R}$ with $\beta \geq 0$ and $\beta^n = \alpha$. If $\alpha = 0$ we can take $\beta = 0$. So suppose $\alpha > 0$. Let

$$S := \{\gamma \in \mathbb{R} \mid \gamma > 0 \quad \text{and} \quad \gamma^n < \alpha\}.$$

Then S is a nonempty subset of $\mathbb{R}$. To see this, let $\alpha_0 := \dfrac{\alpha}{1+\alpha}$. Then $0 < \alpha_0 < 1$ and by lemma 6.2.3, $\alpha_0^n \leq \alpha_0$. Also $\alpha_0 < \alpha$, i.e., $\alpha_0 \in S$. Further $\forall \gamma \in S$ if $0 < \gamma \leq 1$, $\gamma \leq 1 \leq 1+\alpha$. In case $\gamma > 1$, then by lemma 6.2.3, $1 \leq \gamma \leq \gamma^n \leq \alpha \leq 1+\alpha$. Hence $\forall \gamma \in S, \gamma \leq 1+\alpha$. Thus S is a nonempty subset of $\mathbb{R}$ and is bounded above. By the *lub*-property of $\mathbb{R}$, $\beta := lub(S)$ exists. We shall show that $\beta^n = \alpha$. Equivalently we show that both $\beta^n > \alpha$ and $\beta^n < \alpha$ are not possible. First suppose $\beta^n < \alpha$. Consider any $\epsilon > 0$, then $\beta(1+\epsilon) > \beta$. We try to choose $\epsilon > 0$ such that $\beta(1+\epsilon) \in S$, i.e.,

$$\beta^n(1+\epsilon)^n < \alpha.$$

For that we note that if $0 < \epsilon < 1$, then by lemma 6.2.3,

$$\beta^n(1+\epsilon)^n < \beta^n + 3^n\beta^n\epsilon.$$

6.2 Axiomatic definition of reals 243

If we choose $0 < \epsilon < 1$ such that $\beta^n + 3^n \beta^n \epsilon < \alpha$, i.e., $\epsilon < \dfrac{\alpha - \beta^n}{3^n \beta^n}$, which is possible as $(\alpha - \beta^n) > 0$, we will have

$$\begin{aligned} \beta^n(1+\epsilon)^n &< \beta^n + 3^n \beta^n \epsilon \\ &< \beta^n + (\alpha - \beta^n) = \alpha. \end{aligned}$$

Thus for $\beta^n < \alpha$, if we choose $\epsilon := \min\left\{1, \dfrac{\alpha - \beta^n}{3^n \beta^n}\right\}$, then $\beta(1+\epsilon) \in S$ producing a contradiction as $\beta(1+\epsilon) > \beta = lub(S)$. Next, let us assume, if possible, that $\beta^n > \alpha$. Now we try to choose $\epsilon > 0$ such that $\dfrac{\beta}{1+\epsilon} \in S$ and at the same time $\left(\dfrac{\beta}{1+\epsilon}\right)^n > \alpha$. We note that if $0 < \epsilon < 1$, using lemma 6.2.3 we have

$$\frac{\beta^n}{(1+\epsilon)^n} > \frac{\beta^n}{1 + 3^n \epsilon}.$$

Thus for $0 < \epsilon < 1$ if $\dfrac{\beta^n}{1+3^n\epsilon} > \alpha$, i.e., $\dfrac{\beta^n - \alpha}{3^n \alpha} > \epsilon$, then we will have

$$\frac{\beta^n}{(1+\epsilon)^n} > \alpha, \tag{6.1}$$

i.e., $\beta^n > \alpha(1+\epsilon)^n$. Since $\dfrac{\beta}{1+\epsilon} < \beta$, $\dfrac{\beta}{1+\epsilon}$ cannot be $lub(S)$ and hence $\exists \gamma \in S$ such that $\dfrac{\beta}{1+\epsilon} < \gamma$. Then

$$\left(\frac{\beta}{1+\epsilon}\right)^n < \gamma^n < \alpha. \tag{6.2}$$

Thus if we choose $\epsilon := \min\left\{1, \dfrac{\beta^n - \alpha}{3^n \alpha}\right\}$, then it follows from (6.2) that $\left(\dfrac{\beta}{1+\epsilon}\right)^n < \alpha$, which contradicts (6.1). Hence $\beta^n > \alpha$ is also not possible. Thus $\beta^n = \alpha$. This proves the existence part of the theorem.

To prove the uniqueness of β, we note that if $\beta_1 \neq \beta$ is such that $\beta^n = \alpha$, then either $\beta_1 < \beta$ or $\beta_1 > \beta$. But then either $\beta_1^n > \beta^n = \alpha$ or $\beta_1^n < \beta^n = \alpha$. Since neither is possible, $\beta_1 = \beta$. ∎

6.2.5 Definition. For $\alpha \in \mathbb{R}, \alpha \geq 0$, the unique $\beta \in \mathbb{R}$ such that $\beta \geq 0$ and $\beta^n = \alpha$ is called the **positive n^{th} - root** of α and is denoted by $\alpha^{1/n}$ or $\sqrt[n]{\alpha}$. For $n = 2$, we write $\sqrt{\alpha}$ for $\sqrt[2]{\alpha}$.

6.2.6 Exercise. Let $\alpha, \beta \in \mathbb{R}$ with $\alpha \geq 0$ and $\beta \geq 0$. Show that $\forall \, n \in \mathbb{N}, (\alpha\beta)^{1/n} = \alpha^{1/n}\beta^{1/n}$.

6.2.7 Exercise. For $\alpha \in \mathbb{R}$ with $(1 + \alpha) \geq 0$, show that

$$(1+\alpha)^{1/2} \leq 1 + \frac{\alpha}{2}.$$

6.2.8 Definition. An element $\alpha \in \mathbb{R}$ such that α is not a rational is called an **irrational number**.

6.2.9 Example. In section 4.1.2 we had shown that there does not exist any rational number $r \in \mathbb{Q}$ such that $r^2 = 2$. In view of theorem 6.2.4, definitions 6.2.5 and 6.2.8, we can say $\sqrt{2}$ is an irrational number. In fact, by exercise 4.1.4, $\sqrt{p}$ is an irrational number $\forall$ prime p. One can ask, how are irrationals distributed in $\mathbb{R}$? An answer is given by the next theorem.

6.2.10 Exercise. Prove that $\sqrt{2}$ and $\sqrt{6}$ are irrational numbers. What can you say about $\sqrt{2} + \sqrt[3]{2}$ and $\sqrt{2} + \sqrt{3}$?

6.2.11 Theorem. *For α and $\beta \in \mathbb{R}$, the following hold:*

(i) *If α is a rational and β an irrational number then $\alpha + \beta, \alpha\beta$ and α/β (if $\beta \neq 0$) are all irrationals (here $\alpha/\beta := \alpha\beta^{-1}$).*

(ii) *If $\alpha < \beta$, then $\exists$ an irrational γ such that $\alpha < \gamma < \beta$. Thus like rationals, irrationals are also 'dense' in $\mathbb{R}$.*

6.2 Axiomatic definition of reals

Proof. (i) The proof of (i) follows from the fact that $\mathbb{R}\setminus\mathbb{Q}$ is the set of irrational numbers and that $\mathbb{Q}$ is a field, (i.e., closed under addition, multiplication and division by nonzero elements). For example if $(\alpha+\beta) \in \mathbb{Q}$ then $\beta = (\alpha+\beta) - \alpha \in \mathbb{Q}$, not true. Thus $\alpha \in \mathbb{Q}$ and $\beta \in \mathbb{R}\setminus\mathbb{Q}$ implies $(\alpha+\beta) \in \mathbb{R}\setminus\mathbb{Q}$. Similarly, other claims can be proved.

(ii) Let $\alpha, \beta \in \mathbb{R}$. If one of α or β is an irrational and the other is a rational number, then $\gamma = \dfrac{\alpha+\beta}{2}$ will be an irrational and $\alpha < \gamma < \beta$. Suppose both α and β are rational numbers. Then $(\alpha+\sqrt{2})$ and $(\beta+\sqrt{2})$ are both irrational numbers. Thus $\exists$ a rational r such that $(\alpha+\sqrt{2}) < r < (\beta+\sqrt{2})$. Hence $\alpha < (r-\sqrt{2}) < \beta$ and $\gamma := r - \sqrt{2}$ is an irrational. Finally, if both α and β are irrationals, then we first choose a rational r such that $\alpha < r < \beta$ and now choose an irrational γ such that $\alpha < \gamma < r < \beta$. ∎

6.2.12 Notes.

(i) Even though both the rationals and irrationals are 'dense' in $\mathbb{R}$, we shall show in section 6.8 that the 'number' of irrationals is 'much more' than that of rationals.

(ii) We said that numbers such as $\sqrt{2}, \sqrt[3]{3}$, etc., are irrational numbers. An equivalent way of saying this is that $x = \sqrt{2}$ is a solution of the equation $x^2 - 2 = 0$. We say $x \in \mathbb{R}$ is a **real algebraic number** if x satisfies an equation $a_n x^n + a_n x^{n-1} + \cdots + a_1 x + a_0 = 0$, where each $a_i \in \mathbb{Z}$ and $a_n \neq 0$. Note that every rational number p/q is algebraic for it satisfies the equation $qx - p = 0$. The number $\sqrt{2}$ is an algebraic number which is irrational. In theorem 6.3.6 we give example of a real number which is irrational and is not algebraic, see notes 6.3.10(i) and (ii). A criterion for deciding the irrationality of an algebraic number is given in the next theorem.

6.2.13 Theorem. *If p and q are coprime integers and p/q satisfies the equation $a_n x^n + a_{n-1} x^{n-1} + \cdots + a_1 x + a_0 = 0$, where each $a_i \in \mathbb{Z}$, and $a_n \neq 0$, then p divides a_0 and q divides a_n.*

Proof. By the given hypothesis,
$$a_n(p/q)^n + a_{n-1}(p/q)^{n-1} + \cdots + a_1(p/q) + a_0 = 0. \tag{6.3}$$
Thus $a_n p^n = q(-a_{n-1}p^{n-1} - \cdots - a_1 pq^{n-2} - a_0 q^{n-1})$. Since q divides the right-hand side of this equality and p, q are coprime, it follows that q divides a_n. Also from (6.3) we have
$$\begin{aligned} a_0 q^n &= -a_n p^n - a_{n-1}p^{n-1}q - \cdots - a_1 pq^{n-1} \\ &= p(-a_n p^{n-1} - a_{n-1}p^{n-2}q - \cdots - a_1 q^{n-1}). \end{aligned}$$
Since p divides the right hand side of the above equation and p, q are coprime, p must divide a_0. ∎

6.2.14 Corollary. *Let $r \in \mathbb{Q}$ satisfy the equation $x^n + a_{n-1}x^{n-1} + \cdots + a_1 x + a_0 = 0, a_i \in \mathbb{Z}$ for $i = 0, 1, \ldots, n$. Then r must be an integer.*

6.2.15 Exercise. Show that $\sqrt{5 + \sqrt{3}}$ and $\sqrt[3]{\sqrt{2}}$ are irrational numbers.

6.2.16 Exercise. Let α, β and γ be real numbers such that $\gamma > 0$ and $\alpha < \beta$. Show that $\exists$ a rational r such that $\alpha < r\gamma < \beta$.

6.2.17 Exercise. Show that $\forall\, m \in \mathbb{N}$, $\exists$ unique $n \in \mathbb{N}$ such that
$$2^n \leq m < 2^{n+1}.$$

6.3 Sequences of real numbers and their applications

Since $\mathbb{R}$ is an ordered field, using the order on $\mathbb{R}$ we can define $\forall\, \alpha \in \mathbb{R}$ its absolute value, $|\alpha|$, by
$$|\alpha| := \begin{cases} \alpha & \text{if } \alpha \geq 0, \\ -\alpha & \text{if } \alpha \leq 0. \end{cases}$$

6.3 Sequences of real numbers

The absolute value satisfies the triangle inequality and other properties as stated in section 4.8. The notions of sequences, subsequences, Cauchy sequences, convergent sequences in $\mathbb{R}$ can be defined and analysed using this absolute value, as was done for the rationals in sections 4.7, 5.2 and 5.3. We have already proved the following facts about real sequences.

6.3.1 Theorem. *Let $\{\alpha_n\}_{n\geq 1}$, be a sequence of real numbers. Then the following hold:*

(i) *Every sequence $\{\alpha_n\}_{n\geq 1}$ has a subsequence which is either monotonically increasing or decreasing.*

(ii) *If $\{\alpha_n\}_{n\geq 1}$ converges, it converges to a unique real number.*

(iii) *If $\{\alpha_n\}_{n\geq 1}$ is convergent, it is also Cauchy.*

(iv) *If $\{\alpha_n\}_{n\geq 1}$ is Cauchy, it is bounded.*

(v) *If $\{\alpha_n\}_{n\geq 1}$ is monotonically increasing and is bounded above, it is convergent.*

(vi) *If $\{\alpha_n\}_{n\geq 1}$ is monotonically decreasing and is bounded below, it is convergent.*

(vii) *If $\{a_n\}_{n\geq 1}$ is bounded, it has a convergent subsequence. This is called* **Bolzano-Weierstrass property**.

(viii) *If $\{\alpha_n\}_{n\geq 1}$ is Cauchy, it is convergent (Cauchy completeness).*

Proof. We have proved all these results in theorem 5.2.22. Note that only in the proof of (viii) do we require that $\mathbb{R}$ has Archimedean property. ∎

Though proved earlier, we have restated these results for two reasons: one for the sake of completion, to indicate that these are consequences of the *lub*-property of $\mathbb{R}$. And second to impress that "every Cauchy sequence of reals is convergent" is either built in the construction of real number system (if Cantor's method is adopted) or it is a consequence of the least upper bound property (if Dedekind's method is adopted) of the real number system. Also lub and glb of sets can be described in terms of sequences, as shown in the next theorem.

6.3.2 Theorem. *Let S be any nonempty subset of $\mathbb{R}$. The following statements are equivalent for $\alpha \in \mathbb{R}$:*

(i) $\quad \alpha = lub(S)$.

(ii) $\quad \alpha$ *is an upper bound for S and $\exists$ a sequence $\{\alpha_n\}_{n\geq 1}$ of elements of S such that $\{\alpha_n\}_{n\geq 1}$ is monotonically increasing and converges to α.*

Proof. Suppose $\alpha = lub(S)$. Then there exists an $\alpha_1 \in S$ such that $\alpha < \alpha_1 + 1$. For otherwise $\forall\, \alpha_1 \in S$, $\alpha_1 \leq \alpha - 1$, contradicting the fact that α is $lub(S)$. Thus $\exists\, \alpha_1 \in S$ such that $\alpha_1 \leq \alpha < \alpha_1 + 1$. In case $\alpha_1 = \alpha$, we can take $\alpha_n = \alpha_1\ \forall\, n$ and we have the required sequence $\{\alpha_n\}_{n\geq 1}$. In case $\alpha_1 < \alpha$, we can choose $\alpha_2 \in S$ such that $\alpha_1 < \alpha_2 \leq \alpha < \alpha_2 + \frac{1}{2}$. For if not, then $\forall\, \alpha_2 \in S_2$ with $\alpha_2 > \alpha_1$ we will have $(\alpha - \frac{1}{2}) \geq \alpha_2$. But then $\forall\, \gamma \in S$, $\gamma \leq \alpha - 1/2 < \alpha$, which contradicts the fact that $\alpha = lub(S)$. Proceeding this way, by induction, we will have a sequence $\{\alpha_n\}_{n\geq 1}$ of elements of S such that $\forall\, n \geq 1$,

$$\alpha_1 \leq \alpha_2 \leq \ldots \leq \alpha_n \leq \ldots \leq \alpha < \alpha_n + \frac{1}{n}.$$

Now given $\epsilon > 0$, we can choose $n_0 \in \mathbb{N}$ such that $\frac{1}{n_0} < \epsilon$ (note that we are using the Archimedean property of $\mathbb{R}$. Then $\forall\ n \geq n_0$,

$$\alpha - \alpha_n < \frac{1}{n} < \frac{1}{n_0} < \epsilon.$$

6.3 Sequences of real numbers

Hence $\{\alpha_n\}_{n\geq 1}$ converges to α. This proves that (i) implies (ii). Conversely, suppose (ii) holds, i.e., α is an upper bound for S and $\exists$ a sequence $\{\alpha_n\}_{n\geq 1}$ of elements of S such that $\{\alpha_n\}_{n\geq 1}$ is increasing and converges to α. We have to show that $\alpha = lub(S)$. Let $\alpha_0 \in \mathbb{R}$ be any upper bound for S. Suppose $\alpha_0 < \alpha$ and $\epsilon := \alpha - \alpha_0$. Then $\exists$ a positive integer n_0 such that

$$\alpha - \alpha_n < \epsilon \; \forall \, n \geq n_0, \text{ i.e., } \alpha - \alpha_{n_0} < \alpha - \alpha_0.$$

Hence $\alpha_0 < \alpha_{n_0} \in S$, contradicting the fact that α_0 is an upper bound for S. Hence $\alpha \leq \alpha_0$, i.e., $\alpha = lub(S)$. ∎

6.3.3 Exercise. State and prove the result corresponding to theorem 6.3.2 for $glb(S)$.

6.3.4 Exercise. Let $\mathcal{C}_\mathbb{R}$ denote the set of all real sequences which are convergent. Prove the following:

(i) $\{x_n\}_{n\geq 1} \in \mathcal{C}_\mathbb{R}$ iff every subsequence $\{x_{n_k}\}_{k\geq 1}$ of $\{x_n\}_{n\geq 1}$ is convergent to the same point and in that case $\lim_{n\to\infty} x_n = \lim_{k\to\infty} x_{n_k}$.

(ii) If $\{a_n\}_{n\geq 1} \in \mathcal{C}_\mathbb{R}$, then $\{|a_n|\}_{n\geq 1} \in \mathcal{C}_\mathbb{R}$ and $\lim_{n\to\infty} |a_n| = |\lim_{n\to\infty} a_n|$.

(iii) If $\{a_n\}_{n\geq 1}, \{b_n\}_{n\geq 1} \in \mathcal{C}_\mathbb{R}$ and $\alpha, \beta \in \mathbb{R}$, then $\{\alpha a_n + \beta b_n\}_{n\geq 1}$ and $\{a_n b_n\}_{n\geq 1} \in \mathcal{C}_\mathbb{R}$ with

$$\lim_{n\to\infty}(\alpha a_n + \beta b_n) = \alpha(\lim_{n\to\infty} a_n) + \beta(\lim_{n\to\infty} b_n)$$

and

$$\lim_{n\to\infty}(a_n b_n) = \left(\lim_{n\to\infty} a_n\right)\left(\lim_{n\to\infty} b_n\right).$$

Further, if $\lim_{n\to\infty} b_n \neq 0$, $\exists n_0 \in \mathbb{N}$ such that $b_n \neq 0 \; \forall \, n \geq n_0$,

$$\left\{\frac{a_n}{b_n}\right\}_{n\geq 1} \in \mathcal{C}_\mathbb{R} \quad \text{and} \quad \lim_{n\to\infty}\left(\frac{a_n}{b_n}\right) = \frac{\lim_{n\to\infty} a_n}{\lim_{n\to\infty} b_n}.$$

(Hint: Proceed as in theorem 4.7.9.)

6.3.5 Exercise. Let A and B be nonempty subsets of $\mathbb{R}$ which are bounded above. Show that $A + B := \{a + b | a \in A, b \in B\}$ is also bounded above and $\sup(A + B) = \sup(A) + \sup(B)$.

We give next some important applications of the fact that every Cauchy sequence of real numbers is convergent. For any nonnegative integer n, define $n!$, called n-**factorial**, as follows:
$$0! := 1,$$
and
$$n! := n(n-1)(n-2) \cdots 3 \cdot 2 \cdot 1 \quad \text{for } n \geq 1.$$

6.3.6 Theorem. Let $\alpha_n := 1 + 1 + \dfrac{1}{2!} + \dfrac{1}{3!} + \cdots + \dfrac{1}{(n-1)!}$, $n \in \mathbb{N}$. Then the following hold:

(i) The sequence $\{\alpha_n\}_{n \geq 1}$ is monotonically increasing.

(ii) The sequence $\{\alpha_n\}_{n \geq 1}$ is bounded and $2 < \alpha_n < 3 \ \forall\, n > 3$.

(iii) The sequence $\{\alpha_n\}_{n \geq 1}$ is convergent.

(iv) Let $e := \lim\limits_{n \to \infty} \alpha_n$. The number e, called **Euler's number**, is irrational.

Proof. (i) Obviously $\{\alpha_n\}_{n \geq 1}$ is monotonically increasing.

(ii) Clearly $\forall\, n \geq 2$, $2 < \alpha_{n+1}$. It is also easy to check that
$$n! > 2^{n-1} \quad \forall\, n \geq 3.$$
Thus we have $\forall\, n \geq 2$,
$$\begin{aligned}
\alpha_{n+1} &= 1 + 1 + \frac{1}{2!} + \frac{1}{3!} + \cdots + \frac{1}{n!} \\
&< 1 + 1 + \frac{1}{2} + \frac{1}{2^2} + \cdots + \frac{1}{2^{n-1}} \\
&= 1 + \frac{1 - (\frac{1}{2})^n}{1 - \frac{1}{2}} \\
&= 3 - \frac{1}{2^{n-1}} < 3.
\end{aligned}$$

6.3 Sequences of real numbers

Hence $\forall\, n \geq 3$, $2 < \alpha_n < 3$.

(iii) Since $\{\alpha_n\}_{n\geq 1}$ is a monotonically increasing sequence and is bounded above, it is convergent by theorem 6.3.1(v).

(iv) Let $e := \lim_{n\to\infty} \alpha_n$. Note that $e = lub\{\alpha_1, \alpha_2, \ldots\}$ by theorem 6.3.2. Suppose e is a rational number, i.e., $e = q/n$ for some positive integers q and n. Multiplying the numerator and denominator by $(n-1)!$ we may assume that $e = \dfrac{m}{n!}$, for some positive integers m and n, both greater than 2. Let us analyze α_{n+1}. Since

$$\begin{aligned}
\alpha_{n+1} &= 1 + 1 + \frac{1}{2!} + \cdots + \frac{1}{n!}, \\
&= \frac{p}{n!},
\end{aligned} \qquad (6.4)$$

where p is some positive integer. Also for any $k \geq 1$ and $n > 2$ we have

$$\begin{aligned}
& \frac{1}{(n+1)!} + \frac{1}{(n+2)!} + \cdots + \frac{1}{(n+k)!} \\
&= \frac{1}{(n+1)!}\left[1 + \frac{1}{(n+2)} + \frac{1}{(n+2)(n+3)} + \cdots \right. \\
&\qquad\qquad\qquad\qquad \left. + \frac{1}{(n+2)(n+3)\cdots(n+k)}\right] \\
&< \frac{1}{(n+1)!}\left[1 + \frac{1}{(n+2)} + \frac{1}{(n+2)^2} + \cdots + \frac{1}{(n+2)^{k-1}}\right] \\
&< \frac{1}{(n+1)!}\left[1 + \frac{1}{2} + \frac{1}{2^2} + \cdots + \frac{1}{2^{k-1}}\right] \\
&< \frac{2}{(n+1)!} = \frac{2}{(n+1)(n!)} \leq \frac{1}{2(n!)}. \qquad (6.5)
\end{aligned}$$

The last inequality follows as $n > 2$. Thus $\forall\, k \geq 1$ and $n \geq 3$, using (6.4) and (6.5), we have

$$\alpha_{n+k+1} = 1 + 1 + \frac{1}{2!} + \cdots + \frac{1}{(n+k)!}$$

$$= \alpha_{n+1} + \left[\frac{1}{(n+1)!} + \cdots + \frac{1}{(n+k)!}\right]$$
$$< \frac{p}{n!} + \frac{1}{2(n!)} = (p+\frac{1}{2})\frac{1}{n!}.$$

Since this holds $\forall\, k \geq 1$ and $n \geq 3$ and $e = lub\{\alpha_1, \alpha_2, \ldots\}$, we have
$$e \leq (p+\frac{1}{2})\frac{1}{n!} < \frac{p+1}{n!}.$$

Also
$$\alpha_{n+1} := \frac{p}{n!} < e.$$

Hence
$$\frac{p}{n!} < e = \frac{m}{n!} < \frac{p+1}{n!},$$

i.e., $p < m < p+1$. Thus there exists a positive integer between p and $(p+1)$, which is not true. Hence e is irrational. ∎

6.3.7 Exercise (Alternative definition of e).
Let $a_n := \left(1 + \frac{1}{n}\right)^n$ and $b_n := 1 + 1 + \frac{1}{2!} + \cdots + \frac{1}{n!}$. Prove the following:

(i) For every $n \geq 1, 2 < a_n < 3$.

   (Hint: For $2 \leq k \leq n$, $\dfrac{n!}{k!((n-k)!)}\left(\dfrac{1}{n}\right)^k \leq \dfrac{1}{k!} \leq \dfrac{1}{2^{k-1}}$.)

(ii) The sequence $\{a_n\}_{n \geq 1}$ is monotonically increasing.

   (Hint: For $2 \leq k \leq n$,
$$\frac{(n+1)!}{k!((n+1-k)!)}\left(\frac{1}{n+1}\right)^k \geq \frac{n!}{k!((n-k)!)}\left(\frac{1}{n}\right)^k.)$$

(iii) For $1 \leq n < m, a_m \geq b_n \geq a_n$ and hence
$$\lim_{m \to \infty} a_m = \lim_{n \to \infty} b_n = e.$$

6.3 Sequences of real numbers

6.3.8 Example (Definition of π). As an application of the fact that bounded monotone sequences of reals are convergent (theorem 6.3.1), we approximate the area of the unit circle. Recall that in example 4.8.2, in order to compute the area of the unit circle, we constructed regular polygon of 2^{n+1} sides inscribing the circle and denoted its area by A_n. Since $\{A_n\}_{n\geq 1}$ is an increasing sequence and is bounded above, it converges to a real number, denoted by π, which we call the area of the circle. Thus

$$\pi := \lim_{n \to \infty} A_n.$$

Let C denote the perimeter of the unit circle and C_n denote the perimeter of the inscribed regular polygon of 2^{n+1} sides. It is easy to see from figure 6.1 that $\dfrac{A_n}{C_n} = \dfrac{r_n}{2}$, where r_n is the length OA. It is clear geometrically that as $n \to \infty$, $r_n \to 1$ and $C_n \to C$. Thus $\pi = C$, the perimeter of the unit circle.

The value of π can be approximated using elementary geometry and familiarity with trigonometric functions. (Rigorous definition of trigonometric function requires techniques of advance analysis, for example refer to [Ebb].) Let B_n denote the area of the 2^{n+1}-sided regular polygon circumscribing the circle, as shown in figure 6.1 below.

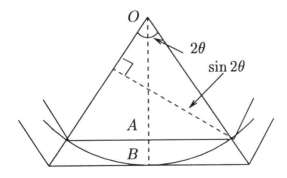

Figure 6.1 : Approximation of π.

Note that $A_n < \pi < B_n \ \forall\ n \geq 1$. The sides of the inscribed and the circumscribed polygon are in the ratio $1 : \cos\theta$, where θ is as shown in figure 6.1. Hence the areas of these polygons will be in the ratio $1 : \cos^2\theta$. Thus

$$A_n < \pi < B_n = \frac{A_n}{\cos^2\theta}.$$

It is also easy to show that $\forall\ n \geq 1$, $A_n = 2^n \sin 2\theta$. Thus

$$2\cos^2\theta = 1 + \sqrt{1 - \sin^2 2\theta}$$
$$= 1 + \sqrt{1 - \left(\frac{A_n}{2^n}\right)^2}.$$

Thus

$$A_n < \pi < \frac{2 A_n}{1 + \sqrt{1 - \left(\frac{A_n}{2^n}\right)^2}}.$$

In particular, since $A_2 = 2$ and $A_3 = 2\sqrt{2}$, we have

$$2 < 2\sqrt{2} < \pi < \frac{8}{\sqrt{2}+1} < 4.$$

This gives approximate values of π provided that we can find $\sqrt{2}$. This is done in the next example.

6.3.9 Example (Approximating square roots). In theorem 4.8.23 we showed that the sequence $x_1 = \frac{3}{2}$ and $x_{n+1} = \frac{1}{2}\left(x_n + \frac{2}{x_n}\right)$ for $n \geq 1$ gives a sequence $\{x_n\}_{n\geq 1}$ of rationals which is monotonically decreasing and is bounded below by 1. Hence it will converge by theorem 6.3.1. If $\lim_{n\to\infty} x_n = \ell$, then $\ell \neq 0$ and $\ell = \frac{1}{2}\left(\ell + \frac{1}{\ell}\right)$, i.e., $\ell^2 = 2$. Hence, $\lim_{n\to\infty} x_n = \sqrt{2}$. The sequence $\{x_n\}_{n\geq 1}$ can be used to approximate $\sqrt{2}$. As shown in the proof of theorem 4.8.23,

$$x_{n+1}^2 - 2 \leq \frac{1}{8^{n-1}}.$$

6.4 Decimal representation of real numbers

Since $x_n \geq \sqrt{2} > 1$, we have $x_n + \sqrt{2} > 2$ and hence

$$x_{n+1} - \sqrt{2} \leq \frac{1}{8^{n-1}(x_{n+1} + \sqrt{2})}$$
$$< \frac{1}{2 \times 8^{n-1}}.$$

Thus given $\epsilon > 0$, we can choose n such that $\frac{1}{2 \times 8^{n-1}} < \epsilon$. Then $x_{n+1} - \sqrt{2} < \epsilon$, i.e., x_{n+1} will approximate $\sqrt{2}$ by error at most ϵ. Similar methods can be used to locate the roots of polynomials.

6.3.10 Notes.

(i) We showed in theorem 6.3.6 that e is an irrational number. It is natural to ask: is e algebraic? It was proved by Charles Hermite (1822–1901) in 1873 that the real number e is not algebraic. For a proof refer to [Lang].

(ii) The first scientific attempt to compute π, (as defined in example 6.3.8) is due to Archimedes (240 B.C.). He showed that $\frac{223}{71} < \pi < \frac{22}{7}$. In the 5th century Tsu Chung-chi, a Chinese mathematician, gave rational approximation for π to be $\frac{355}{113}$. Hindu mathematician Bhaskara in 1150 A.D. gave its value as $\frac{3927}{1250}$. In 1767 Johann Heinrich Lambert (1728–1777) proved that π is an irrational number. The number π is in fact not algebraic. This was proved by Ferdinand Lindermann (1852–1939) in 1882, refer to [Lan] for a proof. A real number which is not algebraic is called a real **transcendental numbers**.

6.4 Decimal representation of real numbers

In section 4.7 we associated with every rational x a unique decimal expansion and showed that this expansion is periodic (theorem

4.7.2). The same process can be carried out for the real numbers. The only concept that we had used in theorem 4.7.2 was that of the integral part of a rational. Since $\forall\, x \in \mathbb{R}$, we have the concept of integral part of x, as given in theorem 6.2.1 (v), for every $x \in \mathbb{R}$ we can associate to it a unique sequence $a_0, a_1, a_2, \ldots$ of integers such that the following hold:

(i) $a_0 \leq x < a_0 + 1$ and $0 \leq a_i \leq 9\; \forall\, n \geq 1$.

(ii) For $n \in \mathbb{N}$ if $r_n := a_0 + \dfrac{a_1}{10} + \dfrac{a_2}{10^2} + \ldots + \dfrac{a_n}{10^n}$, $\{r_n\}_{n \geq 1}$ is a convergent sequence of rationals and $\lim_{m \to \infty} r_m = x$.

(iii) There does not exist any $N \in \mathbb{N}$ such that $a_i = 9\; \forall\, i \geq N$.

The unique sequence $a_0, a_1, a_2, \ldots$ satisfying (i), (ii) and (iii) is called a **decimal representation** of the real number x and we write this as $x = a_0 \cdot a_1 a_2 \ldots$. As mentioned in theorem 4.7.2, if x is a rational, then the sequence has the additional property that it is periodic, i.e., there exist positive integers j and m such that $a_n = a_{n+m}\; \forall\, n \geq j$. We make the following definition.

6.4.1 Definition. A sequence $a_0, a_1, a_2, \ldots$ of integers is called a **decimal** if for every $i \geq 1$, $0 \leq a_i \leq 9$ and there does not exist any $N \in \mathbb{N}$ such that $a_n = 9\; \forall\, n \geq N$. We write this as $a_0 \cdot a_1 a_2 \ldots$. We say the decimal $a_0 \cdot a_1 a_2 \ldots$ is **periodic** if $\exists$ positive integers j, m such that $a_n = a_{n+m}\; \forall\, n \geq j$.

Now we can prove the claim that a real number x has periodic decimal representation iff x is a rational.

6.4.2 Theorem. *Let $a_0 \cdot a_1 a_2 \ldots$ be a decimal. Then there exists a unique real number x such that*

$$x = a_0 \cdot a_1 a_2 \ldots.$$

Further, if the decimal $a_0 \cdot a_1 a_2 \ldots$ is periodic, x is a rational.

6.4 Decimal representation of real numbers

Proof. For every $n \in \mathbb{N}$, let
$$r_n := a_0 + \frac{a_1}{10} + \ldots + \frac{a_n}{10^n}.$$
Then
$$r_{n+1} - r_n = \frac{a_{n+1}}{10^{n+1}} \leq \frac{9}{10^{n+1}} < \frac{1}{10^n}.$$
Thus for every $n > m$,
$$|r_n - r_m| \leq \sum_{j=n}^{m-1} |r_{j+1} - r_j|$$
$$\leq \sum_{j=n}^{m-1} \frac{1}{10^j}$$
$$\leq \frac{1}{10^n} \left(\sum_{j=0}^{m-n-1} \frac{1}{10^j} \right)$$
$$\leq \frac{1}{9 \times 10^{n-1}}.$$

From this, it follows that $\{r_n\}_{n \geq 1}$ is a Cauchy sequence of rationals. By the completeness of $\mathbb{R}$, there exists a real number x such that $\lim_{n \to \infty} r_n = x$. We show that the real number x has the decimal representation $a_0 \cdot a_1 a_2 \ldots$ and that it is a rational if the decimal $a_0 \cdot a_1 a_2 \ldots$ is periodic. Suppose first $a_0 \cdot a_1 a_2 \ldots$ is periodic and j, m are positive integers such that $a_n = a_{n+m} \; \forall \, n \geq j$. For every integer $k \geq 0$, let
$$s_k := \frac{1}{10^{km}} \left(\frac{a_j}{10^j} + \ldots + \frac{a_{m+j-1}}{10^{m+j-1}} \right)$$
and
$$r_{n_k} := r_{j-1} + s_0 + \cdots + s_k = r_{m(k+1)-1}.$$
Then $\{r_{n_k}\}_{k \geq 1}$ is a subsequence of $\{r_n\}_{n \geq 1}$. Since
$$r_{n_k} = r_{j-1} + \left(\frac{a_j}{10^j} + \ldots + \frac{a_{m+j-1}}{10^{m+j-1}} \right)$$
$$\times \left(1 + \frac{1}{10^m} + \frac{1}{10^{2m}} + \cdots + \frac{1}{10^{km}} \right),$$

using theorem 4.8.16(ii), we have

$$\lim_{k\to\infty} r_{n_k} = r_{j-1} + \left(\frac{a_j}{10^j} + \cdots + \frac{a_{m+j-1}}{10^{m+j-1}}\right)\left(\frac{1}{1-\frac{1}{10^m}}\right).$$

Since $\lim_{n\to\infty} r_n = x$, we have $\lim_{n\to\infty} r_{n_k} = \lim_{n\to\infty} r_n = x$. Thus

$$\lim_{n\to\infty} r_n := x = r_{j-1} + \left(\frac{a_j}{10^j} + \cdots + \frac{a_{m+j-1}}{10^{m+j-1}}\right)\left(\frac{10^m}{10^{m-1}}\right).$$

Thus x is a rational. Finally to show that x has the decimal representation $a_0 \cdot a_1 a_2 \ldots$, let x have the decimal representation $b_0 \cdot b_1 b_2 \ldots$. It is easy to check that if $p_n := \frac{b_1}{10} + \cdots + \frac{b_n}{10^n}$, $\lim_{n\to\infty} p_n$ exists and is less than 1 (for $b_0 \cdot b_1 b_2 \ldots$ does not end with repeating 9's). Thus $a_0 = [x] = b_0$. Suppose if possible, there exists $n \in \mathbb{N}$ such that $b_n \neq a_n$. Let N be the smallest such integer and let $b_N < a_N$. Then $b_{N+1} \geq a_{N+1}$ and we get

$$\begin{aligned}
0 &= \lim_{n\to\infty} (r_n - p_n) \\
&= \lim_{n\to\infty} \left[\sum_{i=N}^{n} \left(\frac{a_i}{10^i} - \frac{b_i}{10^i}\right)\right] \\
&= \lim_{n\to\infty} \left[\frac{1}{10^N}(a_N - b_N) + \sum_{i=N+1}^{n} \left(\frac{a_i}{10^i} - \frac{b_i}{10^i}\right)\right] \\
&> \frac{1}{10^N}(a_N - b_N) > 0,
\end{aligned}$$

which is a contradiction, where the last inequality holds because each a_i, b_i is a nonnegative integer and all the b_i's are not equal to 9 after some stage. Thus $a_N \neq b_N$ leads to a contradiction. Hence $a_i = b_i \; \forall \; i$. ∎

6.4.3 Corollary. *Every real number x has a unique decimal representation.*

$$x = a_0 \cdot a_1 a_2 \ldots$$

6.4 Decimal representation of real numbers

and this decimal is periodic iff x is a rational.

Proof. Follows from theorems 4.7.2 and 6.4.2. ∎

6.4.4 Remark. Note that an expression of the type

$$a_0 \cdot a_1 a_2 \ldots a_n 999 \ldots, a_n \neq 9$$

is not a decimal. The reason is that the number represented by this decimal is the same as by the decimal $a_0 \cdot a_1 a_2 \ldots (a_n + 1)00\ldots$. Thus, decimals of the type in which not all the digits are 9 after some stage are in one-to-one correspondence with the real numbers as shown in corollary 6.4.3. Further a decimal corresponds to a rational iff it is periodic. For example the decimal $0 \cdot 40440444044440\ldots$ represents an irrational number. And two real numbers are equal iff they have the same decimal representation.

6.4.5 Note. The decimal representation of a real number can be thought of as a representation with base 10. We can modify the construction of decimal representation by dividing the intervals into n equal parts at each stage, for $n \in \mathbb{N}$ fixed. For example when $n = 2$, if $x \in [0, 1)$, then the corresponding representation is called the **binary expansion**. Thus, the binary expansion of a real number will look like $a_1 a_2 \ldots a_n \cdot d_1 d_2 \ldots$ where each a_i and $d_i = 0$ or 1, and not all the d_i's are 1 after some stage.

6.4.6 Exercise. Consider the interval $[0, 1]$. Let $F_1 := [0, 1]$. Remove the middle-open interval $\left(\frac{1}{3}, \frac{2}{3}\right)$ from F_1 and denote the remaining set by F_2, i.e., $F_2 := \left[0, \frac{1}{3}\right] \cup \left[\frac{2}{3}, 1\right]$. From each of the closed intervals in F_2, delete the open middle-third part of each, i.e., delete $\left(\frac{1}{9}, \frac{2}{9}\right)$ from $\left[0, \frac{1}{3}\right]$ and $\left(\frac{7}{9}, \frac{8}{9}\right)$ from $\left[\frac{2}{3}, 1\right]$. Let F_3 denote the remaining set, i.e.,

$$F_3 := \left[0, \frac{1}{9}\right] \cup \left[\frac{2}{9}, \frac{1}{3}\right] \cup \left[\frac{2}{3}, \frac{7}{9}\right] \cup \left[\frac{8}{9}, 1\right].$$

Continue this process of deleting the open middle-third of each closed interval left from the previous stage to obtain a sequence $\{F_n\}_{n\geq 1}$ of subsets of $[0,1]$. Let $F := \bigcap_{n=1}^{\infty} F_n$. The set F is the set of those points which are left after the deletion process has been completed. The set F is called **Cantor's ternary set**. Show that for $x \in [0,1], x \in F$ iff x has ternary expansion $\cdot d_1 d_2 \ldots$ where each $d_i = 0$ or 2. We shall return to this set in the next section.

6.4.7 Note. The geometric idea of decimal representations of points on the line can itself be used to give a construction of the real number system, refer to [Bur].

6.5 Special subsets of $\mathbb{R}$

In view of the discussions of sections 5.1, 5.2 and 5.3, we can represent the set of real numbers geometrically by the points on a horizontal line, which from now on will be called the **real line**. There are some special subsets of $\mathbb{R}$ which play an important role in analyzing various properties of $\mathbb{R}$. We describe these subsets next.

6.5.1 Definition. A set $I \subseteq \mathbb{R}$ is called an **interval** if $x, y \in I$ and $x < z < y$ imply $z \in I$.

Suppose I is an interval and as a set is bounded above. Then by the lub-property of real numbers, $b := lub(I)$ exists and is called the **right end-point** of I. Similarly, if I is bounded below, then $a := glb(I)$ exists and is called the **left end- point** of I. If I is bounded above but not below and $b := lub(I)$, we denote I by $(-\infty, b)$ if $b \notin I$, and by $(-\infty, b]$ if $b \in I$. Similarly, if I is bounded below but not above and $a := glb(I)$, we denote I by $(a, +\infty)$ if $a \notin I$ and by $[a, +\infty)$ if $a \in I$. If I is bounded both above and below with $a := glb(I)$ and $b := lub(I)$, we denote I by (a, b) if $a \notin I, b \notin I$; by $(a, b]$ if $a \notin I$ and $b \in I$; by $[a, b)$ if $a \in I$ and $b \notin I$;

6.5 Special subsets of ℝ

and by $[a, b]$ if both $a, b \in I$. In case $a \in I$ and $b \in I$, the interval $I = [a, b]$ is called the **closed interval**. If $a, b \notin I$, $I = (a, b)$ is called an **open interval**.

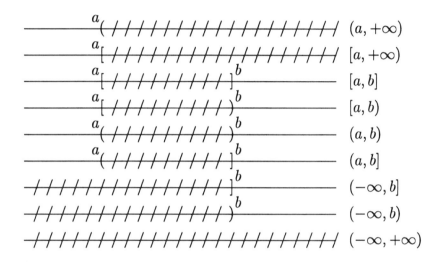

Figure 6.2 : Intervals in ℝ.

Geometrically, (a, b) is the set of all points on the right of a and on the left of b, and so on for other intervals. Analytically, $(a, b) = \{x \in \mathbb{R} | a < x < b\}$, $[a, b] = \{x \in \mathbb{R} | a \leq x \leq b\}$, and so on.

6.5.2 Exercise. Prove the following statements:

(i) If I and J are intervals, $I \cap J$ is an interval. If $I \cap J \neq \emptyset$ then it is an open interval if both I and J are open intervals, and it is a closed interval if both I and J are closed.

(ii) Show that the intersection of any family of intervals is again an interval.

(iii) If I and J are intervals and $I \cap J \neq \emptyset$, $I \cup J$ is an interval.

(iv) If $\{I_n\}_{n\geq 1}$ is a sequence of intervals such that $I_n \subseteq I_{n-1}$ $\forall\, n$, $\bigcup_{n=1}^{\infty} I_n$ is an interval.

6.5.3 Exercise. Prove the following statements:

(i) $(a-b) = \bigcup_{n=1}^{\infty} \left[a+\dfrac{1}{n}, b-\dfrac{1}{n}\right]$, where the union is over those $n \in \mathbb{N}$ such that $\dfrac{b-a}{2} > \dfrac{1}{n}$.

(ii) $[a,b] = \bigcap_{n=1}^{\infty} (a-\dfrac{1}{n}, b+\dfrac{1}{n})$.

6.5.4 Exercise. Prove the following statements:

(i) If $\alpha \in (a,b)$ and $a < b$, $\exists\, \epsilon > 0$ such that $(\alpha-\epsilon, \alpha+\epsilon) \subset (a,b)$.

(ii) A sequence $\{a_n\}_{n\geq 1}$ of real numbers converges to $a \in \mathbb{R}$ iff for every open interval I with $a \in I$, there exists a positive integer n_0 such that $a_n \in I$ $\forall\, n \geq n_0$.

(iii) If $\{a_n\}_{n\geq 1}$ is a sequence of real numbers such that $a_n \in [a,b]$ $\forall\, n \geq 1$ and $\{a_n\}_{n\geq 1}$ converges to $a \in \mathbb{R}$, then $a \in [a,b]$.

(iv) Analyze the claim of (iv) when $[a,b]$ is replaced by an arbitrary interval.

In view of exercise 6.5.4(iii) above, closed intervals have the property that they contain the limits of convergent sequence of their elements. This motivates the next definition.

6.5.5 Definition. A nonempty subset A of $\mathbb{R}$ is said to be **closed** if for every convergent sequence $\{a_n\}_{n\geq 1}$ in A, $\lim_{n\to\infty} a_n \in A$. We declare the empty set $\emptyset$ to be a closed subset of $\mathbb{R}$.

6.5 Special subsets of $\mathbb{R}$

6.5.6 Examples.

(i) Clearly $\mathbb{R}$ is a closed subset of itself.

(ii) For every $a \in \mathbb{R}$, $\{a\}$ is a closed subset of $\mathbb{R}$. This is because a sequence $\{a_n\}_{n\geq 1}$ in $\{a\}$ means $a_n = a \ \forall \ n$, and hence $\lim_{n\to\infty} a_n := a$.

(iii) For $a, b \in \mathbb{R}$, $a \leq b$, each of the intervals $[a, +\infty), (-\infty, b]$ and $[a, b]$ is a closed subset of $\mathbb{R}$. None of the intervals $(a, b], [a, b)$ and (a, b) is a closed subset of $\mathbb{R}$. For example, $\{b - \frac{1}{n}\}_{n\geq 1} \in [a, b)$ and $\lim_{n\to\infty}(b - \frac{1}{n}) = b$ but $b \notin [a, b)$.

6.5.7 Exercise. Let $A \subseteq \mathbb{R}$ be a closed set. Let $\{a_n\}_{n\geq 1}$ be a sequence of real numbers such that $\lim_{n\to 0} a_n = a \in \mathbb{R}$. If there exists a subsequence $\{a_{n_k}\}_{k\geq 1}$ of $\{a_n\}_{n\geq 1}$ such that $a_{n_k} \in A \ \forall \ k \geq 1$, show that $a \in A$.

6.5.8 Exercise. Let $A \subseteq \mathbb{R}$ be a closed set. Show that $|A| := \{|x| \mid x \in A\}$ is also a closed set.

6.5.9 Proposition. *The closed subset of $\mathbb{R}$ has the following properties:*

(i) If A and B are closed subsets of $\mathbb{R}$, so is the set $A \cup B$.

(ii) If $\{C_\alpha\}_{\alpha\in S}$ is any collection of closed subsets of $\mathbb{R}$, then $C := \bigcap_{\alpha\in S} C_\alpha$ is also a closed subset of $\mathbb{R}$.

Proof. (i) Let $\{x_n\}_{n\geq 1}$ be a sequence in $A \cup B$ and let $\{x_n\}_{n\geq 1}$ converge to x. Suppose $\forall \ n \geq 1$, there exists $k_n \in \mathbb{N}$ such that $x_{k_n} \in A$. Then we can choose $k_n \in \mathbb{N}$ such that $k_1 < k_2 < k_3 < $ with $x_{k_n} \in A \ \forall \ n$. Then $\{x_{k_n}\}_{n\geq 1}$ is a subsequence of $\{x_n\}_{n\geq 1}$ of elements of A and it converges to x. Thus $x \in A$. In case the above is not true, $\exists \ k \in \mathbb{N}$ such that $x_m \in B \ \forall \ m > k$. Let $k_n := k+n$. Then $\{x_{k_n}\}_{n\geq 1}$ is subsequence of sequence $\{x_n\}_{n\geq 1}$ of elements of B and converges to x. Then $x \in B$. In either case, $x \in A \cup B$.

(ii) Let $\{x_n\}_{n\geq 1}$ be a sequence of elements of C and let it converge to x. Then $\forall\,\alpha \in S, \{x_n\}_{n\geq 1}$ is a sequence of elements of C_α and hence $x \in C_\alpha$. Thus $x \in C$. ∎

6.5.10 Exercise. Prove the following statements:

(i) The union of a finite number of closed subsets of $\mathbb{R}$ is a closed set. Hence we can deduce that every finite subset of $\mathbb{R}$ is a closed set.

(ii) A subset A of $\mathbb{R}$ is said to be **bounded** if there exist real numbers $a, b \in \mathbb{R}$ such that $A \subseteq [a, b]$. If A is a subset of $\mathbb{R}$ which is closed and bounded, $lub(A) \in A$ and $glb(A) \in A$.

Let A be any subset of $\mathbb{R}$. Clearly, A need not be a closed subset of $\mathbb{R}$. Supposing we attach to A all those $x \in \mathbb{R}$ which are the limits of sequences in A. We should expect that new set to be closed. This motivates our next definition.

6.5.11 Definition. Let $A \subseteq \mathbb{R}$ and let

$$\overline{A} := \{x \in \mathbb{R} \mid x = \lim_{n\to\infty} a_n, \text{ for some sequences } \{a_n\}_{n\geq 1} \text{ in } A\}.$$

The set $\overline{A}$ is called the **closure** of the set A.

6.5.12 Theorem. *For $A \subseteq \mathbb{R}$, the following hold:*

(i) $A \subseteq \overline{A}$.

(ii) $x \in \overline{A}$ iff $\forall$ open interval I with $x \in I, I \cap A \neq \emptyset$.

(iii) A is closed iff $\overline{A} = A$.

(iv) $\overline{A}$ is a closed subset of A and if C is any closed subset of $\mathbb{R}$ such that $A \subseteq C$, then $C \subseteq \overline{A}$. In other words, $\overline{A}$ is the smallest closed subset of $\mathbb{R}$ which includes A.

6.5 Special subsets of $\mathbb{R}$

Proof. (i) For $a \in A$, consider the constant sequence $\{x_n\}_{n\geq 1}$, $x_n = a \ \forall \ n$. Then $a = \lim_{n\to\infty} x_n$ and hence $a \in \overline{A}$.

(ii) Let $x \in \overline{A}$ and I be any open interval such that $x \in I$. If $x \in A$, then clearly $I \cap A \neq \emptyset$. Let $x \notin A$ and let there exist a sequence $\{x_n\}_{n\geq 1}$ of elements of A converging to x. But then $x_n \in I \ \forall \ n \geq n_0$, for some $n_0 \in \mathbb{N}$ (see exercise 6.5.4). Hence $I \cap A \neq \emptyset$. Conversely, let $x \in \mathbb{R}$ be such that for every interval I with $x \in I$, we have $I \cap A \neq \emptyset$. If $x \in A$, then $x \in \overline{A}$ by (i). If $x \notin A$, then $\forall \ n$, choose $x_n \in (x-1/n, x+1/n) \cap A$. Then $\{x_n\}_{n\geq 1}$ is a sequence of elements of A, and clearly $\{x_n\}_{n\geq 1}$ converges to x. Hence $x \in \overline{A}$.

(iii) By (i), $A \subseteq \overline{A}$ and if A is closed, clearly $\overline{A} \subseteq A$. Thus $\overline{A} = A$, if A is closed. Conversely, let $\overline{A} = A$ and $a = \lim_{n\to\infty} a_n$ where $a_n \in A \ \forall \ n$. Then $a \in \overline{A}$, by definition, and hence $a \in A$. Thus A is closed.

(iv) We first show that $\overline{A}$ is closed. Let $z \in \mathbb{R}$ be a limit of a sequence in $\overline{A}$. Let I be an open interval such that $z \in I$. Then by (ii), $I \cap \overline{A}$ being nonempty. Let $x \in I \cap \overline{A}$. Since $x \in \overline{A}$ and $x \in I, I \cap A \neq \emptyset$. Hence $z \in \overline{A}$, by (ii) again. Thus $\overline{(\overline{A})} = \overline{A}$ and hence $\overline{A}$ is closed by (iii). Next, let C be any closed subset of $\mathbb{R}$ and $A \subseteq C$. Let $x \in \overline{A}$ and I be any open interval such that $x \in I$. Then $\emptyset \neq I \cap A \subseteq I \cap C$. Hence by (ii), $x \in \overline{C} = C$. ∎

6.5.13 Exercise. Let A be any nonempty subset of $\mathbb{R}$. Show that $x \in \overline{A}$ iff $\forall \ \epsilon > 0, \ \exists \ a \in A$ such that $|x - a| < \epsilon$.

6.5.14 Examples.

(i) If $A = (0,1)$ or $[0,1)$ or $(0,1]$, then $\overline{A} = [0,1]$.

(ii) For $\mathbb{Q}$, the set of rationals, $\overline{\mathbb{Q}} = \mathbb{R}$. Also $\overline{\mathbb{R} \setminus \mathbb{Q}} = \mathbb{R}$.

6.5.15 Exercise. Let A, B be subsets of $\mathbb{R}$. Show that $\overline{A \cup B} = \overline{A} \cup \overline{B}$ but $\overline{A \cap B} = \overline{A} \cap \overline{B}$ need not be true.

6.5.16 Definition. Let $A \subseteq \mathbb{R}$ be a nonempty subset of $\mathbb{R}$.

(i) A point $x \in \mathbb{R}$ is called a **limit point** of A if $\forall$ open interval I such that $x \in I, (I \setminus \{x\}) \cap A \neq \emptyset$.

(ii) A point $x \in \mathbb{R}$ is called an **interior point** of A if there exists an open interval I such that $x \in I \subseteq A$. We denote the set of all interior points of A by A°.

6.5.17 Exercise. For $\emptyset \neq A \subseteq \mathbb{R}$ and $x \in \mathbb{R}$, prove the following statements:

(i) x is a limit point of A iff $\forall \ r > 0, ((x-r, x+r) \setminus \{x\}) \cap A \neq \emptyset$.

(ii) x is a limit point of A iff $\forall \ r > 0, (x-r, x+r) \cap A$ is an infinite set.

(iii) x is a limit point of A iff there exists a sequence $\{x_n\}_{n \geq 1}$ in $A \setminus \{x\}$ converging to x. (In proving (iii) one requires the axiom of choice.)

(iv) x is an interior point of A iff $(x-r, x+r) \subseteq A$ for some $r > 0$ such that $(x-r, x+r) \subseteq A$.

(v) If $x \notin A$, then $x \notin A^\circ$.

6.5.18 Examples.

(i) For the set $(0,1)$, every element of $(0,1)$ is an interior point of it and every element of $[0,1]$ is a limit point of it. For $(0,1]$, the set of interior points is $(0,1)$.

(ii) The set $\left\{\frac{1}{2}, \frac{1}{3}, \frac{1}{4}, \ldots\right\}$ has no interior point and 0 is the only limit point of it.

(iii) The set $\mathbb{Q}$ of rationals has no interior point and every real number is a limit point of $\mathbb{Q}$.

(iv) If A is finite set, A has no limit points.

6.5 Special subsets of ℝ

In view of exercise 6.5.13, $x \in \overline{A}$ iff $\forall \ \epsilon > 0 \ \exists \ a \in A$ such that $|x - a| < \epsilon$. Thus $x \notin \overline{A}$ iff $\exists \ \epsilon > 0$ such that $\forall \ a \in A, |x - a| \geq \epsilon$, i.e., $(x - a, x + a) \cap A = \emptyset$, i.e., $x \in (\mathbb{R} \setminus A)^\circ$. Thus we have the following theorem.

6.5.19 Theorem. *For any nonempty subset A of $\mathbb{R}$, the following hold:*

(i) $\mathbb{R} \setminus \overline{A} = (\mathbb{R} \setminus A)^\circ$.

(ii) $\overline{A} = \mathbb{R} \setminus ((\mathbb{R} \setminus A)^\circ)$.

(iii) $\overline{(\mathbb{R} \setminus A)} = \mathbb{R} \setminus A^\circ$.

(iv) $A^\circ = \mathbb{R} \setminus (\overline{\mathbb{R} \setminus A})$.

Proof. The proof of (i) is already given above. (ii), (iii) and (iv) follow by changing A to $\mathbb{R} \setminus A$. ∎

6.5.20 Corollary. *For any subset A of $\mathbb{R}$, the following hold:*

(i) $A = A^\circ$ *iff* $\mathbb{R} \setminus A$ *is closed.*

(ii) A *is closed iff* $(\mathbb{R} \setminus A^\circ) = \mathbb{R} \setminus A$.

Proof. Follows from theorem 6.5.19, since A is closed iff $A = \overline{A}$, by theorem 6.5.12. ∎

Corollary 6.5.20 motivates the following definition.

6.5.21 Definition. A subset A of $\mathbb{R}$ is said to be **open** if $A = A^\circ$.

6.5.22 Examples.

(i) In view of corollary 6.5.20, A is open iff $\mathbb{R} \setminus A$ is a closed set.

(ii) The empty set, $\emptyset$, is an open set and for $A \neq \emptyset$, A is open iff $\forall \ x \in A$ there exists $\epsilon > 0$ such that $(x - \epsilon, x + \epsilon) \subseteq A$.

(iii) Every open interval is an open set.

(iv) If A and B are both open, then $A \cap B$ is an open set, for intersection of any two open intervals is an open interval.

(v) If $\{A_\alpha\}_{\alpha \in I}$ is any collection of open sets, then $A := \bigcup_{\alpha \in I} A_\alpha$ is also an open set, for if $x \in A$, then $x \in A_\alpha$ and hence there exists $a > 0$ such that $(x - \epsilon, x + \epsilon) \subseteq A_\alpha \subseteq \bigcup_{\alpha \in I} A_\alpha$.

6.5.23 Exercise. For $x, \delta \in \mathbb{R}$ with $\delta > 0$, show that the set $\{y \in \mathbb{R} \mid |x - y| > \delta\}$ is open.

We describe next an arbitrary open set in terms of open intervals.

6.5.24 Theorem. *Let U be any nonempty open set. Then there exists a collection of pairwise disjoint open intervals $\{I_\alpha\}_{\alpha \in I}$ such that $U = \bigcup_{\alpha \in I} I_\alpha$.*

Proof. Define a relation $\sim$ on U as follows: we say $x \sim y$ if $[x, y]$ or $[y, x] \subseteq U$. It is easy to check that this is an equivalence relation. Let I_x denote the equivalence class containing $x \in U$. We claim that I_x is an open interval for every x. Clearly if $y, z \in I_x$, say $y < z$, and $w \in \mathbb{R}$ is such that $y < w < z$, clearly $[x, w] \subseteq [x, z] \subseteq U$. Thus $w \in U$, i.e., I_x is an interval. Next, if $y \in I_x$, say $y > x$, then $[x, y] \subseteq U$ and since $y \in U$, there exists $\epsilon > 0$ such that $(y - \epsilon, y + \epsilon) \subseteq U$. Without loss of generality let $y - \epsilon < x$. Then $\forall\ z \in (y - \epsilon, y + \epsilon)$, $[x, z] \subseteq [x, y] \cup (y - \epsilon, y + \epsilon) \subseteq U$, i.e., $z \in I_x$.

Figure 6.3

Thus $y \in (y - \epsilon, y + \epsilon) \in I_x$. Hence I_x is an open interval. Let $\mathcal{C}$

denote the set of all equivalence classes under the relation $\sim$. Then elements of $\mathcal{C}$ are pairwise disjoint open intervals and $U = \bigcup_{I \in \mathcal{C}} I$. ∎

6.5.25 Note. For a refinement of theorem 6.5.24, see theorem 6.8.5.

6.6 Heine-Borel property and compact subsets of $\mathbb{R}$

In section 6.5 we saw that if $\{a_n\}_{n \geq 1}$ is a sequence of elements of a set $A \subseteq \mathbb{R}$ and $\{a_n\}_{n \geq 1}$ converges to $a \in \mathbb{R}$, $a \in A$ iff A is a closed set. But given a sequence $\{a_n\}_{n \geq 1}$ of elements of A, $\{a_n\}_{n \geq 1}$ may not converge at all. In fact $\{a_n\}_{n \geq 1}$ need not have any convergent subsequence. For example the sequence $\{n\}_{n \geq 1}$ in $\mathbb{N}$ has no convergent subsequence. This motivates our next definition.

6.6.1 Definition. A set $A \subseteq \mathbb{R}$ is said to be **compact** if every sequence in A has a convergent subsequence, converging to a point in A.

6.6.2 Example.

(i) For $a \leq b$, the interval $[a, b]$ is a compact set. Clearly, every sequence $\{a_n\}_{n \geq 1}$ in $[a, b]$ is bounded and hence has a convergent subsequence, by the Bolzano-Weierstrass property (see 6.3.1 (vii)). Further the limit of the subsequence will be in $[a, b]$ for $[a, b]$ is a closed set.

(ii) If $A \subseteq \mathbb{R}$ is compact, then clearly it is closed. Thus $(0, 1]$ is not a compact set. The set $\mathbb{R}$ itself, though closed, is not compact for $\{n\}_{n \geq 1}$ has no convergent subsequence.

6.6.3 Theorem. *A subset A of $\mathbb{R}$ is compact iff it is closed and bounded.*

Proof. Suppose A is compact. Clearly it is closed. Suppose A is not bounded, i.e., $\forall\, n \in \mathbb{N}$, $\exists\, a_n \in A$ such that $|a_n| > n$. Consider the sequence $\{a_n\}_{n\geq 1}$. Then $\{a_n\}_{n\geq 1}$ cannot have a convergent subsequence, for that subsequence will have to be bounded, but $\{a_n\}_{n\geq 1}$ is not bounded. Hence A is also bounded. Conversely, let A be a closed and bounded set. Let $\{a_n\}_{n\geq 1}$ be any sequence of elements of A. Then $\{a_n\}_{n\geq 1}$ is a bounded sequence and hence by Bolzano-Weierstrass property (theorem 6.3.1 (vii)), $\{a_n\}_{n\geq 1}$ has a convergent subsequence. Since A is closed, the subsequence has to converge in A. Hence A is compact. ∎

6.6.4 Corollary. *Let A be a nonempty compact subset of $\mathbb{R}$. Then $glb(A), lub(A)$ exist and are elements of A.*

Proof. Since A is compact, it is closed and bounded. Thus $glb(A)$ and $lub(A)$ exist. By theorem 6.3.2. and exercise 6.3.3, there exist sequences $\{x_n\}_{n\geq 1}$ and $\{y_n\}_{n\geq 1}$ in A such that $glb(A) = \lim_{n\to\infty} x_n$ and $lub(A) = \lim_{n\to\infty} y_n$. Since A is closed set, it follows that $glb(A), lub(A) \in A$. ∎

6.6.5 Definition. Let A be a subset of $\mathbb{R}$. An **open cover** of A is a collection $\{G_\alpha\}_{\alpha \in I}$ of open subsets of $\mathbb{R}$ such that $A \subseteq \bigcup_{\alpha \in I} G_\alpha$.

6.6.6 Examples.

(i) Consider the interval $(0,1)$. Then the collection $\{(0, 1-1/n)\}_{n\geq 1}$ forms an open cover of $(0,1)$.

(ii) Consider $[1,\infty)$. Then $\{(0,\infty)\}$ is an open cover of $[1,\infty)$. So are the collections $\{(n-1, n+1)\}_{n\in\mathbb{N}}$ and $\{(r-1, r+1) | r \in \mathbb{Q}\}$.

We describe next an important property of closed bounded intervals.

6.6 Heine-Borel property and compact sets

6.6.7 Theorem (Heine-Borel). *Let $I = [a,b]$ and let $\mathcal{S}$ be a collection of nonempty open subsets of $\mathbb{R}$ such that $I \subseteq \bigcup_{J \in \mathcal{S}} J$. Then $\exists\, J_1, J_2, \ldots, J_n \in \mathcal{S}$ such that $I \subseteq \bigcup_{i=1}^{n} J_i$.*

Proof. Let A be the set of those $x \in [a,b]$ such that $[a,x] \subseteq \bigcup_{i=1}^{m} J_i$, for some $m \in \mathbb{N}$ and $J_1, J_2, \ldots, J_m \in \mathcal{S}$. We first show that $A \neq \emptyset$. Since $a \in [a,b] \subseteq \bigcup_{J \in \mathcal{S}} J$, $\exists\, J \in \mathcal{S}$ such that $a \in J$. Thus $[a,a] \subseteq J \in \mathcal{S}$, showing that $a \in A$. Also, A is bounded above by b. Thus by the *lub*-property of the real numbers, $\alpha := lub(A)$ exists. Clearly, we have $\alpha \leq b$. Suppose $\alpha < b$. Since $\alpha \in [a,b]$, $\exists\, J \in \mathcal{S}$ such that $\alpha \in J$. Since J is an open set, we can choose an open interval (c,d) such that $\alpha \in (c,d) \subseteq J$. Without any loss of generality, we may assume that $a \leq c \leq d < b$. Since $\alpha \in (c,d)$, $\exists\, \beta \in (c,d)$ such that $\beta > \alpha$. Also, $\exists\, x \in (c,d)$ such that $x \in A$, for if not then $\forall\, x \in A, x < c$ and hence c will be an upper bound for A with $c < \alpha$, contradicting the fact that $\alpha = lub(A)$. Hence $a < c < x < \alpha < \beta < d < b$. Since $x \in A$, $\exists\, k \in \mathbb{N}$ and $J_1, J_2, \ldots, J_k \in \mathcal{S}$ such that $[a,x] \subseteq \bigcup_{i=1}^{k} J_i$. But then

$$[a, \beta] \subseteq \left(\bigcup_{i=1}^{k} J_i\right) \bigcup (J).$$

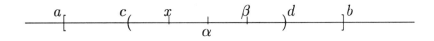

Figure 6.4

Thus $\beta \in A$, which is not possible as $\beta \in A$ and $\beta > \alpha = lub(A)$.

Hence $\alpha < b$ is not true, i.e., $\alpha = b$. This proves the theorem. ∎

6.6.8 Note. Another way of stating theorem 6.6.7 is the following: whenever a family $\mathcal{S}$ of open sets 'covers' a closed bounded interval I, finite number of members of $\mathcal{S}$ which 'cover' I.

6.6.9 Exercise. Let $\mathcal{U} := \{(0, 1 - 1/n) | n \in \mathbb{N}\}$. Show that $\mathcal{U}$ is a covering of $(0,1)$ by open intervals but $(0,1)$ is not covered by any finite number of members of $\mathcal{U}$. Why does this not contradict theorem 6.6.7?

6.6.10 Exercise. Let $\mathcal{U} := \{(0, n) | n \in \mathbb{N}\}$. Show that $\mathcal{U}$ is a covering of $(0, \infty)$ by open intervals but $(0, \infty)$ is not covered by any finite number of members of $\mathcal{U}$.

6.6.11 Theorem. *For a subset A of $\mathbb{R}$, the following are equivalent:*

(i) *A is compact.*

(ii) *A is closed and bounded.*

(iii) *If $\mathcal{G} = \{G_\alpha\}_{\alpha \in I}$ is any open cover of A, there exist finitely many elements $G_{\alpha_1}, \ldots, G_{\alpha_n}$ of $\mathcal{G}$ such that $A \subseteq \bigcup_{i=1}^{n} G_{\alpha_i}$.*

Proof. We have already proved in theorem 6.6.3 that (i) ⇔ (ii). We show that (ii) ⇔ (iii). Suppose A is a closed bounded set and $\mathcal{G} = \{G_\alpha\}_{\alpha \in I}$ is an open cover of A. Since A is bounded, $\exists\ n \in \mathbb{N}$ such that $A \subseteq [-n, n]$. Also A being closed, $\mathbb{R} \setminus A$ is an open set. Thus $\mathcal{G} \cup \{\mathbb{R} \setminus A\}$ is an open covering of $[-n, n]$. By theorem 6.6.7, $\exists\ G_{\alpha_1}, \ldots, G_{\alpha_m}$ in $\mathcal{G}$ such that $[-n, n] \subseteq \bigcup_{i=1}^{m} G_{\alpha_i}$. Let $\mathcal{G}_1 := \{G_{\alpha_1}, \ldots, G_{\alpha_m}\}$. Then $\mathcal{G}_1 \setminus \{\mathbb{R} \setminus A\} \subseteq \mathcal{G}$ is a finite sub-cover of A, i.e., $\mathcal{G}_1 \setminus \{\mathbb{R} \setminus A\}$ has finite number of elements and $A \subseteq \bigcup_{G \in \mathcal{G}_1} G$. Thus (ii) ⇒ (iii). Finally, suppose (iii) holds. We shall prove that A is closed and bounded. Consider the open covering $\{(-n,n)\}_{n\geq 1}$ of A. By the given hypothesis, $\exists\, n \in \mathbb{N}$ such that $A \subseteq \bigcup_{k=1}^{n}(-k,k) = (-n,n)$. Hence A is bounded. To show that A is closed, we shall prove that $\mathbb{R} \setminus A$ is open. Let $x \in \mathbb{R} \setminus A$. Let $U_n := \{y \in \mathbb{R} \mid |x-y| > \frac{1}{n}\}$, $n \in \mathbb{N}$. Then each U_n is an open set (by exercise 6.5.23) and $\mathbb{R} \setminus \{x\} = \bigcup_{n=1}^{\infty} U_n$. Since $x \notin A$, we have $A \subseteq (\mathbb{R} \setminus \{x\}) \subseteq \bigcup_{n=1}^{\infty} U_n$. Thus $\{U_n \mid n \geq 1\}$ forms a covering of A and by the given hypothesis, there exists $m \in \mathbb{N}$ such that $A \subseteq \bigcup_{n=1}^{m} U_n = U_m$. Hence $(x - \frac{1}{m}, x + \frac{1}{m}) \cap A = \emptyset$, i.e.,

$$\left(x - \frac{1}{m}, x + \frac{1}{m}\right) \subseteq \mathbb{R} \setminus A.$$

Thus $\mathbb{R} \setminus A$ is an open set. Hence by example 6.5.22(i), A is closed, i.e., (iii) $\Rightarrow$ (iii). ∎

6.6.12 Note. In theorem 6.6.7 we showed that the completeness, i.e., the *lub*-property, of the ordered field $\mathbb{R}$ implied the Heine-Borel property. The same argument applies in any ordered field which is complete. In fact the converse is also true, i.e., if an ordered field has the Heine-Borel property, it is complete. For details refer to [Men].

6.7 Nested interval property

To describe another property of intervals, we have the following definition. For a bounded interval I with end points a and b such that $a < b$, we write $\lambda(I) := b - a$, called the **length** of I.

6.7.1 Definition. Let $\{I_n\}_{n\geq 1}$ be a sequence of bounded intervals. The sequence $\{I_n\}_{n\geq 1}$ is said to be **nested** if the following hold:

(i) $\{I_n\}_{n\geq 1}$ is a decreasing sequence of intervals, i.e., $I_{n+1} \subseteq I_n$ $\forall\, n = 1, 2, \ldots$.

(ii) The sequence $\{\lambda(I_n)\}_{n\geq 1}$ converges to zero.

6.7.2 Examples.

(i) Let $I_n := (0, \frac{1}{n})$, $n = 1, 2, \ldots$. Then $\{I_n\}_{n\geq 1}$ is a nested sequence.

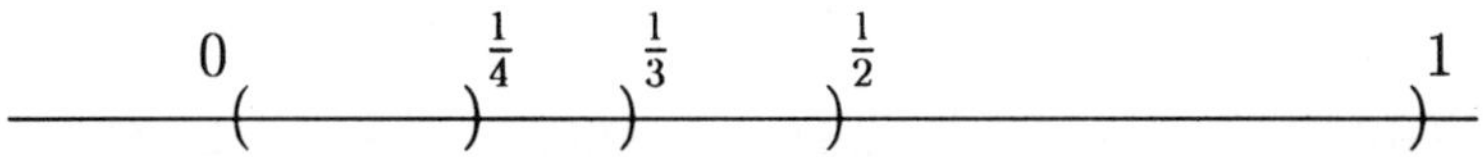

Figure 6.5 : Nested sequence $\left\{\left(0, \frac{1}{n}\right)\right\}_{n\geq 1}$.

(ii) The sequence $\{I_n\}_{n\geq 1}$ where $I_n = (n, n + \frac{1}{n})$, $n = 1, 2, \ldots$, is not nested even though $\{\lambda(I_n)\}_{n\geq 1}$ converges to zero.

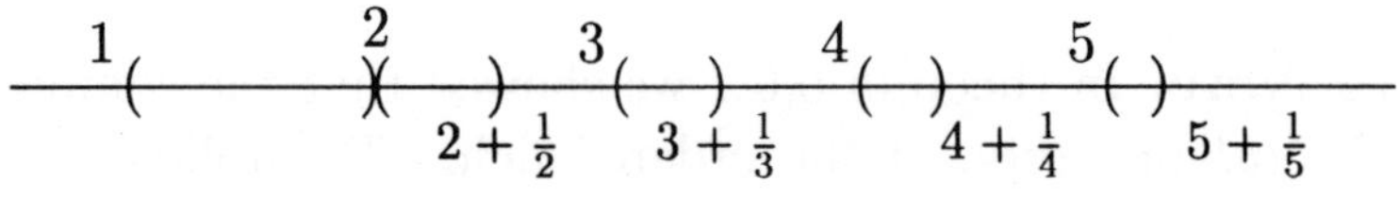

Figure 6.6 : $\left\{\left(n, n + \frac{1}{n}\right)\right\}_{n\geq 1}$ is not nested.

We describe next an important property of a nested sequence of closed intervals.

6.7.3 Theorem (Nested interval property). Let $\{I_n\}_{n\geq 1}$ be a sequence of intervals such that

(i) I_n is a closed interval $\forall\, n$.

(ii) $\{I_n\}_{n\geq 1}$ is a nested sequence.

6.7. Nested interval property

Then $\exists$ unique $a \in \mathbb{R}$ such that $\bigcap_{n=1}^{\infty} I_n = \{a\}$.

Proof. Let $I_n := [a_n, b_n], n = 1, 2, \ldots$. Since the sequence $\{I_n\}_{n \geq 1}$ is decreasing, we have

$$b_1 \geq b_2 \geq b_3 \geq \cdots \geq b_n \geq b_{n+1} \geq \cdots \geq a_{n+1} \geq a_n \geq \cdots a_2 \geq a_1.$$

Thus $\{b_n\}_{n \geq 1}$ is a monotonically decreasing sequence which is bounded below and $\{a_n\}_{n \geq 1}$ is a monotonically increasing sequence which is bounded above. Thus, by theorem 6.3.1, $\{a_n\}_{n \geq 1}$ and $\{b_n\}_{n \geq 1}$ are both convergent to say a and b, respectively. Clearly $a \leq b$. Consider the interval $[a, b] = \{x \in \mathbb{R} \mid a \leq x \leq b\}$. Clearly $[a, b] \neq \emptyset$ and $[a, b] \subseteq [a_n, b_n] \; \forall \; n$. Thus $[a, b] \subseteq \bigcap_{n=1}^{\infty} [a_n, b_n]$. Let $\epsilon = b - a \geq 0$. Since $\epsilon < \lambda(I_n) \; \forall \; n$ and $\{\lambda\{I_n\}\}_{n \geq 1}$ converges to zero, we have $\epsilon = 0$, i.e., $a = b$. Suppose $\exists \, c \in \bigcap_{n=1}^{\infty} I_n, c \neq a$. Then either $c < a$ or $c > a$. If $c < a$, given $\delta = a - c$ we can choose $n_0 \in \mathbb{N}$ such that $a - a_n < \delta = a - c$, i.e., $c < a_n \leq a$. Hence $c \notin [a_n, b_n]$ for $n \geq n_0$. Similarly if $c > a$, we can show that $c \notin [a_n, b_n]$ for some $n \geq n_0$. Thus $\bigcap_{n=1}^{\infty} I_n = \{a\}$. Uniqueness of a is obvious. ∎

6.7.4 Exercise. In the proof of theorem 6.7.3 where exactly do you think the fact that each I_n is a closed interval has been used? Do you think the theorem remains true if I_n's are not necessarily closed?

6.7.5 Exercise. Consider $I_n = (0, 1/n), n = 1, \ldots$. Then $\{I_n\}_{n \geq 1}$ is a sequence of nested intervals and $\bigcap_{n=1}^{\infty} I_n = \emptyset$.

6.7.6 Exercise. Let $a_0 = 1, a_1 = 2$ and for $n \geq 2$,

$$a_n := \frac{1}{2}\left(a_{n-1} + \frac{2}{a_{n-1}}\right) \quad \text{and} \quad b_n := \frac{2}{a_n}.$$

Show that $\{[a_n, b_n]\}_{n\geq 1}$ is a nested sequence. Find $\{x\}$ such that $\{x\} = \bigcap_{n=1}^{\infty} [a_n b_n]$.

6.7.7 Exercise. For each $n \geq 1$, let

$$a_n := \left(1 + \frac{1}{n}\right)^n \quad \text{and} \quad b_n := \left(1 + \frac{1}{n}\right)^{n+1}.$$

Show that $\{[a_n, b_n]\}_{n\geq 1}$ is a nested sequence. Compute $\bigcap_{n=1}^{\infty} [a_n, b_n]$. (Hint: see exercise 6.3.7.)

6.7.8 Notes.

(i) By the nested interval property, each nested sequence of closed intervals locates a point in the line and each point on the line can be located by a nested sequence. In fact, sequences of nested intervals with rational end points are enough to determine points on the line. This idea can be used to construct the real number system. For details, refer to [Cou].

(ii) Since the concept of intervals, open intervals and closed intervals can be defined in any ordered field, the proofs of theorems 6.2.1, 6.3.1 and 6.7.3 extend to any ordered field. Thus in an ordered field F, the *lub*-property implies that F is Archimedean and has the nested interval property. The converse is also true, i.e., if an ordered field is Archimedean and has the nested interval property, then it has the *lub* property. For details refer to [Men].

6.8 Uncountability of $\mathbb{R}$ and the continuum hypothesis

The aim of this section is to compare the 'number' of elements in $\mathbb{Q}$, $\mathbb{R} \setminus \mathbb{Q}$ and $\mathbb{R}$. Recall that in chapter 2 we had defined the notion of countable sets and showed that $\mathbb{N}$ and $\mathbb{N} \times \mathbb{N}$ are countably

6.8. Uncountability of $\mathbb{R}$ and the continuum hypothesis

infinite sets. Other examples of countably infinite sets are $\mathbb{Z}$ and $\mathbb{Q}$.

6.8.1 Theorem. *The sets $\mathbb{Z}$ and $\mathbb{Q}$ are countably infinite.*

Proof. That $\mathbb{Z}$ is countably infinite follows from the fact that

$$\mathbb{Z} = \{\ldots -3, -2, -1\} \cup \{0\} \cup \{1, 2, 3, \ldots\}$$

and theorem 2.4.23 (iii). To show that $\mathbb{Q}$ is countably infinite, we note that

$$\mathbb{Q} = \mathbb{Q}^+ \cup \{0\} \cup \mathbb{Q}^-,$$

where $\mathbb{Q}^+ := \left\{\dfrac{m}{n} \mid m, n \in \mathbb{Z}, m > 0, n > 0\right\}$ and $\mathbb{Q}^- := \{-r \mid r \in \mathbb{Q}^+\}$. In view of theorem 2.4.23 (iii) to show that $\mathbb{Q}$ is countably infinite, it is enough to show that $\mathbb{Q}^+$ is countably infinite. Since $\mathbb{Q}^+$ is not finite (easy to see) and $\mathbb{Q}^+ \subseteq \mathbb{N} \times \mathbb{N}$, it follows from corollary 2.4.21 and theorem 2.4.19 (v) that $\mathbb{Q}^+$ is countably infinite. ∎

6.8.2 Remark. A more geometric way of counting the elements of $\mathbb{Q}^+$ is to arrange its elements in rows, n^{th}-row having elements of the form m/n, and count them diagonally:

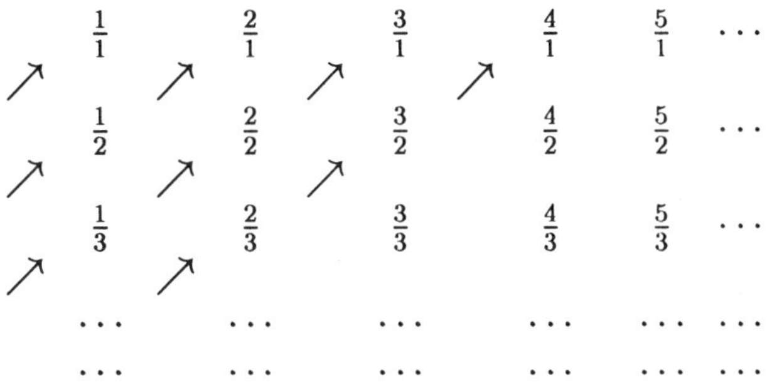

Figure 6.7 : Countability of $\mathbb{Q}$.

6.8.3 Example. The set of all real algebraic numbers (see note 6.2.12(ii) for definition) is countable. For any equation of the form

$$a_n x^n + a_{n-1} x^{n-1} + \cdots + a_1 x + a_0 = 0,$$

where $n \in \mathbb{N}$ and each $a_i \in \mathbb{Z}$ with $a_n \neq 0$, the number $N = n + |a_n| + |a_{n-1}| + \cdots + |a_0|$ is called the **index** of the equation. Note that $N \geq 2$ for every such equation. Clearly, there are only finitely many equations of a given index. Also, each such equation can have at most n solutions (this will follow from corollary 7.6.10). Thus A_N, the set of real solutions of $a_n x^n + a_{n-1} x^{n-1} + \cdots + a_1 x + a_0 = 0$ of index N, is a finite set $\forall\ N$. Hence $\bigcup_{N=2}^{\infty} A_N$, set of all real algebraic numbers is at most a countable set, by corollary 2.4.25.

6.8.4 Proposition. *Let $\mathcal{C}$ be any collection of pairwise disjoint intervals. Then $\mathcal{C}$ is countable.*

Proof. For each $I \in \mathcal{C}$, choose a rational $\xi_I \in I$ (using axiom of choice) and construct the set $A := \{\xi_I : I \in \mathcal{C}\}$. Then A is a countable set, by theorem 2.4.19 (v), and $I \longmapsto \xi_I$ is a one-one map. Thus $\mathcal{C}$ is countable by proposition 2.4.24. ■

6.8.5 Theorem. *If U is any nonempty open set in $\mathbb{R}$ then U is a union of countable number of pairwise disjoint open intervals.*

Proof. Follows from theorem 6.5.24 and proposition 6.8.4. ■

Next we show that $\mathbb{R}$ is not countable. Recall that a set is said to be uncountable if it is not countable.

6.8.6 Theorem. *The set $\mathbb{R}$ is uncountable.*

Proof. Since $\mathbb{N} \subseteq \mathbb{R}$, $\mathbb{R}$ is not countably finite, by theorem 2.4.19 (iv). Let $S \subset \mathbb{R}$ be any countably infinite set. To show that $\mathbb{R}$ is uncountable, it is enough to show that $S \neq \mathbb{R}$. Since S is

6.8. Uncountability of ℝ and the continuum hypothesis

countably infinite, there exists a one-one, onto map $f : \mathbb{N} \longrightarrow S$. Let $x_n := f(n) \ \forall \ n$. Then $S = \{x_1, x_2, \ldots\}$. Choose any closed interval I_1 such that $x_1 \notin I$ and $\lambda(I_1) = 1$, for example take $I_1 = [x_1 + 1, x_1 + 2]$. Next choose a closed interval $I_2 \subset I_1$ such that $x_2 \notin I_2$ and $\lambda(I_2) < \dfrac{1}{2}$. For example, divide I_1 into three equal subintervals. Then X_1 does not belong to at least one of these subintervals and it has length $\dfrac{1}{3}$. Proceeding this way, having defined $I_1, I_2, \ldots, I_{n-1}$, choose a closed interval $I_n \subset I_{n-1}$ such that $x_n \notin I_n$ and $\lambda(I_n) < \dfrac{1}{2^n}$. By induction, we get a sequence $\{I_n\}_{n \geq 1}$ of closed intervals such that $I_{n+1} \subset I_n \ \forall \ n$ and $\{\lambda(I_n)\}_{n \geq 1}$ converges to zero. Thus by the nested interval property (theorem 6.7.3), $\bigcap_{n=1}^{\infty} I_n = \{x\}$. Clearly $\forall \ n, x \in I_n$ and since $x_n \notin I_n, x \neq x_n \ \forall \ n$. Thus $x \notin S$, which is a contradiction. Hence $S \neq \mathbb{R}$. ∎

Aliter. We can also prove that ℝ is uncountable using the decimal representations. As noted earlier, ℝ is not finite. So suppose ℝ is countably infinite, i.e., we can enumerate the elements of ℝ, say $\mathbb{R} = \{x_1, x_2, \ldots\}$. Let us write the decimal expansion of each x_n:

$$\begin{aligned} x_1 &= a_1 \cdot a_{11} a_{12} a_{13} \cdots a_{1n} \cdots \\ x_2 &= a_2 \cdot a_{21} a_{22} a_{23} \cdots a_{2n} \cdots \\ &\vdots \\ x_n &= a_n \cdot a_{n1} a_{n2} a_{n3} \cdots a_{nn} \cdots, \end{aligned}$$

and so on, where $\forall \ i$ and $j, 0 \leq a_{ij} \leq 9$. Consider a real number y whose decimal representation is given by

$$y = 0 \cdot y_1 y_2 y_3 \cdots y_n \cdots,$$

where for each n

$$y_n := \begin{cases} 0 & \text{if } 9 \geq a_n \geq 5, \\ 6 & \text{if } 0 \leq a_n < 5. \end{cases}$$

Then $y \neq x_n$ for any n, because y differs from x_n in the nth-decimal places. Thus $y \in \mathbb{R} = \{x_1, x_2, \ldots\}$ and $y \neq x_n$ for any n, a contradiction. Hence $\mathbb{R}$ is uncountable. ∎

6.8.7 Corollary. *Irrationals form an uncountable subset of* $\mathbb{R}$.

Proof. Since $\mathbb{R} = (\mathbb{R} \setminus \mathbb{Q}) \cup \mathbb{Q}$ is uncountable and $\mathbb{Q}$ is countably infinite, the set of irrationals, $\mathbb{R} \setminus \mathbb{Q}$, cannot be countably infinite because of theorem 2.4.23 (iii). ∎

6.8.8 Exercise. Let X denote the set of all sequences $\{x_n\}_{n \geq 1}$ such that $x_n = 0$ or $1 \ \forall \ n$. Show that X is uncountable.

6.8.9 Exercise. Show that real transcendental numbers exist and form an uncountable subset of $\mathbb{R}$. (Hint: example 6.8.3.)

6.8.10 Notes.

(i) Exercise 6.8.9 demonstrates the existence of transcendental numbers without actually producing one. This was proved by Cantor in 1874 and it was a sensational discovery, for it is quite hard to prove that a given number is transcendental.

(ii) Saying that $\mathbb{Q}$ is countably infinite is essentially saying that $\mathbb{Q}$ has as many elements as $\mathbb{N}$ has (looks surprising !). On the other hand, $\mathbb{R}$ is uncountable is the same as saying that $\mathbb{R}$ has much greater number of elements than $\mathbb{Q}$ has. In some sense $\mathbb{N}, \mathbb{Q}, \mathbb{Z}, \mathbb{R}$ and $\mathbb{R} \setminus \mathbb{Q}$, all have infinite number of elements. The 'infinity' of $\mathbb{N}$ and $\mathbb{Q}$ is much 'smaller' than the 'infinity' of $\mathbb{R}$ and $(\mathbb{R} \setminus \mathbb{Q})$.

Though we have shown that $\mathbb{R}$ is an infinite set and is not countable, we haven't 'counted' the number of elements in $\mathbb{R}$. For that we make the following definition.

6.8.11 Definition. Given two sets A and B, we say A is **equivalent** to B if there exists a one-one, onto map $f : A \to B$. We write this as $A \sim B$.

6.8. Uncountability of $\mathbb{R}$ and the continuum hypothesis

6.8.12 Examples.

(i) For $n \in \mathbb{N}$, let $[1, n] := \{1, 2, \ldots, n\} \subset \mathbb{N}$. Then $[1, n] \sim [1, m]$ iff $n = m$. Obviously, if $n = m$ then $[1, n] \sim [1, m]$, the required map being the identity map. Conversely, suppose $[1, n] \sim [1, m]$ but $n \neq m$, say $m < n$. Let $\phi : [1, n] \to [1, m]$ be the given bijective map and let $\eta : [1, m] \to [1, n]$ be defined by $\eta(k) := k \ \forall \, k \in [1, m]$. Then η is one-one but not onto and hence the map $\eta \circ \phi : [1, n] \to [1, n]$ is a one-one map which is not onto, contradicting theorem 2.4.11. Hence $m < n$ is not true. Similarly, $m > n$ is also not true.

(ii) $\mathbb{N}$ is not equivalent to $[1, n]$ for any $n \in \mathbb{N}$, for otherwise $\mathbb{N}$ will be both a finite as well as an infinite set.

(iii) Clearly any two of the sets $\mathbb{N}, \mathbb{Z}^+, \mathbb{Z}^-, \mathbb{Q}$ and $\mathbb{N} \times \mathbb{N}$ are equivalent to one another.

(iv) The open intervals $(0, 1)$ and $(0, a)$ are equivalent for $a > 0$. For example $f : (0, 1) \to (0, a), x \mapsto ax$ is a one-one, onto map.

(v) Consider the map $x \mapsto \dfrac{x}{|x| + 1}$ from $\mathbb{R}$ into $(0, 1)$. It is easy to show that this is a one-one, onto map. Hence $(0, 1) \sim \mathbb{R}$. Thus $(0, 1)$ is uncountable.

6.8.13 Exercise. Prove the following statements:

(i) Let X be any set. Then $\sim$ is an equivalence relation on X, i.e., $\forall \, A, B, C \in X, A \sim A; A \sim B \Rightarrow B \sim A; A \sim B$ and $B \sim C \Rightarrow A \sim C$.

(ii) Let $a, b \in \mathbb{R}$ be such that $a < b$. Show that $(a, b) \sim (0, 1)$ and $[a, b] \sim [0, 1]$.

(iii) Show that Cantor's ternary set F is equivalent to $[0, 1]$.

(Hint: For $x \in F$ consider its ternary expansion, $x = \cdot a_1 a_2 a_3 \ldots$ and define $\phi(x)$ to be the real number with binary expansion $\cdot b_1 b_2 \ldots$, where $b_i := a_i/2 \ \forall \, i$.)

(iv) Show that $(0,1) \sim (0,1]$. (Hint: consider $f(x) := x$ if $x \neq \frac{1}{n}$ and $f(\frac{1}{n}) := \frac{1}{n-1}$.)

An important theorem which allows one to check the equivalence of two sets is the following.

6.8.14 Theorem (Cantor-Bernstein-Schröder). *If X and Y are two nonempty sets such that each is equivalent to a subset of the other then they are equivalent.*

Proof. Let $X_0 := X \setminus \text{Range}(g)$. Then X_0 is the portion of X which keeps g from being onto. Let $X_{n+1} := g(f(X_n)) \; \forall \; n \geq 0$ and $Y_n := f(X_n) \; \forall \; n \geq 0$. Then $X_{n+1} = g(Y_n) \; \forall \; n \geq 0$. Note that if $x \in X$ and $x \notin X_n$ for any $n \geq 0$, then $x \in \text{Range}(g)$. We define $h : X \to Y$ as follows:

$$h(x) := \begin{cases} f(x) & \text{if } x \in X_n \text{ for some } n, \\ g^{-1}(x) & \text{otherwise}. \end{cases}$$

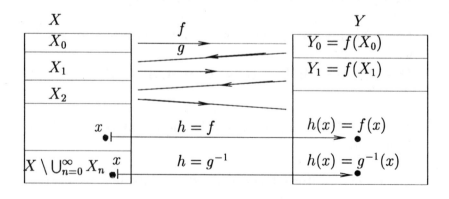

Figure 6.8 : Definition of h.

6.8. Uncountability of $\mathbb{R}$ and the continuum hypothesis

We show that h is one-one and onto. Suppose $x, x' \in X$ and $x \neq x'$. If x, x' both belong to $\bigcup_{n=0}^{\infty} X_n$ or $X \setminus \bigcup_{n=0}^{\infty} X_n$, then it follows that $h(x) \neq h(x')$, as both f, g are one-one. The other possibility is that one of x, x' is in $\bigcup_{n=0}^{\infty} X_n$ and the other in $X \setminus \bigcup_{n=0}^{\infty} X_n$. Let $x \in X_m$ and $x' \notin \bigcup_{n=0}^{\infty} X_n$. Then $h(x) = f(x) \in f(X_m) = Y_m$. But $h(x') = g^{-1}(x') \notin Y_n$, for if $g^{-1}(x') \in Y_m$, then $x' \in g(Y_m) = X_{m+1} \subseteq \bigcup_{n=0}^{\infty} X_n$, which is not true. Hence h is one-one. To show that h is onto, observe that $h(X_n) = f(X_n) = Y_n \; \forall \; n$. Thus $Y_n \subseteq \text{Range}(h) \; \forall \; n$. Let $y \in Y \setminus \bigcup_{n=0}^{\infty} Y_n$. Then $g(y) \notin X \setminus \text{Range}(g) = X_0$. Also $g(y) \notin X_{n+1} = g(Y_n)$ for otherwise $y \in Y_n$. Thus $g(y) \notin \bigcup_{n=0}^{\infty} X_n$ and hence $h(g(y)) = g^{-1}(g(y)) = y$. Thus $Y = \text{Range}(h)$, i.e., the map h is also onto. ∎

As an application of this theorem we have the following:

6.8.15 Theorem. *Let X be any subset of $\mathbb{R}$ such that X includes a nonempty open interval. Then $X \sim \mathbb{R}$.*

Proof. Let $(a, b) \subseteq X$. Since $X \sim X \subseteq \mathbb{R}$ and $\mathbb{R} \sim (a, b) \subseteq X$, by theorem 6.8.12 we have $X \sim \mathbb{R}$. ∎

6.8.16 Corollary. *Any two intervals in $\mathbb{R}$ are equivalent.*

Proof. Follows from theorem 6.8.15. ∎

6.8.17 Definition. Let X be any nonempty set.

(i) If X is a finite set, we define its **cardinality** to be the number of elements in X and denote it by the corresponding natural number.

(ii) If X is countably infinite, i.e., $X \sim \mathbb{N}$, we say X has **cardinality aleph-nought** and denote it by $\aleph_0$.

(iii) If X is any infinite set such that $X \sim \mathbb{R}$, then we say that X has **cardinality** that of the **continuum** and denote it by c.

Let $\mathcal{C}$ denote the set $\{1, 2, \ldots, \aleph_0, c\}$. Elements of $\mathcal{C}$ are called **cardinal numbers**. Note that like $1, 2, \ldots$, etc. $\aleph_0$ and c are just symbols. We can compare these cardinal numbers as follows.

6.8.18 Definition. Let α and β be two cardinal numbers. We say α is **less than or equal to** β, and write $\alpha \leq \beta$, if there exists a set X with cardinality β such that X includes a set A with cardinality α. We say $\alpha = \beta$ if there exist sets X and Y such that $X \sim Y$ with X having cardinality α and Y having cardinality β. We write $\alpha < \beta$ if $\alpha \leq \beta$ and $\alpha \neq \beta$.

6.8.19 Examples.

(i) Clearly if $n \leq m$ in $\mathbb{N}$, then $n \leq m$ as cardinal numbers also.

(ii) Clearly $n \leq \aleph_0 \ \forall \ n \in \mathbb{N}$. In fact $n < \aleph_0 \ \forall \ n \in \mathbb{N}$, since $[1, n]$ is not equivalent to $\mathbb{N}$ for any $n \in \mathbb{N}$ by example 6.8.12(ii).

(iii) It follows from theorem 6.8.6 that $\aleph_0 < c$.

The above examples say that $1 < 2 < \cdots < \aleph_0 < c$. The inequality $\aleph_0 < c$ means that the infinite of $\aleph_0$ is smaller than that of c. To see the smallness of $\aleph_0$ compared to that of c, let us consider the sets $A_n := \left\{ \dfrac{m}{n} \mid m \in \mathbb{N} \text{ and } gcd(m, n) = 1 \right\}, n \in \mathbb{N}$ and clearly each A_n has cardinality $\aleph_0$ and $A_n \cap A_r = \emptyset$ if $n \neq r$. Since $\bigcup_{n=1}^{\infty} A_n = \mathbb{Q}^+$ the cardinality of $\bigcup_{n=1}^{\infty} A_n$ is $\aleph_0$. Thus, even though we have $\aleph_0$ copies of disjoint sets each having $\aleph_0$ elements, we still obtain a set with cardinality $\aleph_0 < c$. The cardinal numbers which correspond to finite sets are called **finite cardinals** and those which

6.8. Uncountability of $\mathbb{R}$ and the continuum hypothesis

correspond to infinite sets are called **infinite cardinals**. Thus, $\aleph_0$ and c are infinite cardinals. The cardinal $\aleph_0$ is the smallest infinite cardinal, i.e., $\aleph_0$ is the cardinality of an infinite set, namely $\mathbb{N}$ and since every infinite set contains a copy of $\mathbb{N}$, we have $\aleph_0$ less than or equal to the 'cardinality' of that set. (Note that we have not defined the cardinality of an arbitrary set. Intuitively, it is the 'number' of elements in the set and we can say that two sets have the same cardinality if they are equivalent. This can be taken as an additional axiom in ZF-theory.) We can call c an uncountable infinite cardinal. Cantor conjectured that there exists no set A with the following two properties: (i) A includes a copy of $\mathbb{N}$ but A is not equivalent to $\mathbb{N}$ and (ii) a copy of A is included in $\mathbb{R}$ but A is not equivalent to $\mathbb{R}$. In the terminology of cardinalities, this implies that there is no cardinal number between $\aleph_0$ and c. This is called the **Continuum Hypothesis**. As we have mentioned in chapter 1, Cohen showed in 1963 that the continuum hypothesis is neither provable nor disprovable from the axioms of set theory. Thus in the Zermelo-Fraenkel axiomatic set theory, the question 'how many real numbers are there?' remains unresolved. We close this section with an answer to the following question: does there exist sets with 'cardinality' bigger than that of $\mathbb{R}$? More precisely, can we construct a set X such that X includes a copy of $\mathbb{R}$ but X is not equivalent to $\mathbb{R}$? Cantor gave a method of constructing such sets. Recall that for a set $X, \mathcal{P}(X)$ denotes the set of all subsets of X.

6.8.20 Theorem (Cantor). *Let X be any nonempty set. Then there exists a one-one mapping of X into $\mathcal{P}(X)$ but X is not equivalent to $\mathcal{P}(X)$.*

Proof. The map $x \longmapsto \{x\}, x \in X$, is clearly a one-one map from X into $\mathcal{P}(X)$. Hence $\mathcal{P}(X)$ includes a copy of X. Suppose there exists an onto map $P : X \to \mathcal{P}(X), x \longmapsto P(x)$. Let $x \in X$ be arbitrary. Since $P(x)$ is a subset of X, either $x \in P(x)$ or $x \notin P(x)$. Let $U := \{x \in X | x \notin P(x)\}$. Since U is a subset of X,

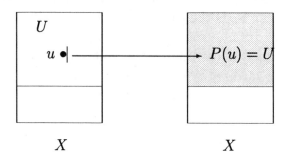

Figure 6.9

i.e., $U \in \mathcal{P}(X)$ $\exists\, u \in X$ such that $P(u) = U$. Now two possibilities arise. Either $u \in P(u) = U$, in which case, by definition of U, $u \notin U$; or $u \notin P(u) = U$, in which case, by definition of U again, $u \in U$. Thus either possibility leads to a contradiction. Hence there does not exist any onto map from X to $\mathcal{P}(X)$. Thus $X \sim P(X)$ is not true. ∎

6.8.21 Note. The above argument is similar to that of Russell's paradox. Russell's paradox demolished Cantor's set theory whereas the same argument in Zermelo-Fraenkel axiomatic set theory becomes constructive in yielding sets of 'bigger cardinalities'.

6.8.22 Definition. For a set X having cardinality α, we say $\mathcal{P}(X)$ has **cardinality** 2^α.

Here 2^α is just a symbolic notation justified by our next theorem.

6.8.23 Theorem. *For $\alpha \in \mathcal{C}$, the following hold:*

(i) $\alpha < 2^\alpha$.

(ii) *If α is a finite cardinal, then 2^α is in fact the natural number 2^α.*

(iii) If $\alpha = \aleph_0$, then $2^{\aleph_0} = c$.

Proof. (i) That $\alpha < 2^\alpha$ follows from theorem 6.8.20.

(ii) Let α be a finite cardinal, i.e., $\alpha \in \mathbb{N}$. Let X be a set with cardinality α. We have to show that $\mathcal{P}(X)$ has 2^α number of elements. Note that there is only one subset of X with no elements, i.e., the empty set and that there is only one subset of X with α elements, namely X itself. In general, if $1 < n < \alpha$, to construct a subset of X having n elements we have to choose n elements of X and there are $\dfrac{\alpha!}{(n!)((\alpha-n)!)} := \binom{\alpha}{n}$ ways of doing that. Hence, $\mathcal{P}(X)$ has $1 + \binom{\alpha}{2} + \cdots + \binom{\alpha}{n} + \cdots + \binom{\alpha}{\alpha} = (1+1)^\alpha = 2^\alpha$ elements.

(iii) Let $\alpha = \aleph_0$. Consider $\mathbb{N}$ the set of natural numbers and $I = (0,1)$. We shall show that $\mathcal{P}(\mathbb{N}) \sim I$. For that, let $A \in \mathcal{P}(\mathbb{N})$ and consider $p_A \in I$ to be the point with decimal expansion $p(A) = \cdot x_1 x_2 x_3 \ldots$ where $x_i = 1$ if $i \in A$ and $x_i = 0$ if $i \notin A$. Then p is clearly a one-one function from $\mathcal{P}(\mathbb{N})$ to I. Also for any $x \in (0,1)$, consider its decimal expansion $x = \cdot a_1 a_2 a_3 \ldots$. Let $s(x) := \{n \in \mathbb{N} | a_n = 1\}$. Then $s(x) \in \mathcal{P}(\mathbb{N})$ and clearly the map $x \longmapsto s(x)$ is a one-one function from I into $\mathcal{P}(X)$. Hence, by theorem 6.8.14, $I \sim \mathcal{P}(\mathbb{N})$. It follows from example 6.8.12 (v) and exercise 6.8.13 (i) that $\mathcal{P}(\mathbb{N}) \sim \mathbb{R}$, i.e., $2^{\aleph_0} = c$. ∎

6.8.24 Note. Examples 6.8.19 and theorems 6.8.20 and 6.8.23 tell us that there are infinite number of infinite sets with cardinal numbers:

$$1 < 2 < 3 < \cdots < \aleph_0 < 2^{\aleph_0} = c < 2^c < 2^{2^c} < \cdots.$$

6.9 Bolzano-Weierstrass property

In theorem 6.3.1 we proved that every bounded sequence $\{x_n\}_{n\geq 1}$ of real numbers has a convergent subsequence. Thus, if we write $A :=$

$\{x_n | n \geq 1\}$ then A is a bounded set, i.e., there exists a nonnegative real number M such that $|a| \leq M \ \forall \ a \in A$. And A has a limit point if A is not finite, namely the limit of the convergent subsequence. This property holds for every bounded infinite subset of $\mathbb{R}$. Suppose A is any bounded infinite subset of $\mathbb{R}$. Then by theorem 2.4.26, A has a countably infinite subset B. If we enumerate elements of B as a sequence, then by theorem 6.3.1, B has a limit and hence B has a limit point, i.e., A has a limit point. This is known as the Bolzano-Weierstrass property of real numbers. We give below another proof of this using the nested interval property, theorem 6.7.3.

6.9.1 Definition. A subset S of $\mathbb{R}$ is said to be **bounded** if there exists $n \in \mathbb{N}$ such that $|x| \leq n \ \ \forall \, x \in S$.

6.9.2 Theorem (Bolzano-Weierstrass property). *Let S be any bounded infinite subset of $\mathbb{R}$. Then S has at least one limit point.*

Proof. We gave a proof of this just before definition 6.9.1. Here is another proof which uses the nested interval property. Since S is a bounded set, there exists a closed interval $I_0 = [a, b]$ such that $S \subseteq I_0$. Divide I_0 into two closed intervals, $I_1^1 := \left[a, \dfrac{a+b}{2}\right]$ and $I_1^2 := \left[\dfrac{a+b}{2}, b\right]$. Then at least one of the sets $I_1^1 \cap S$ and $I_1^2 \cap S$ is infinite for otherwise S will be finite by theorem 2.4.23. Let I_1 be either of the sets I_1^1 or I_1^2 such that $I \cap S$ is infinite. Having constructed closed intervals $I_1 \supset I_2 \supset \cdots \supset I_k$ such that $\lambda(I_k) = \dfrac{(b-a)}{2^k}$ and $I_k \cap S$ is infinite, divide I_k into two equal closed intervals, say I_k^2, and I_k^1, and choose I_{k+1} to be one of the intervals I_k^1 or I_k^2 whose intersection with S is infinite. Then by induction, we get a sequence $\{I_k\}_{k \geq 1}$ of closed intervals such that

(i) $\quad I_k \supset I_{k+1} \ \forall \ k.$

6.9 Bolzano-Weierstrass property

(ii) $\lambda(I_k) = \dfrac{b-a}{2^k}$.

(iii) $I_k \cap S$ is infinite $\forall\, k \geq 1$.

Using the nested interval property (theorem 6.7.3), properties (i) and (ii) imply that $\bigcap_{k=1}^{\infty} I_k = \{x\}$ for some $x \in \mathbb{R}$. We claim that x is a limit point of S. Let J be any open interval such that $x \in J := (c, d)$. Since $x \in I_k \ \forall\, k$, and $\lambda(I_k) = \dfrac{b-a}{2^k}$, we can choose k sufficiently large such that $\dfrac{b-a}{2^k} < \min\{x - c, d - x\}$.

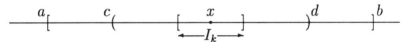

Figure 6.10

Then $I_k \subset J$ and $I_k \cap S$ is an infinite set, by (iii). Thus the set $(I_k \setminus \{x\}) \cap S \neq \emptyset$ and hence $(J \setminus \{x\}) \cap S \neq \emptyset$, i.e., x is a limit point of S. ∎

6.9.3 Notes.

(i) In the proof of theorem 6.9.2 we used the nested interval property of the real number system. One can also deduce this using the Heine-Borel theorem as follows: Let S be any bounded infinite set in $\mathbb{R}$. Since S is bounded, there exists $a, b \in \mathbb{R}$ such that $S \subseteq [a, b]$. Suppose S has no limit point. Then each $x \in [a, b]$ is not a limit point of S and hence there exists an open interval I_x such that $x \in I_x$ and $(I_x \setminus \{x\}) \cap S = \emptyset$. Then the collection $\{I_x | x \in S\}$ forms an open covering of S. By Heine-Borel property (theorem 6.6.7), there exist $x_1, \ldots, x_n \in S$ such that $S \subseteq \bigcup_{j=1}^{n} I_{x_j}$. But then

$S = \bigcup_{j=1}^{n} (I_{x_j} S) \subseteq \{x_1, \ldots, x_n\}$, showing that S is a finite set, not true. Hence S has a limit point. The above arguments work in any ordered field.

(ii) A closer look at the proof of theorem 6.9.2 will tell us that this theorem holds in any ordered field which has the nested interval property. Thus in view of note 6.7.8 (ii), in every ordered field F, completeness implies Bolzano-Weierstrass property. The converse is also true. For details refer to [Men].

(iii) In view of (i) and (ii) above, in an ordered field F completeness, Bolzano-Weierstrass property, and the nested interval property are all equivalent to each other.

6.10 Connected subsets of $\mathbb{R}$

In this section we give another description of the continuity of points on the real line or the property that any two points of $\mathbb{R}$ are not 'separated'. We make this more precise.

6.10.1 Definition.

(i) Let $x, y \in \mathbb{R}$. We say that x is **separated** from y if there exists an open interval I such that $y \in I$ but $x \notin I$.

(ii) Let A be any nonempty subset of $\mathbb{R}$ and $x \in \mathbb{R}$. We say that A is **separated** from x if for every $a \in A$ there exists some open interval I such that $a \in I$ but $x \notin I$.

(iii) Let A and B be nonempty subsets of $\mathbb{R}$. We say that B is **separated** from A if for every $b \in B$ there exists an open interval I such that $b \in I$ and $I \cap A = \emptyset$.

6.10.2 Examples.

(i) Clearly for $x, y \in \mathbb{R}$, x is separated from y iff $x \neq y$. Also $\{x\}$ is separated from $\{y\}$ iff x is separated from y.

(ii) Every $x \in \mathbb{R}$ is separated from $A := (x, y)$, for every $y > x$. Also x is separated from every $a \in (x, y)$. In general, if a point x is separated from a set A, then x is separated from every $a \in A$. However, this need not imply $\{x\}$ is separated from A. For example, let A be the set of all irrationals in $\mathbb{R}$. Then every rational r is separated from every $a \in A$ but $\{r\}$ is not separated from A.

(iii) The intervals $(0,1)$ and $(1,2)$ are separated from each other. Clearly if a set A is separated from another set B then $A \cap B = \emptyset$. However, the converse need not be true. For example consider $A = \mathbb{Q}$, the set of all rationals in $\mathbb{R}$, and $B = \mathbb{R} \setminus \mathbb{Q}$, the set of all irrationals in $\mathbb{R}$. Then $A \cap B = \emptyset$ but neither is A separated from B nor is B separated from A.

6.10.3 Definition. Let S be a subset of $\mathbb{R}$.

(i) A pair of nonempty sets A and B is called a **partition** of S if $A \cap B = \emptyset$ and $A \cup B = S$. We denote this by (A, B).

(ii) A partition (A, B) of S is called a **separation** of S if A and B are separated from each other.

(iii) A subset S of $\mathbb{R}$ is said to be **connected** if there does not exist any separation of S. A set which is not connected is said to be **disconnected**.

6.10.4 Examples.

(i) Clearly $(0,1) \cup (1,2)$ is a disconnected set.

(ii) $((0,1], (1,2))$ is a partition of $(0,2)$ but is not a separation of $(0,1)$.

6.10.5 Theorem. *A subset S of $\mathbb{R}$ is connected iff it is an interval.*

Proof. Suppose S is a connected set. If $S = \emptyset$ or $S = \{a\}$ for some $a \in \mathbb{R}$, clearly S is an interval. So, suppose $x, y \in S$ with $x < y$. Suppose $\exists\, z \in \mathbb{R}$ such that $x < z < y$ and $z \notin S$. Let $A := S \cap (-\infty, z)$ and $B := S \cap (z, +\infty)$. Then (A, B) is a partition of S. Let $a \in A$. Then $a < z$, $a \in (a-1, \frac{a+z}{2})$ and $(a-1, \frac{a+z}{2}) \cap B = \emptyset$. Similarly $\forall\, b \in B, b \in \left(\frac{b+z}{2}, b+1\right)$ and $\left(\frac{b+z}{2}, b+1\right) \cap A = \emptyset$. Thus (A, B) is a separation of S, which is not possible as S is connected. Hence $\forall\, z \in \mathbb{R}$ if $x < z < y$, then $z \in S$, i.e., S is an interval.

Figure 6.11

Conversely, let S be any interval. We shall show that S is connected. Suppose, if possible there exists a separation (A, B) of S. Let $a \in A$ and $b \in B$. Consider the point $\frac{a+b}{2}$. Without loss of generality, let us assume that $a < b$. Then $a < \frac{a+b}{2} < b$ and hence $\frac{a+b}{2} \in S$. Either $\frac{a+b}{2} \in A$ or $\frac{a+b}{2} \in B$. In the earlier case, let $a_1 := \frac{a+b}{a}$ and $b_1 := b$. In the latter case, let $a_1 := a$ and $b_1 := \frac{a+b}{2}$. Again consider the mid-point of $[a_1, b_1]$, i.e., $\frac{a_1 + b_1}{2}$. Once again, since $\frac{a_1 + b_1}{2} \in S$, either $\frac{a_1 + b_1}{2} \in A$ or $\frac{a_1 + b_1}{2} \in B$. If $\frac{a_1 + b_1}{2} \in A$, let $a_2 := \frac{a_1 + b_1}{2}$ and $b_2 := b_1$. If $\frac{a_1 + b_1}{2} \in B$, let $a_2 := a_1$ and $b_2 := \frac{a_1 + b_1}{2}$. Proceeding similarly, we can construct a sequence $\{a_n\}_{n \geq 1}$ of elements of A and a sequence $\{b_n\}_{n \geq 1}$ of element of B such that $\forall\, n \geq 1$, $a_1 \leq a_2 \leq \ldots \leq a_{n-1} \leq a_n \leq \ldots \leq b_n \leq b_{n-1} \leq \ldots \leq b_2 \leq b_1$, and $b_n - a_n = \frac{b-a}{2^n}$. Thus by the completeness of $\mathbb{R}$, both $\lim\limits_{n \to \infty} a_n$ and $\lim\limits_{n \to \infty} b_n$ exist. Using the fact that $b_n - a_n = \dfrac{b-a}{2^n}$, it is easy to see that $\lim\limits_{n \to \infty} a_n = \lim\limits_{n \to \infty} b_n$. Let $x := \lim\limits_{n \to \infty} a_n = \lim\limits_{n \to \infty} b_n$. Now either $x \in A$ or $x \in B$. If $x \in A$, as A is separated from B, there exists an open interval I such that $x \in I$ and $I \cap B = \emptyset$. But that is not possible as $x = \lim\limits_{n \to \infty} b_n$ and hence $b_n \in I$ for all $n \geq n_0$ for some $n_0 \in \mathbb{N}$. Thus $x \notin A$. Similarly $x \notin B$. Hence S has no separation, i.e., S is connected. ∎

6.10.6 Corollary. *If U is a subset of $\mathbb{R}$ which is both open and closed, then either $U = \emptyset$ or $U = \mathbb{R}$.*

Proof. Suppose there exists a nonempty set $U \subsetneq \mathbb{R}$. Then $(U, \mathbb{R} \setminus U)$ is a separation of $\mathbb{R}$, not possible by theorem 6.10.5. Hence $U = \mathbb{R}$. ∎

6.10.7 Exercise. Let A be a nonempty proper subset of $\mathbb{R}$. Show that either A contains a limit point of $\mathbb{R} \setminus A$ or $\mathbb{R} \setminus A$ contains a limit point of A.

6.10.8 Note. The concept of connectedness of sets can be defined in any ordered field, we only need the concept of intervals for that. Theorem 6.10.5 remains true (with the same proof as given above) in any complete ordered field. Thus, if an ordered field F is complete, its connected subsets are precisely the intervals (note that $\emptyset, F$ and $\{a\}$ for $a \in F$ are all intervals in our definition). The converse of this is also true, i.e., if in an ordered field the connected sets are precisely the intervals, then F is complete. For details refer to [Men].

6.11 Extreme values of functions on $\mathbb{R}$

Recall that we had defined the motion of a function in section 1.3. In this section we consider functions which are defined on subsets of the real line and take values in the real line. Let $f : D \subseteq \mathbb{R} \to \mathbb{R}$ be function. We would like to answer the question: when does

$f(D)$, the range of f, have a maximum and/or minimum? The problem of finding maximum/minimum of functions arise in many practical problems in diverse fields. To analyze this problem, let us make the following definition.

6.11.1 Definition. Let D be a subset of $\mathbb{R}$ and $f : D \to \mathbb{R}$ be a function. We say f has a **maximum** in D if there exists a point $x_{\max} \in D$ such that $f(x_{\max}) \geq f(x) \ \forall \ x \in D$. The value $f(x_{\max})$ is called the **maximum value** of f. Similarly, we say f has a **minimum** in D if there exists a point $x_{\min} \in D$ such that $f(x_{\min}) \leq f(x) \ \forall \ x \in D$. The value $f(x_{\min})$ is called the **minimum value** of f.

In order for a function $f : D \to \mathbb{R}$ to have a maximum, clearly $f(D)$, the range of f, has to be bounded above with $lub(f(D))$ to be the maximum value. But then, we have to find a point $x_{\max} \in D$ such that $f(x_{\max}) = lub(f(D))$. One way of doing this is the following: by theorem 6.3.2, we can find a sequence $\{x_n\}_{n \geq 1}$ in D such that $\{f(x_n)\}_{n \geq 1}$ converges to $lub(f(D))$. Suppose $\{x_n\}_{n \geq 1}$ is convergent or at least has a subsequence $\{x_{n_k}\}_{k \geq 1}$ that is convergent to some point $x \in D$. Then $\{f(x_{n_k})\}_{k \geq 1}$ being a subsequence of $\{f(x_n)\}_{n \geq 1}$ will also converge to $lub(f(D))$. Finally, if we can ensure that $\{f(x_{n_k})\}_{k \geq 1}$ also converges to $f(x)$, by the uniqueness of limit we will have $f(x) = lub(f(D))$. Thus, for $f : D \to \mathbb{R}$ to have maximum in D, we require the following:

(i) The set $f(D)$ is bounded.

(ii) Every sequence $\{x_n\}_{n \geq 1}$ in D has a subsequence convergent in D, which is nothing but the compactness of D.

(iii) The function f has the property that whenever a sequence $\{u_k\}_{k \geq 1}$ in D is convergent to a point $u \in D$, then $\{f(u_k)\}_{k \geq 1}$ is convergent to $f(u)$.

We show that properties (ii) and (iii) are enough to prove that $f : D \to \mathbb{R}$ has a maximum and minimum in D.

6.11 Extreme values of functions on $\mathbb{R}$

6.11.2 Theorem (Weierstrass). *Let D be a nonempty compact subset of $\mathbb{R}$ and $f : D \to \mathbb{R}$ be a function with the property that whenever a sequence $\{u_k\}_{k \geq 1}$ in D is convergent to a point $u \in D$, then $\{f(u_k)\}_{k \geq 1}$ converges to $f(u)$. Then $f(D)$ is a compact subset of $\mathbb{R}$ and f has a maximum and minimum in D.*

Proof. We first show that $f(D)$ is a bounded set. Suppose $f(D)$ is not a bounded set. Then $\forall\, n \in \mathbb{N}$, there exists a point $u_n \in D$ such that $|f(u_n)| > n$. Since D is compact, the sequence $\{u_n\}_{n \geq 1}$ has a subsequence, say $\{u_{n_k}\}_{k \geq 1}$, convergent in D. Let $\lim_{k \to \infty} u_{n_k} = u \in D$. But then, by the given hypothesis, $\{f(u_{n_k})\}_{k \geq 1}$ is convergent to $f(u)$, which is not possible as $|f(u_{n_k})| > n_k \geq k$, $\forall\, k \in \mathbb{N}$. Thus $f(D)$ is bounded. Thus by the completeness of $\mathbb{R}$, $\alpha := lub\,(f(D))$ and $\beta := glb\,(f(D))$ exist. By theorem 6.3.2, there exists a sequence $\{x_n\}_{n \geq 1}$ in D such that $\{f(x_n)\}_{n \geq 1}$ converges to α. Since D is compact, $\{x_n\}_{n \geq 1}$ has a subsequence $\{x_{n_k}\}_{k \geq 1}$ convergent in D. Let $x_{\max} := \lim_{k \to \infty} x_{n_k}$. By the given property of f, $\{f(x_{n_k})\}_{k \geq 1}$ converges to $f(x_{\max})$. Being a subsequence of $\{f(x_n)\}_{n \geq 1}$, $\{f(x_{n_k})\}_{k \geq 1}$ also converges to α. Hence, by the uniqueness of the limit, $\alpha = f(x_{\max})$. This proves the existence of maximum of f. The existence of minimum can be proved similarly. Finally, to show that $f(D)$ is compact, we only have to show that $f(D)$ is also a closed subset of $\mathbb{R}$ (see theorem 6.6.11). Let $\{y_n\}_{n \geq 1}$ be a sequence in $f(D)$, convergent to some point $y \in \mathbb{R}$. We have to show that $y \in f(D)$. Choose $z_n \in D$ such that $f(z_n) = y_n$. Since D is compact, $\{z_n\}_{n \geq 1}$ has a convergent subsequence, say $\{z_{n_k}\}_{k \geq 1}$ with $\lim_{k \to \infty} (z_{n_k}) = z \in D$. But then, by the given property of f, $\lim_{k \to \infty} f(z_{n_k}) = f(z)$. Since $\{f(z_{n_k})\}_{k \geq 1}$ is a subsequence of $\{f(z_n)\}_{n \geq 1}$, $y = \lim_{n \to \infty} f(z_{n_k}) = f(z)$. Hence $y \in f(D)$, i.e., $f(D)$ is closed. ∎

The special property of functions required in the above theorem plays an important role in analysis. We separate it out as a definition.

6.11.3 Definition. Let D be a subset of $\mathbb{R}$ and $f : D \to \mathbb{R}$. We say that f is **continuous at a point** $u \in D$ if for every sequence $\{u_k\}_{k \geq 1}$ in D converging to u, the sequence $\{f(u_k)\}_{k \geq 1}$ converges to $f(u)$. We say that f is **continuous** on D if f is continuous at every point in D. If f is not continuous at a point $u \in D$, we say f is **discontinuous at u** (or f has a **discontinuity** at u or u is a **point of discontinuity** for f).

6.11.4 Note. To show that a function $f : D \to \mathbb{R}$ is continuous at $u \in D$, one has to show that for every sequence $\{u_k\}_{k \geq 1}$ in D converging to u, the sequence $\{f(u_k)\}_{k \geq 1}$ converges to $f(u)$. However, to show that f is discontinuous at u, it is enough to produce only one sequence $\{u_k\}_{k \geq 1}$ in D converging to u such that $\{f(u_k)\}_{k \geq 1}$ is not convergent to $f(u)$.

6.11.5 Examples.

(i) Let $f : D \to \mathbb{R}$ be the identity function, i.e., $f(x) = x \ \forall \, x \in D$, where D is any subset of $\mathbb{R}$. Then f is obviously continuous on D.

(ii) Let $c \in \mathbb{R}$ and D be any subset of $\mathbb{R}$. The **constant function**, i.e., $f(x) = c \ \forall \, x \in D$ is continuous on D.

(iii) Let $a \in \mathbb{R}$ and $f : \mathbb{R} \to \mathbb{R}$ be defined by

$$f(x) := \begin{cases} 1 & \text{for } x > a, \\ 0 & \text{for } x \leq a. \end{cases}$$

Then f is continuous at every point $x \neq a$ and is discontinuous at $x = a$. For example, the sequence $\{a + \frac{1}{n}\}_{n \geq 1}$ converges to a but $\{f(a + \frac{1}{n})\}_{n \geq 1}$ is the constant sequence converging to $1 \neq f(a)$.

(iv) Let f, g and h be the following functions:

$$f : \mathbb{Q} \to \mathbb{R}, \ f(x) := 1 \ \forall \, x \in \mathbb{Q};$$
$$g : \mathbb{R} \setminus \mathbb{Q} \to \mathbb{R}, \ g(x) := 0 \ \forall \, x \in \mathbb{R} \setminus \mathbb{Q};$$
$$h : \mathbb{R} \to \mathbb{R},$$

$$h(x) := \begin{cases} 1, & \text{if } x \in \mathbb{Q}, \\ 0, & \text{if } x \in \mathbb{R} \setminus \mathbb{Q}. \end{cases}$$

6.11 Extreme values of functions on $\mathbb{R}$

Clearly f and g are functions continuous at every point but h is not continuous at any point. For example, for $x \in \mathbb{Q}$ choose a sequence $\{x_n\}_{n \geq 1}$ of irrationals in $\mathbb{R}$ converging to x. Then $h(x) = 0 \ \forall \ n$ and hence $\{h(x_n)\}_{n \geq 1}$ does not converge to $h(x) = 1$. Thus h is not continuous at any $x \in \mathbb{Q}$.

6.11.6 Proposition. *Let D be any subset of $\mathbb{R}$ and $f : D \to \mathbb{R}$ have the property that for every sequence $\{u_k\}_{k \geq 1}$ in D converging to $u \in D$, $\{f(u_k)\}_{k \geq 1}$ is convergent. Then $f\{(u_k)\}_{k \geq 1}$ converges to $f(u)$, i.e., f is continuous at u.*

Proof. Let $\{u_k\}_{k \geq 1}$ be any sequence in D converging to $u \in D$. The hypothesis says that $\{f(u_k)\}_{k \geq 1}$ is convergent. We have to show that $\{f(u_k)\}_{k \geq 1}$ is convergent to $f(u)$. Consider the sequence $\{x_n\}_{n \geq 1}$, defined by

$$x_n := \begin{cases} u_k & \text{if } n = 2k-1, k \geq 1, \\ u & \text{if } n = 2k, k \geq 1. \end{cases}$$

Clearly the sequence $\{x_n\}_{n \geq 1}$ is convergent to u and by the given hypothesis, $\{f(x_n)\}_{n \geq 1}$ is convergent. But $\{f(x_{2k-1})\}_{k \geq 1}$ is a subsequence of $\{f(x_n)\}_{n \geq 1}$ and is convergent to $f(u)$. Thus it follows that $\{f(x_n)\}_{n \geq 1}$ converges to $f(u)$. Finally, $\{f(u_k) = f(x_{2k-1})\}_{k \geq 1}$ being a subsequence of $\{f(x_n)\}_{n \geq 1}$ is also convergent to $f(u)$. ∎

6.11.7 Notes.

(i) If $u \in D$ is not a limit point of D, then the only sequence $\{u_k\}_{k \geq 1}$ of elements of D that can converge to u is the constant sequence, i.e., $u_k = u \ \forall \ k$. Thus $f(u_n) = u \ \forall \ u$ and hence $f(u_n) \to f(u)$. Thus f is continuous at such points $u \in D$. The nontrivial situation arises only when $u \in \mathbb{R}$ is a limit point of D. In that case there exist sequences of distinct elements of D converging to u.

(ii) Continuity of f at $u \in D$ implies that for every sequence $\{u_n\}_{n \geq 1}$ in D with $\{u_n\}_{n \geq 1}$ converging to u, $\{f(u_k)\}_{k \geq 1}$ converges to the same value, namely $f(u)$. In general, it can happen that for every sequence $\{u_k\}_{k\geq 1}$ in D converging to u, a limit point of D, $\{f(u_k)\}_{k\geq 1}$ converges to the same value, say $\ell \in \mathbb{R}$. In that case we can call ℓ as the 'value expected of the function f' at u. This is because via whichever sequence we 'approaches' the point u in D, the image sequence 'approaches' the same value ℓ. Continuity of f at u means that $\ell = f(u)$, i.e., f takes the value it is expected to take at u. In a sense, there is continuity in the behaviour of the function f at the point u. The 'value expected' of f at u, whenever it exists, is called the **limit** of f at u and is denoted by $\lim_{x\to u} f(x)$. Proposition 6.11.6 says that f is continuous at a limit point $u \in D$ iff the limit of f at u exists and is equal to the value of the function at u, i.e., $\lim_{x\to u} f(x) = f(u)$.

6.11.8 Example. Let D be any nonempty subset of $\mathbb{R}$ and let $f : \mathbb{R} \to \mathbb{R}$ be defined by

$$f(x) := \begin{cases} 1 & \text{if } x \in D, \\ 0 & \text{if } x \notin D. \end{cases}$$

This is called the **indicator function** of the set D. For example, the function h defined in example 6.11.5(iv) is the indicator function of the set of rationals in $\mathbb{R}$. Suppose there exists a point $x \notin D$ which is a limit point of D. Then, by exercise 6.5.17, there exists a sequence $\{x_n\}_{n\geq 1}$ of elements of D such that $\{x_n\}_{n\geq 1}$ converges to x. Since $f(x_n) = 1 \; \forall \; x$ and $f(x) = 1, f$ becomes discontinuous at x. Similarly, if there exists $x \in D$ such that x is a limit point of $\mathbb{R} \setminus D$, f can be shown to be discontinuous at x. Now in view of exercise 6.10.7, it follows that, f has at least one point of discontinuity if $D \neq \emptyset$ and $D \neq \mathbb{R}$.

6.11.9 Exercise. Let $f : \mathbb{R} \to \mathbb{R}$ be defined by

$$f(x) := \begin{cases} x & \text{if } x \text{ is irrational}, \\ 0 & \text{if } x \text{ is rational}. \end{cases}$$

Show that f is continuous only at $x = 0$.

6.11 Extreme values of functions on ℝ

6.11.10 Exercise. Let $f_i : \mathbb{R} \to \mathbb{R}$, $i = 1, 2$, be defined by

$$f_1(x) := \begin{cases} x & \text{if } x \text{ is irrational,} \\ 1 - x & \text{if } x \text{ is rational,} \end{cases}$$

and

$$f_2(x) := [x], \text{ the greatest integer } \leq x, \ x \in \mathbb{R}.$$

Analyze f_1 and f_2 for points of continuity.

6.11.11 Exercise. Let D be a subset of $\mathbb{R}$ and $f_i : D \to \mathbb{R}$, $i = 1, 2$, be functions which are continuous at a point $a \in D$. Prove the following:

(i) For $\alpha \in \mathbb{R}$, αf_1 and $|f_1|$ are continuous at $a \in D$, where

$$\begin{aligned}(\alpha f_1)(x) &:= \alpha(f(x)) \quad \forall\, x \in D; \\ |f_1|(x) &:= |f(x)| \quad \forall\, x \in D.\end{aligned}$$

(ii) The functions $f_1 + f_2$ and $f_1 f_2$ are both continuous at $a \in D$, where

$$\begin{aligned}(f_1 + f_2)(x) &:= f_1(x) + f_2(x), \quad x \in D; \\ (f_1 f_2)(x) &:= f_1(x) f_2(x), \quad x \in D.\end{aligned}$$

(iii) If $f_2(x) \neq 0 \ \forall\, x \in D$. The functions f_1/f_2 is continuous at $a \in D$, where

$$(f_1/f_2)(x) := f_1(x)/f_2(x), \quad x \in D.$$

(Hint: exercise 6.3.4.)

6.11.12 Exercise. Let D be a subset of $\mathbb{R}$ and $f_1, f_2 : D \to \mathbb{R}$ be functions. Define $\forall\, x \in D$

$$(f_1 \vee f_2)(x) := \max\{f_1(x), f_2(x)\}$$

and
$$(f_1 \wedge f_2)(x) := \min\{f_1(x), f_2(x)\}.$$

Show that
$$f_1 \vee f_2 = \frac{1}{2}(f_1 + f_2 + |f_1 - f_2|)$$

and
$$f_1 \wedge f_2 = \frac{1}{2}(f_1 + f_2 - |f_1 - f_2|).$$

Using exercise 6.11.11, show that both $f_1 \wedge f_2$ and $f_1 \vee f_2$ are continuous at a.

We can restate theorem 6.11.2 in the terminology of continuous functions as follows:

6.11.13 Theorem (Extreme value). *Let D be a compact subset of $\mathbb{R}$ and $f : D \to \mathbb{R}$ be a continuous function. Then $f(D)$ is a compact subset of $\mathbb{R}$ and f has a maximum and a minimum in D.*

In particular, the above theorem tells us that if $D = [a, b]$, a closed bounded interval, then $f(D)$ is a compact set and $f(D) \subseteq [f(x_{\min}), f(x_{\max})]$. One can say much more: $f[a, b]$ is in fact a closed bounded interval, as proved in the next theorem.

6.11.14 Lemma (Location of zeros). *Let I be an interval and $f : I \to \mathbb{R}$ be continuous. Let $a, b \in I$ be such that $a < b$ and $f(a) < 0 < f(b)$. Then there exists a number $c \in (a, b)$ such that $f(c) = 0$.*

First proof. Let $A := \{x \in [a, b] | f(x) \geq 0\}$. The idea is to show that $glb(A)$ exists and is the required point. Clearly $a \notin A$ and $b \in A$. Thus A is a nonempty subset of $\mathbb{R}$ and is bounded below by a. Thus $glb(A)$ exists. Let $c \in [a, b], c := glb(A)$. By theorem 6.3.1, there exists a sequence $\{x_n\}_{n \geq 1}$ in A such that $\lim_{n \to \infty} f(x_n) = c$ and hence $f(c) \geq 0$. Thus $c \in A$ and $c \neq a$, i.e., $a < c$. We choose any sequence $\{y_n\}_{n \geq 1}$ such that $a < y_n < c$ and $\lim_{n \to \infty} y_n = c$. By

6.11 Extreme values of functions on $\mathbb{R}$ 301

continuity of f again, $\lim_{n\to\infty} f(y_n) = f(c)$. Since $y_n \notin A \ \forall \, n, f(y_n) < 0$. Thus $f(c) \leq 0$, consequently $f(c) = 0$. ∎

Second proof (bisection method). Let $I_1 = [a, b]$. We know that $f(a) < 0$ and $f(b) > 0$. Consider the point $\dfrac{a+b}{2} \in [a, b]$. If $f(\dfrac{a+b}{2}) = 0$, we take $c = \dfrac{a+b}{2}$ and the proof is complete. If $f(\dfrac{a+b}{2}) > 0$, we define $a_2 := a, b_2 := \dfrac{a+b}{2}$. In case $f(\dfrac{a+b}{2}) < 0$, we define $a_2 := \dfrac{a+b}{2}$ and $b_2 := b$. In either case, we get $I_2 := [a_2, b_2] \subset [a, b]$ with $f(a_2) < 0 < f(b_2)$. We continue this bisection process. Suppose that the intervals $I_1, \ldots, I_n := [a_n, b_n]$ with $f(a_n) < 0 < f(b_n)$ have been obtained. Consider the point $\dfrac{a_n + b_n}{2}$. If $f(\dfrac{a_n + b_n}{2}) = 0$, the proof is complete. In case $f(\dfrac{a_n + b_n}{2}) > 0$, we define $a_{n+1} := a_n$ and $b_n := \dfrac{a_n + b_n}{2}$. In case $f(\dfrac{a_n + b_n}{2}) < 0$, we define $a_{n+1} := \dfrac{a_n + b_n}{2}$ and $b_n := b_n$. In either case, we get an interval $I_{n+1} := [a_{n+1}, b_{n+1}] \subset [a, b]$ with $f(a_{n+1}) < 0 < f(b_{n+1})$. Thus, either for some n we will have the point $\dfrac{a_n + b_n}{2} \in [a, b]$ such that $f(\dfrac{a_n + b_n}{2}) = 0$ or we will get a decreasing sequence of closed intervals $I_n = \{[a_n, b_n]\}_{n \geq 1}$. Since $b_n - a_n = (b-a)/2^{n-1}$, the sequence $\{[a_n, b_n]\}_{n \geq 1}$ is a nested sequence of closed intervals. Thus, by theorem 6.7.3, there exists a point $c \in [a, b]$ such that $\{c\} = \bigcap_{n=1}^{\infty} I_n$. Since $a_n \leq c \leq b_n$, $0 \leq c - a_n \leq (b_n - a_n) = (b-a)/2^{n-1}$. Thus $\lim_{n\to\infty} b_n = \lim_{n\to\infty} a_n = c$. Since f is continuous we have $f(c) = \lim_{n\to\infty} f(b_n) = \lim_{n\to\infty} f(a_n)$. Since $f(a_n) < 0 < f(b_n) \ \forall \, n$, we have $\lim_{n\to\infty} f(a_n) \leq 0 \leq \lim_{n\to\infty} f(b_n)$. Hence $f(c) = 0$. ∎

6.11.15 Theorem (Intermediate value). *Let I be an interval and $f : I \to \mathbb{R}$ be continuous on I. Let $a, b \in I$ and $\alpha \in \mathbb{R}$ be such that $f(a) < \alpha < f(b)$. Then there exists a point $c \in I$ between a and b such that $f(c) = \alpha$.*

Proof: Assume first that $a < b$. Consider the function $g(x) := f(x) - \alpha, x \in I$. Then g is a continuous function on I and $g(a) < 0 < g(b)$. Thus, by lemma 6.11.14, there exists $c \in (a, b)$ such that $0 = g(c) = f(c) - \alpha$. Hence $f(c) = \alpha$. In case $b < a$, consider the function $h(x) := \alpha - f(x)$ and apply lemma 6.11.14. ∎

6.11.16 Corollary. *Let I be an interval and $f : I \to \mathbb{R}$ be a continuous function. Then $f(I)$ is an interval. If I is both closed and bounded, then $f(I)$ is also closed and bounded.*

Proof: Let I be any interval. Let $\alpha, \beta \in f(I)$ and $\gamma \in \mathbb{R}$ be such that $\alpha < \gamma < \beta$. Let $a, b \in I$ with $f(a) = \alpha$ and $f(b) = \beta$. By theorem 6.11.15, there exists c between a and b such that $f(c) = \gamma$. Since I is an interval, $c \in I$. Hence $\gamma = f(c) \in f(I)$, i.e., $f(I)$ is an interval.

Next suppose that I is both closed and bounded, say $I = [a, b]$. Then, by theorem 6.11.13, there exists $x_{\max}$ and $x_{\min} \in I$ such that $f(x_{\min}) = \inf f(I)$ and $f(x_{\max}) = \sup f(I)$. Clearly $f(I) \subseteq [f(x_{\min}), f(x_{\max})]$. Let $x \in [f(x_{\min}), f(x_{\max})]$. Then, by theorem 6.11.15, there exists c between $x_{\min}$ and $x_{\max}$, and hence in I, such that $f(c) = \gamma$. Thus $f(I) = [f(x_{\min}), f(x_{\max})]$. ∎

6.11.17 Corollary. *Let $f : D \to \mathbb{R}$ be any continuous function and $A \subseteq D$ be any connected subset of $\mathbb{R}$. Then $f(A)$ is also a connected subset of $\mathbb{R}$.*

Proof. Follows from theorem 6.10.5 and corollary 6.11.16. ∎

6.11.18 Exercise. *Let $I \subseteq \mathbb{R}$ be an interval and $f : I \to \mathbb{R}$ be a continuous function. Prove or disprove the following:*

6.11 Extreme values of functions on $\mathbb{R}$ 303

(i) If I is an open interval, then so is $f(I)$.

(ii) If I is a closed interval, then so is $f(I)$.

(iii) If I is a bounded interval, then so is $f(I)$.

As another application of lemma 6.11.14, we show that every real polynomial of odd degree has at least one real root.

6.11.19 Lemma. *Let $a_0, a_1, \ldots, a_n$ be real numbers, $a_n \neq 0$ and $\phi(x) = \dfrac{a_0}{x^n} + \dfrac{a_1}{x^{n-1}} + \cdots + a_n$, for $x \neq 0$. Then there exists a positive number K such that $\phi(x)$ has the same sign as a_n for $|x| \geq K$.*

Proof: Clearly

$$|\phi(x) - a_n| \leq \frac{|a_0|}{|x^n|} + \cdots + \frac{|a_{n-1}|}{|x|}.$$

Thus for any positive real number K, if $|x| \geq K$ then

$$|\phi(x) - a_n| \leq \frac{(|a_0| + \cdots + |a_{n-1}|)}{K}.$$

We choose $K > 1$, large enough, such that

$$\frac{|a_0| + \cdots + |a_{n-1}|}{K} < |a_n|.$$

Then for $|x| > K$,

$$|\phi(x) - a_n| < |a_n|. \tag{6.6}$$

Now suppose $\phi(x) > 0$ and $a_n < 0$. Then $\phi(x) - a_n > 0$ and from (6.6) we have

$$\phi(x) - a_n < -a_n, \quad \text{i.e.,} \quad \phi(x) < 0,$$

which is a contradiction. Similarly if $\phi(x) < 0$ and $a_n > 0$, then $a_n - \phi(x) > 0$ and (6.6) gives

$$a_n - \phi(x) < a_n, \quad \text{i.e.,} \quad \phi(x) > 0,$$

which is again a contradiction. Hence $\phi(x)$ and a_n have the same sign. ∎

6.11.20 Theorem (Roots of real polynomials). *Every polynomial of odd degree with real coefficients has at least one real root.*

Proof: Let $p(x)$ be a polynomial of odd degree with real coefficients, say $p(x) = a_0 + a_1 x + \cdots + a_n x^n$, $a_n \neq 0$. Let

$$q(x) := \frac{a_0}{x^n} + \cdots + \frac{a_{n-1}}{x} + a_n.$$

By lemma 6.11.19, there exists a positive real number K such that for $|x| > K$, $q(x)$ and a_n have the same sign. Now $p(x) = q(x)x^n$. In case $a_n > 0$, for $x > K$ we have $p(x) > 0$ and for $x < -K$, n being odd, we have $p(x) < 0$. Thus by lemma 6.11.14, $p(x) = 0$ for some x with $|x| \geq K$. Similarly, if $a_n < 0$, then $p(x) < 0$ for $x > K$ and $p(x) > 0$ for $x < -K$. Once again $p(x) = 0$ for some x with $|x| \leq K$. ∎

We close this section with some equivalent ways of describing the continuity of functions.

6.11.21 Theorem. *Let D be a nonempty subset of $\mathbb{R}$, $u \in D$ and $f : D \to \mathbb{R}$ be a function. Then the following are equivalent:*

(i) f is continuous at $u \in D$.

(ii) For every real number $\epsilon > 0$, there exists a real number $\delta > 0$ such that $\forall x \in D$ if $|x - u| < \delta$ then $|f(x) - f(u)| < \epsilon$.

Proof: To prove that the statement (i) implies (ii) we shall prove that if (ii) does not hold, then (i) also does not hold. Saying that (ii) does not hold means that there exists a real number $\epsilon > 0$ such that for every real number $\delta > 0$, we can find a point $x \in D$ such that $|x - u| < \delta$ but $|f(x) - f(u)| \geq \epsilon$. We start with $\delta = \dfrac{1}{n}, n \in \mathbb{N}$ and choose $x_n \in D$ such that $|x_n - u| < \dfrac{1}{n}$ but $|f(x_n) - f(u)| \geq \epsilon$. This means that the sequence $\{x_n\}_{n \geq 1}$ converges to u but $\{f(x_n)\}_{n \geq 1}$ does not converge to $f(u)$. Hence (i) does not hold. Thus (i) implies (ii). Conversely, let (ii) hold and let $\{x_n\}_{n \geq 1}$ be a sequence of elements of D converging to u. Let $\epsilon > 0$ be arbitrary. Then by (ii), there exists $\delta > 0$ such that whenever $x \in D$ with $|x - u| < \delta$, we have $|f(x) - f(u)| < \epsilon$. Since $\{x_n\}_{n \geq 1}$ converges to a, we can choose $n_0 \in \mathbb{N}$ such that $|x_n - u| < \delta \ \forall \ n \geq n_0$. Then $|f(x_n) - f(u)| < \epsilon \ \forall \ n \geq n_0$. Hence the sequence $\{f(x_n)\}_{n \geq 1}$ converges to $f(u)$. ∎

6.11.22 Remark. For a function $f : D \to \mathbb{R}$, continuity of f at a point u is a 'local' property, i.e., it depends upon the values of f at points near the point u only. The condition (ii) in theorem 6.11.21 says that given any 'degree of closeness', the values of f at points 'close' to you are 'close' to $f(u)$, the value of function at u. We can make this more precise in terms of 'neighborhoods'.

6.11.23 Definition. Let $u \in \mathbb{R}$ and A be a subset of $\mathbb{R}$ such that $u \in A$. The set A is called a **neighborhood** of u if u is an interior point of A, i.e., there exists an open interval I such that $u \in I \subseteq A$.

For example an open set $A \subseteq \mathbb{R}$ is a neighbourhood of each of its elements. We can restate theorem 6.11.21 as follows:

6.11.24 Theorem. *Let D be a subset of $\mathbb{R}$, and $f : D \to \mathbb{R}$ be a function. For $u \in D$, the following statements are equivalent:*

(i) f is continuous at u.

(ii) Given any neighbourhood V of $f(u)$, there exists a neighbourhood U of u such that $f(U \cap D) \subseteq V$.

Proof: Suppose f is continuous at $u \in D$ and V is a neighbourhood of $f(u)$. By definition, there exists an open interval I such that $f(u) \in I \subseteq V$. Choose $\epsilon > 0$ such that $f(u) \in (f(u) - \epsilon, f(u) +$

ϵ) =: $I_\epsilon \subseteq I \subseteq V$. By continuity of f at u (theorem 6.11.21), there exists $\delta > 0$ such that for $x \in D$ with $|x - u| < \delta$, $f(x) \in I_\epsilon$. Let $U = (u - \delta, u + \delta)$. Then U is a neighbourhood of u and $f(U \cap D) \subseteq V$. Thus (i) implies (ii). To prove (ii) $\Rightarrow$ (i), let $\epsilon > 0$ be given. Then $V := (f(u) - \epsilon, f(u) + \epsilon)$ is a neighbourhood of $f(u)$ and by (ii), there exists a neighbourhood U of u such that $f(U \cap D) \subseteq V$. Choose $\delta > 0$ such that $u \in (u - \delta, u + \delta) \subseteq U$. Then $\forall\, x \in D \cap (u - \delta, u + \delta), f(x) \in V$, i.e., for $x \in D$ with $|u - x| < \delta, |f(u) - f(x)| < \delta$. Hence (i) holds. ∎

Continuity of a function at every point in the domain of f can also be described in terms of open sets as follows:

6.11.25 Theorem. *Let D be a subset of $\mathbb{R}$ and $f : D \to \mathbb{R}$. Then the following statements are equivalent:*

(i) f is continuous at every point of D.

(ii) For every open subset V of $\mathbb{R}$, there exists an open set U in $\mathbb{R}$ such that $U \cap D \subseteq f^{-1}(V) := \{x \in D\,|\,f(x) \in V\}$.

Proof. Suppose (i) holds and let V be an open subset of $\mathbb{R}$. In case $f^{-1}(V) = \emptyset$, we take $U = \emptyset$. If $f^{-1}(V) \neq \emptyset$, for every $u \in f^{-1}(V)$, $f(u) \in V$ and V, being an open set, is a neighbourhood of $f(u)$. Thus, by theorem 6.11.24, there exists a neighbourhood $U(u)$ of u such that $f(U(u) \cap D) \subseteq V$. Let $U := \bigcup_{u \in U} U(u)$. Then U is an open subset of $\mathbb{R}$, by example 6.5.22, and $f(U \cap D) \subseteq V$, i.e., $U \cap D \subseteq f^{-1}(V)$. Since $f^{-1}(V) \subseteq U$, we have $U \cap D \subseteq f^{-1}(V)$. Thus (i) $\Rightarrow$ (ii) holds. Conversely, let (ii) hold and $u \in D$ be arbitrary. Let V be a neighborhood of $f(u)$. Let I be an open interval such that $f(u) \in I \subset V$. Then by (ii), there exists an open set U such that $U \cap D = f^{-1}(I)$, i.e., $f(U \cap D) \subseteq I \subseteq V$. Since $f(u) \in I$, it follows that $u \in f^{-1}(I) = U \cap D$. Thus U is a neighborhood of u such that $f(U \cap D) \subseteq V$. Hence, f is continuous at u, by theorem 6.11.24. ∎

6.11.26 Corollary. *Let* $f : \mathbb{R} \to \mathbb{R}$. *Then* f *is continuous at every point iff* $f^{-1}(V)$ *is an open set for every open set* V *of* $\mathbb{R}$.

Proof. Follows from theorem 6.11.25 with $D = \mathbb{R}$. ∎

Continuity property of functions in the form given by theorem 6.11.25 and corollary 6.11.26 finds generalization to 'topological spaces'. As an application of corollary 6.11.26, we give an alternative proof of corollary 6.11.17, which avoids the use of theorem 6.10.5 and corollary 6.11.16.

6.11.27 Theorem. *Let* $C \subseteq \mathbb{R}$ *be connected and* $f : C \to \mathbb{R}$ *be continuous. Then* $f(C)$ *is a connected subset of* $\mathbb{R}$.

Proof. Suppose $f(C)$ is not connected and let (A, B) be a separation of it. Since $f(C) = A \cup B$ and $A \cap B = \emptyset$, we have $C = f^{-1}(A) \cup f^{-1}(B)$ and $f^{-1}(A) \cap f^{-1}(B) = \emptyset$. Let $a \in f^{-1}(A)$. Then $f(a) \in A$. Since A is separated from B, there exists an open interval I such that $f(a) \in I$ and $I \cap B = \emptyset$. But then by continuity of f (theorem 6.11.26) there exists an open interval J such that $a \in J$ and $f(J \cap C) \subseteq I$. Thus $f(J \cap C) \cap B = \emptyset$ and hence $(J \cap C) \cap f^{-1}(B) = \emptyset$, i.e., $J \cap f^{-1}(B) = \emptyset$. Thus $f^{-1}(A)$ is separated from $f(B)$. Similarly, $f^{-1}(B)$ is separated from A. Thus $(f^{-1}(A), f^{-1}(B))$ is a separation of C, i.e., C is not separated, which is a contradiction. Hence $f(C)$ is connected. ∎

6.12 Abstractions: metric spaces

We saw in the previous sections that in trying to understand the behaviour of sequences and continuous functions on the real line, the notion of absolute value played an important role. Absolute value of real numbers provided us the notion of distance, $|x - y|$, between any two points $x, y \in \mathbb{R}$ and this enabled us to analyze the notion of limits of sequences. The notions of closed sets and compact sets were defined in terms of limits of sequence of their elements. The notion of continuity of function at a point also involves limits of sequences. It is possible to generalize these concepts to sets on which there is the notion of 'distance'- an abstraction of the absolute value function and the notion of distance in analytical geometry. The aim of this section is to define this notion and extend theorem 6.11.13 to such sets.

6.12.1 Definition. A **metric** on a set X is a function $d : X \times X \to \mathbb{R}$ that satisfies the following properties : $\forall\, x, y, z \in X$

(i) $d(x, y) \geq 0$.

(ii) $d(x, y) = 0$ if and only if $x = y$.

(iii) $d(x, y) = d(y, x)$.

(iv) $d(x, y) \leq d(x, z) + d(z, x)$.

A set X together with a metric d on it is called a **metric space** and is denoted by the pair (X, d). The property (i) says that the 'distance' between any two points is always nonnegative and (ii) says that it is zero iff the points coincide. Property (iii) says that the distance between x and y is the same as between y and x, called the **symmetry** of d. Finally, (iv) is called the **triangle inequality**, geometrically it refers to the fact that the sum of two sides of a triangle is always bigger than or equal to the third side.

6.12.2 Examples.

(i) For $x, y \in \mathbb{R}, d(x, y) := |x - y|$ is a metric on $\mathbb{R}$. For any $\alpha \in \mathbb{R}, \alpha \geq 0$, we can define for $x, y \in \mathbb{R}$,

$$d_\alpha(x, y) := \alpha |x - y|.$$

Then each d_α is a metric on $\mathbb{R}$.

6.12 Abstractions: metric spaces

(ii) Consider the set $\mathbb{R}^n$, the n-fold Cartesian product of $\mathbb{R}$ with itself. For $\boldsymbol{x} = (x_1, \ldots, x_n), \boldsymbol{y} = (y_1, \ldots, y_n) \in \mathbb{R}^n$, let

$$d_1(\boldsymbol{x}, \boldsymbol{y}) := \sum_{i=1}^{n} |x_i - y_i|;$$
$$d_\infty(\boldsymbol{x}, \boldsymbol{y}) := \max\{|x_i - y_i| \mid i = 1, 2, \ldots, n\}.$$

It is easy to check that d_1 and d_∞ are both metrics on $\mathbb{R}^n$. Another metric on $\mathbb{R}^n$ is given as follows:

$$d_2(\boldsymbol{x}, \boldsymbol{y}) := \left(\sum_{i=1}^{n} |x_i - y_i|^2\right)^{1/2}.$$

The only non-trivial property for d_2 to be a metric is the triangle-inequality. To prove that, we first note that for $\boldsymbol{x}, \boldsymbol{y} \in \mathbb{R}^n$, we have

$$\begin{aligned}
0 &\leq \sum_{i,j=1}^{n} (x_i y_j - x_j y_i)^2 \\
&= \sum_{i,j=1}^{n} \left(x_i^2 y_j^2 + x_j^2 y_i^2 - 2x_i x_j y_i y_j\right) \\
&= \sum_{i,j=1}^{n} x_i^2 y_j^2 + \sum_{i,j=1}^{n} x_j^2 y_i^2 - 2 \sum_{i,j=1}^{n} x_i x_j y_i y_j \\
&= 2\left(\sum_{i=1}^{n} x_i^2\right)\left(\sum_{j=1}^{n} y_j^2\right) - 2\left(\sum_{i=1}^{n} x_i y_i\right)^2.
\end{aligned}$$

Thus

$$\left|\sum_{i=1}^{n} x_i y_i\right| \leq \left(\sum_{i=1}^{n} x_i^2\right)^{1/2} \left(\sum_{j=1}^{n} y_j^2\right)^{1/2}.$$

This is called **Cauchy's inequality**. Let us write for $\boldsymbol{x}, \boldsymbol{y} \in \mathbb{R}^n$

$$\langle \boldsymbol{x}, \boldsymbol{y} \rangle := \sum_{i=1}^{n} x_i y_i \quad \text{and} \quad |\boldsymbol{x}| := (\langle \boldsymbol{x}, \boldsymbol{x} \rangle)^{1/2}.$$

Then Cauchy's inequality can be written as

$$|\langle x, y \rangle| \leq |x|\,|y|.$$

The real number $\langle x, y \rangle$ is called the **inner-product** of x with y and $|x|$ is called the **length** or the **norm** of x. For $x = (x_1, \ldots, x_n)$, $y = (y_1, \ldots, y_n) \in \mathbb{R}^n$ and $\alpha \in \mathbb{R}$, let

$$\begin{aligned} x + y &:= (x_1 + y_1, x_2 + y_2, \ldots, x_n + y_n); \\ \alpha x &:= (\alpha x_1, \ldots, \alpha x_n). \end{aligned}$$

Then the following properties are easy to verify: for $x, y, z \in \mathbb{R}^n$ and $\alpha \in \mathbb{R}$,

$$\begin{aligned} \langle x + y, z \rangle &= \langle x, z \rangle + \langle y, z \rangle, \\ \langle x, y \rangle &= \langle y, x \rangle, \\ \langle \alpha x, y \rangle &= \alpha \langle x, y \rangle. \end{aligned}$$

Using these properties and the Cauchy's inequality, we get

$$\begin{aligned} |x + y|^2 &= \langle x + y, x + y \rangle \\ &= \langle x, x \rangle + \langle y, y \rangle + 2\langle x, y \rangle \\ &\leq |x|^2 + |y|^2 + 2|x|\,|y| \\ &= (|x| + |y|)^2. \end{aligned}$$

Thus $|x + y| \leq |x| + |y|$. Since $d_2(x, y) = |x - y|$, for $x, y, z \in \mathbb{R}^n$ we have

$$\begin{aligned} d_2(x, y) &= |x - y| \\ &= |(x - z) + (z - y)| \\ &\leq |x - z| + |z - y| \\ &= d_2(x, z) + d_2(z, y). \end{aligned}$$

Hence d_2 is a metric on $\mathbb{R}^n$. This is called the **Euclidean-metric** on $\mathbb{R}^n$.

6.12 Abstractions: metric spaces

(iii) Let $C[0,1]$ denote the set of all real valued continuous functions on the interval $[0,1]$. For $f, g \in [0,1]$, define

$$\rho_\infty(f, g) := \sup\{|f(x) - g(x)| \mid x \in [0,1]\}.$$

Since $x \longmapsto |f(x) - g(x)|$ is a continuous function, by exercise 6.11.11, $\rho_\infty(f, g)$ is well-defined by theorem 6.11.13. It is easy to check that it is a metric on $C[0, 1]$. The metric ρ_∞ is called the **uniform metric** on $C[0, 1]$. Another metric on $C[0, 1]$ is given by

$$\rho_1(f, g) := \int_0^1 |f(x) - g(x)| dx,$$

where the right-hand side is the Riemann-integral of $|f - g|$. Using properties of the integral one can show that ρ_1 is a metric on $C[0, 1]$. It is called the L_1-**metric**. The metrics ρ_1 and ρ_∞ are in some sense generalizations of the corresponding metrics on $\mathbb{R}^n$, as defined in the previous example (one can visualize a function f as a 'vector' whose xth component is $f(x)$).

(iv) Let $\mathcal{B}$ denote the set of all real sequences which are bounded. For $\{x_n\}_{n \geq 1}, \{y_n\}_{n \geq 1}$ in $\mathcal{B}$, define

$$d(\{x_n\}_{n \geq 1}, \{y_n\}_{n \geq 1}) := \sup\{|x_n - y_n| \mid n \geq 1\}.$$

Using exercise 6.3.5, it can be shown that d is a metric on $\mathcal{B}$.

6.12.3 Exercise. Which of the following are metric on X?

(i) $X = \mathbb{R}^2, d((x_1, y_1), (x_2, y_2)) := |x_1 - y_1|.$

(ii) X is any nonempty set and for $x, y \in X$,

$$d(x, y) := \begin{cases} 0 & \text{if } x = y, \\ 1 & \text{if } x \neq y. \end{cases}$$

(iii) $X = (0, \infty)$ and $d(x, y) := x/y$, for $x, y \in X$.

Next we analyze sequences in a metric space. Recall, a sequence in (X, d) is a function $f : \mathbb{N} \to X$, with domain $\mathbb{N}$ and taking values in X. As usual, we denote the sequence $f : \mathbb{N} \to X$ with $f(n) := x_n$, by $\{x_n\}_{n \geq 1}$.

6.12.4 Definition. Let $\{x_n\}_{n \geq 1}$ be a sequence in a metric space (X, d).

(i) The sequence $\{x_n\}_{n \geq 1}$ is said to be **Cauchy** if $\forall \, \epsilon > 0$ there exists some $n_0 \in \mathbb{N}$ such that $d(x_n, x_m) < \epsilon \; \forall \, n, m \geq n_0$.

(ii) The sequence $\{x_n\}_{n \geq 1}$ is said to be **convergent** to a point $x_0 \in X$ if for every $\epsilon > 0$ there exists $n_0 \in \mathbb{N}$ such that $d(x_n, x_0) < \epsilon \; \forall \, n \geq n_0$.

(iii) The metric space (X, d) is said to be **complete** if every Cauchy sequence in (X, d) is convergent to a point in X.

6.12.5 Examples.

(i) Theorem 6.3.1(viii) tells us that $\mathbb{R}$ with the usual metric, $d(x, y) := |x - y|$, is a complete metric space.

(ii) Consider $\mathbb{R}^2$ with the metric d_1, as in example 6.12.2(ii). Then $(\mathbb{R}^2, d_1)$ is a complete metric space. To see this, note that if $\{\boldsymbol{a}_n\}_{n \geq 1}$ is a Cauchy sequence in $(\mathbb{R}^2, d_1)$ and $\boldsymbol{a}_n := (x_n, y_n) \; \forall \, n$, then both $\{x_n\}_{n \geq 1}$ and $\{y_n\}_{n \geq 1}$ are Cauchy sequences in $\mathbb{R}$ for

$$d_1(\boldsymbol{a}_n, \boldsymbol{a}_m) = |x_n - x_m| + |y_n - y_m| \; \forall \, n, m \geq 1.$$

Thus $\lim\limits_{n \to \infty} x_n =: x_0$ and $\lim\limits_{n \to \infty} y_n =: y_0$ exist. Let $\boldsymbol{a}_0 := (x_0, y_0)$. Since

$$d_1(\boldsymbol{a}_n, \boldsymbol{a}_0) = |x_n - x_0| + |y_n - y_0|,$$

it follows that $\{\boldsymbol{a}_n\}_{n \geq 1}$ converges to $\boldsymbol{a}_0 \in \mathbb{R}^2$. A similar argument will show that $\mathbb{R}^2$ is complete under the d_2-metric.

(iii) Let (X, d) be any metric space and $\{x_n\}_{n \geq 1}$ be a sequence in X converging to a point $x \in X$. Then $\{x_n\}_{n \geq 1}$ is also Cauchy. To see this, let a real number $\epsilon > 0$ be given. Choose $n_0 \in \mathbb{N}$ such

6.12 Abstractions: metric spaces

that $d(x_n, x) < \epsilon/2 \; \forall \; n \geq n_0$. Then $\forall \; n, m \geq n_0$, using triangle inequality, we have

$$\begin{aligned} d(x_n, x_m) &\leq d(x_n, x) + d(x_m, x) \\ &< \epsilon/2 + \epsilon/2 = \epsilon. \end{aligned}$$

Hence $\{x_n\}_{n \geq 1}$ is a Cauchy sequence. The converse need not be true. For example consider $\mathbb{Q}$, the set of rationals in $\mathbb{R}$ with the metric $d(r, s) := |r - s|$. Then as shown in example 6.3.9, the sequence $\{r_n\}_{n \geq 1}$, where $r_1 := \dfrac{3}{2}$ and $r_n := \dfrac{r_{n-1}}{2} + \dfrac{1}{r_{n-1}}$ for every $n \geq 2$, converges to $\sqrt{2} \notin \mathbb{Q}$.

(iv) Consider the metric spaces $(C[0, 1], \rho_\infty)$ and $(C[0, 1], \rho_1)$ as described in example 6.12.2(iii). One can show using theorems about 'uniform convergence' and integrals proved in books on Real Analysis, that $(C[0, 1], \rho_\infty)$ is a complete metric space whereas $(C[0, 1], \rho_1)$ is not complete.

The concepts of open/closed intervals in $\mathbb{R}$ can be generalized to open/closed 'balls' in metric spaces.

6.12.6 Definition. Let (X, d) be a metric space. For $x_0 \in X$ and any real number $r > 0$, the set
$$B(x_0, r) := \{x \in X | d(x_0, x) < r\}$$

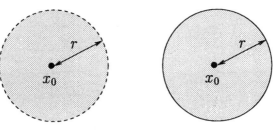

Open ball $B(x_0, r)$ Closed ball $\overline{B}(x_0, r)$

Figure 6.12

is called the **open ball** in X with **center** x_0 and **radius** r. The set
$$\overline{B}(x_0, r) := \{x \in X | d(x_0, x) \leq r\}$$
is called the **closed ball** in X with **center** x_0 and **radius** r.

The notions of interior points of a set, limit points of a set, closure of a set, open subsets and closed subsets of a metric space (X, d) can be defined in terms of open balls and sequences in (X, d) as was done on $\mathbb{R}$. For example, for a subset U of (X, d), a point $u \in U$ is called an **interior point** of U if there exists an open ball B with center at u such that $u \in B \subseteq U$. A subset U is said to be **open** if every point of U is an interior point. The results about open and closed sets of $\mathbb{R}$ as proved in sections 6.5 hold with same proofs for any metric space (X, d). For example, as in $\mathbb{R}$, in any metric space (X, d) every open ball $B(x, r)$ is an open set. To see this, let $y \in B(x, r)$. Then $d(y, x) = r' < r$. Let $\delta := \dfrac{r - r'}{2}$. Then $\forall z \in B(y, \delta)$
$$d(x, z) \leq d(x, y) + d(y, z) = r' + \delta < r.$$

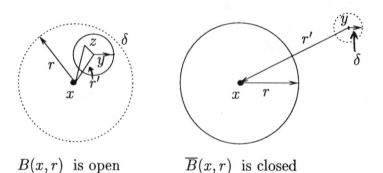

$B(x, r)$ is open $\overline{B}(x, r)$ is closed

Figure 6.13

Thus $y \in B(y, \delta) \subseteq B(x, r)$. Similarly, every closed ball $\overline{B}(x, r)$ is a closed set, for if $y \notin \overline{B}(x, r)$, then $d(x, y) = r' > r$. Clearly for $\delta = \dfrac{r' - r}{2}$, $y \in B(y, \delta) \subseteq X \setminus \overline{B}(x, r)$. Hence $X \setminus \overline{B}(x, r)$ is an open set, i.e. $\overline{B}(x, r)$ is a closed set.

The nested interval property of real numbers (theorem 6.6.3) has the following generalization.

6.12.7 Theorem (Cantor's intersection). *Let (X, d) be a complete metric space and $\{\overline{B}(x_n, r_n)\}_{n \geq 1}$ be any sequence of closed balls with $\overline{B}(x_{n+1}, r_{n+1}) \subseteq \overline{B}(x_n r_n)$ $\forall$ n and $\lim\limits_{n \to \infty} r_n = 0$. Then $\bigcap\limits_{n=1}^{\infty} \overline{B}(x_n, r_n)$ consists of a single point.*

Proof. Suppose $\{\overline{B}(x_n, r_n)\}_{n \geq 1}$ is a sequence of closed balls with the given properties. Intuitively, $\{x_n\}_{n \geq 1}$ should be a Cauchy sequence. To see this, let $m, n, n_0 \in \mathbb{N}$ with $m > n \geq n_0$. Since $\overline{B}(x_m, r_m) \subseteq \overline{B}(x_n, r_n) \subseteq \overline{B}(x_{n_0}, r_{n_0})$, the points $x_n, x_m \in \overline{B}(x_{n_0}, r_{n_0})$ and hence

$$d(x_n, x_m) \leq d(x_n, x_{n_0}) + d(x_m, x_{n_0}) \leq 2r_{n_0}.$$

Now let $\epsilon > 0$ be given. We choose n_0 such that $2r_{n_0} < \epsilon$. Then for every $m, n \geq n_0$, $d(x_n, x_m) \leq 2r_{n_0} < \epsilon$. Hence $\{x_n\}_{n \geq 1}$ is a Cauchy sequence. As (X, d) is complete, $\{x_n\}_{n \geq 1}$ is convergent to some $x \in X$. Since each $\overline{B}(x_n, r_n)$ is a closed set and includes all but finitely many terms of the sequence $\{x_n\}_{n \geq 1}$, $x \in \overline{B}(x_n, r_n)$ $\forall$ n and hence $x \in \bigcap\limits_{n=1}^{\infty} \overline{B}(x_n, r_n)$. In case $y \in \bigcap\limits_{n=1}^{\infty} \overline{B}(x_n, r_n)$ then $d(x, y) \leq 2r_n$ $\forall$ n and hence, $d(x, y) = 0$, i.e., $x = y$. ∎

6.12.8 Remark. Theorem 6.12.7 has a converse: let (X, d) be a metric space such that for every decreasing sequence $\{\overline{B}(x_n, r_n)\}_{n \geq 1}$ of closed balls in X with $\lim\limits_{n \to \infty} r_n = 0$, $\bigcap\limits_{n=1}^{\infty} \overline{B}(x_n, r_n)$ consists of a singleton, then (X, d) is complete. In fact this is a generalization of the fact (as mentioned in note 6.7.8(ii) that if an ordered field has the nested interval property then it is complete).

The notion of compactness for subsets of a metric space (X, d) can also be defined as was done for subsets of $\mathbb{R}$ in section 6.6. A subset A of a metric space (X, d) is said to be **compact** if every sequence of elements of A has a subsequence convergent to a point in A. It can be shown that $A \subseteq X$ is compact iff every covering of A by open subsets has a finite sub-cover (see [Cop-1] for a proof). We will not need this in our future discussions.

6.12.9 Exercise. Let (X, d) be a metric space and A be a compact subset of (X, d). Prove the following statements:

(i) A is a closed and bounded, (i.e., there exists a positive number k such that $d(x, y) \leq k \ \forall \ x, y \in A$).

(ii) For any $x, y \in A$, let $d_A(x, y) := d(x, y)$. Show that (A, d_A) is a complete metric space. The metric d_A is called the **restriction** of the metric d to A.

6.12.10 Examples.

(i) Let A be a compact subset of a metric (X, d) and B be a closed subset of X, $B \subseteq A$. Then B itself is compact. To see this, let $\{x_n\}_{n \geq 1}$ be any sequence in $B \subseteq A$. By compactness of A, there exists a subsequence $\{x_{n_k}\}_{k \geq 1}$ of $\{x_n\}_{n \geq 1}$ convergent in A say to a. Since $x_{n_k} \in B \ \forall \ k$ and B is closed, $a \in B$. Hence B is compact.

(ii) Consider the metric space $(\mathbb{R}^2, d_2)$, where d is the Euclidean metric on $\mathbb{R}^2$. Let A and B be arbitrary compact subsets of $\mathbb{R}$ with the usual metric. We show that $A \times B \subseteq \mathbb{R}^2$ is a compact subset of $(\mathbb{R}^2, d_2)$. Let $\{(x_n, y_n)\}_{n \geq 1}$ be a sequence in $A \times B$. Then $\{x_n\}_{n \geq 1}$ is a sequence of points of A and $\{y_n\}_{n \geq 1}$ is a sequence of point of B. By compactness of A and B, $\{x_n\}_{n \geq 1}$ contains a subsequence $\{x_{n_k}\}_{k \geq 1}$ converging to a point x in A and the sequence $\{y_{n_k}\}_{k \geq 1}$ contains a subsequence $\{y_{n_{k_\ell}}\}_{\ell \geq 1}$ which converges to a point y in

6.12 Abstractions: metric spaces

B. Note that the corresponding subsequence $\{x_{n_{k_\ell}}\}_{\ell \geq 1}$ of $\{x_n\}_{n \geq 1}$ is convergent to x. Since

$$d_2\left((x_{n_{k_\ell}}, y_{n_{k_\ell}}), (x, y)\right) = \sqrt{|x_{n_{k_\ell}} - x|^2 + |y_{n_{k_\ell}} - y|^2},$$

it follows that $\{(x_{n_{k_\ell}}, y_{n_{k_\ell}})\}_{\ell \geq 1}$ is convergent to $(x, y) \in A \times B$. Hence $A \times B$ is compact. In particular, for closed bounded intervals $I, J \subseteq \mathbb{R}$, the set $I \times J$ is compact in $\mathbb{R}^2$.

(iii) Consider the closed ball $\overline{B}(\boldsymbol{x}, r) \subset \mathbb{R}^2$ for $\boldsymbol{x} \in \mathbb{R}^2$ and $r > 0$. Then $\overline{B}(\boldsymbol{x}, r)$ is a closed set and $\overline{B}(\boldsymbol{x}, r) \subseteq I \times J \subset \mathbb{R}^2$, where $I = [x_0 - r, x_0 + r]$ and $J = [y_0 - r, y_0 + r]$ if $\boldsymbol{x} = (x_0, y_0)$. Using examples (i) and (ii) above, it follows that $\overline{B}(\boldsymbol{x}, r)$ is a compact subset of $(\mathbb{R}^2, d_2)$.

Next we extend the notion of continuity to functions between metric spaces.

6.12.11 Definition. Let (X, d) and (Y, ρ) be metric spaces and $x_0 \in X$.

(i) A function $f : X \to Y$ is said to be **continuous** at a point x_0 if for every sequence $\{x_n\}_{n \geq 1}$ in (X, d) converging to $x_0 \in X$, the sequence $\{f(x_n)\}_{n \geq 1}$ converges to $f(x_0)$ in the metric space (Y, ρ).

(ii) We say that f is **continuous everywhere** on X if f is continuous at every point in X.

6.12.12 Examples.

(i) Let (X, d) be any metric space. The identity map $Id_X : X \to X$, $Id_X(x) := x$, is trivially continuous. Another trivial example is the **constant function,** $f : X \to Y$ with $f(x) := y_0 \in Y \ \forall \, x \in X, y_0 \in Y$ fixed; where (X, d) and (Y, ρ) are arbitrary metric spaces.

(ii) Consider the metric space $(\mathbb{R}^2, d_2)$, d_2 being the Euclidean metric. Consider the function $f : \mathbb{R}^2 \to \mathbb{R}$, $f(\boldsymbol{x}) := |\boldsymbol{x}| = d_2(\boldsymbol{x}, 0) \ \forall \, \boldsymbol{x} \in$

$\mathbb{R}^2$. Using the triangle inequality, it is easy to see that

$$|d_2(\boldsymbol{x},\boldsymbol{0}) - d_2(\boldsymbol{y},\boldsymbol{0})| \leq |d_2(\boldsymbol{x},\boldsymbol{y})| \ \forall\ \boldsymbol{x},\boldsymbol{y} \in \mathbb{R}^2.$$

From this it follows that f is a continuous function.

6.12.13 Exercise. Let (X,d) be a metric space and let F be any field on which a metric ρ is defined. Let $f, g : X \to F$ be functions which are continuous at $x_0 \in X$ and let $\alpha, \beta \in F$. Define $\forall\ x \in X$

$$\begin{aligned}(f+g)(x) &:= f(x) + g(x); \\ (fg)(x) &:= f(x)g(x); \\ (\alpha f)(x) &:= \alpha f(x).\end{aligned}$$

Show that all these functions are continuous at $x_0 \in X$.

We finally prove an extension of theorem 6.11.13 for metric spaces.

6.12.14 Theorem. *Let (X,d) be a metric space and $A \subseteq X$ be a compact set. Let $f : X \to \mathbb{R}$ be a continuous function. Then the following hold:*

(i) $f(A)$ is a compact subset of $\mathbb{R}$.

(ii) There exist points $x_{\max}$ and $x_{\min}$ in A such that $f(x_{\max}) \geq f(x)\ \forall\ x \in A$ and $f(x_{\min}) \leq f(x)$ for every $x \in A$.

Proof. Let $\{f(x_n)\}_{n \geq 1}$ be any sequence in $f(A)$. Consider the sequence $\{x_n\}_{n \geq 1}$ in A. Since A is compact, $\{x_n\}_{n \geq 1}$ has a subsequence $\{x_{n_k}\}_{k \geq 1}$ converging to a point in A. Let it converge to $x \in A$. Then by the continuity of f, $\{f(x_{n_k})\}_{k \geq 1}$ converges to $f(x)$. Hence $f(A)$ is a compact subset of $\mathbb{R}$. This proves (i). Since $f(A)$ is compact, by theorem 6.11.13, $\sup(f(A))$ and $\inf(f(A))$ are attained, say at $x_{\max}$ and $x_{\min}$, respectively. Then $f(x_{\max}) \geq f(x)\ \forall\ x \in A$ and $f(x_{\min}) \leq f(x)\ \forall\ x \in A$. ∎

Chapter 7

COMPLEX NUMBERS

7.1 Historical comments

We have seen in the previous chapters that the need to find solutions of some algebraic equations led to the construction of the field of rationals and the need to have a complete ordered field led to the construction of the field of real numbers. The field of real numbers, though topologically and order complete, is not 'algebraically complete', i.e., not every polynomial with real coefficients has a root in $\mathbb{R}$. For example, there is no real number x such that $x^2 + 1 = 0$, because $x^2 + 1$ is always positive.

The Greek Mathematician Diophantos (3rd Century A.D.) thought that the equation $x^2 + 1 = 0$ is not solvable, for at that time even the negative numbers were not recognized. Even as late as the sixteenth century, mathematicians had reservations about negative numbers. The first algebraist to accept negative numbers was Thomas Harriot (1560–1621), but he did not accept negative roots for equations. At about the same time Geronimo Cardano (1501–76) in his work on solutions of algebraic equations introduced the roots of negative numbers and called numbers like $a + b\sqrt{-1}$ as 'formal numbers'. Rafael Bombelli (1526–72) introduced the fundamental rules of computation for these 'formal numbers.' Albert Girard (1595–1632) accepted these numbers as formal solutions of equations. René Descartes (1596–1650) called these numbers 'imaginary numbers'. Isaac Newton (1643–1727) did not attach much significance to these numbers. Gottfried Wilhelm Leibniz (1646–1716) worked formally with these numbers, not caring about their nature. He was led to 'imaginary numbers' in order to express a polynomial with real coefficients as a product of linear and/or quadratic factors with real coefficients, which he needed in the method of partial fractions for integration. Leibniz did not believe that every polynomial with real coefficients can be factored into a product of linear and quadratic factors with real coefficients. Leonhard Euler (1707–83) affirmed without proof that a polynomial of arbitrary degree can be so expressed. He also accepted imaginary numbers and used them intuitively in his calculations. However, he had great difficulty in defining them. A geometric representation of 'imaginary numbers' as points in the plane and as directed line segments were obtained by Caspar Wessel (1745–1818) and Jean Robert Argand (1768–1822). The work of Carl Friedrich Gauss (1777–1855) was very effective in bringing about the acceptance of 'imaginary numbers'. He called them 'complex numbers' and introduced the symbol i for $\sqrt{-1}$. The key to the problem of decomposing a real polynomial into linear and quadratic factors with real coefficients was to show that every polynomial had at least one real or complex root. The proof of this fact came to be known as the 'fundamental theorem of algebra'. Gauss used the geometric representation of complex numbers to give a proof of the fundamental theorem of algebra in his doctoral thesis. (For a detailed discussion on complex numbers and various proof of fundamental theorem of algebra, refer to [Ebb].) He later gave three more proofs of this theorem. Gauss's work succeeded in bringing acceptability to complex numbers and bringing them on par with the real numbers. Argand in 1814 published a proof of the fundamental theorem of algebra, assuming the minimum of a continuous function. Finally, the construction of complex numbers (as presented in section 7.2) assuming the existence of real numbers was given by Sir William Rowan Hamilton (1805–65) in 1835.

7.1 Historical comments

Carl Friedrich Gauss (1777–1855)

Gauss was the son of a mason in Burnswick (Germany) and went to school there. Impressed with his intelligence, the director of his school called him to the attention of Duke Karl Wilhelm who agreed to sponsor his high school studies. Gauss was a child prodigy. In his teens he invented the method of least squares and gave a method of constructing a 17-sided polygon. In 1795, he went to the University of Göttingen and in 1798 to the University of Helmstedt where he obtained his doctorate. From there he returned to Burnswick and later in 1807 joined as a professor of astronomy and director of the observatory at Göttingen, where he remained till the end of his life. He contributed to many branches of mathematics, astronomy and physics. However, he did not publish his discoveries. His contemporaries acknowledged and appreciated his genius and called him the "Prince of Mathematicians."

William Rowan Hamilton (1805–65)

Hamilton was born in Dublin. By the age of ten he could read Latin, Greek, Hebrew, Italian, French, Arabic, Sanskrit and Persian. In

1823, he entered Trinity College in Dublin. In 1827, while still an undergraduate, he wrote a paper in optics which was published in the Transactions of the Royal Irish Academy. In the same year he was appointed as a Professor of Astronomy at Trinity College with the title of Royal Astronomer of Ireland. Other than his important work in optics and mechanics, he gave the construction of complex numbers and invented 'Quaternions'. Along with metaphysics, mathematics and physics, he was also interested in poetry and general literature.

7.2 Construction of complex numbers

Let $\mathbb{C}$ denote $\mathbb{R} \times \mathbb{R}$, the set of all ordered pairs (x, y) for $x, y \in \mathbb{R}$.

7.2.1 Definition. For $(x_1, y_1), (x_2, y_2) \in \mathbb{C}$, let

$$(x_1, y_1) \oplus (x_2 + y_2) := (x_1 + x_2, y_1 + y_2)$$

and

$$(x_1, y_1) \otimes (x_2, y_2) := (x_1 x_2 - y_1 y_2, x_1 y_2 + x_2 y_1).$$

The binary operation $\oplus$ is called **addition** and $\otimes$ is called **multiplication** on $\mathbb{C}$.

7.2 Construction of complex numbers

7.2.2 Theorem. $(\mathbb{C}, \oplus, \otimes)$ *is a field.*

Proof. It is easy to verify that $(\mathbb{C}, \oplus)$ is a commutative group with $(0,0)$ as the additive identity element and $(-x,-y)$ being the additive inverse of $(x,y) \in \mathbb{C}$. Also it is easy to verify that $\otimes$ is associative, commutative and distributive over $\oplus$. The element $(1,0)$ is the identity for $\otimes$, i.e., $(x,y) \otimes (1,0) = (x,y)$. For $Z = (x,y) \in \mathbb{C}$ and $(x,y) \neq (0,0)$, let $Z^{-1} := (\frac{x}{x^2+y^2}, \frac{-y}{x^2+y^2})$. Then

$$\begin{aligned} Z \otimes Z^{-1} &= (x,y) \otimes (\frac{x}{x^2+y^2}, \frac{-y}{x^2+y^2}) \\ &= (\frac{x^2}{x^2+y^2} + \frac{y^2}{x^2+y^2}, \frac{-xy}{x^2+y^2} + \frac{xy}{x^2+y^2}) \\ &= (1,0). \end{aligned}$$

Similarly, $Z^{-1} \otimes Z = (1,0)$. Thus $(\mathbb{C} \setminus \{(0,0)\}, \otimes)$ is also a commutative group. Hence $(\mathbb{C}, \oplus, \otimes)$ is a field. ■

7.2.3 Note. The binary operation of multiplication '$\otimes$' in $\mathbb{C}$ as defined above appears to be artificial. However, it is motivated by the following considerations. Our aim is to give mathematical meaning to numbers of the form $a \pm \sqrt{-b}$ where $a, b \in \mathbb{R}, b \geq 0$. We can write $\sqrt{-b} = \sqrt{-1(b)}$. In case we expect $\sqrt{-b}$ to exist and to have properties similar to that of real numbers, we should be able to write $\sqrt{-b} = (\sqrt{-1})\sqrt{b}$. Note that $\sqrt{b}$ is a real number as $b \geq 0$. Suppose we denote $\sqrt{-1}$ by i. Then i is a quantity which can be 'multiplied' with itself to give -1, i.e., $i^2 = -1$ and can be multiplied with real numbers and added to real numbers with the usual arithmetic rules. Thus $a + \sqrt{-b} := a \pm i\sqrt{b}$, $a, b \in \mathbb{R}$. If we denote $\pm\sqrt{b}$ by c, then we have 'numbers' of the form $a + ic$ with $a, c \in \mathbb{R}$. Now we can add and multiply them with the usual arithmetical rules to get

$$\begin{aligned} (a_1 + ic_1) + (a_2 + ic_2) &= (a_1 + a_2) + i(c_1 + c_2). \quad (7.1) \\ (a_1 + ic_1) \times (a_2 + ic_2) &= a_1 a_2 + ic_1 a_2 + i a_1 c_2 + i^2 c_1 c_2 \end{aligned}$$

$$= (a_1a_2 - c_1c_2) + i(a_1c_2 + a_2c_1). \quad (7.2)$$

Thus, to make $a + ic$ rigorous, we write $(a + ic)$ as the ordered pair (a, c). Then (7.1) and (7.2) give the binary operations $\oplus$ and $\otimes$ as defined in 7.2.1. For example, for $Z = (a, b) = a + ib$, we have

$$\begin{aligned} Z^{-1} &= \frac{1}{a + ib} \\ &= \frac{1}{a + ib} \times \frac{a - ib}{a - ib} \\ &= \frac{a - ib}{a^2 + b^2} = \frac{a}{a^2 + b^2} - i\frac{b}{a^2 + b^2}. \end{aligned}$$

Thus for $Z = (a, b)$, $Z^{-1} := \left(\frac{a}{a^2 + b^2}, \frac{-b}{a^2 + b^2}\right)$ is the multiplicative inverse. We shall show that $\oplus$ and $\otimes$ extends the addition and multiplication of real numbers.

7.2.4 Definition. The elements of the field $(\mathbb{C}, \oplus, \otimes)$ as given by theorem 7.2.2 are called **complex numbers**. The element $(0, 1) \in \mathbb{C}$ is called the **imaginary unit** and is denoted by i.

7.2.5 Theorem. *The field $\mathbb{C}$ is a field extension of $\mathbb{R}$.*

Proof. Consider the set $\mathbb{R} \times \{0\} \subseteq \mathbb{C}$. For $(x_1, 0), (x_2, 0) \in \mathbb{R} \times \{0\}$,

$$(x_1, 0) \oplus (x_2, 0) := (x_1 + x_2, 0) \quad \text{and} \quad (x_1, 0) \otimes (x_2, 0) = (x_1 x_2, 0).$$

Further, it is easy to check that $(\mathbb{R} \times \{0\}, \oplus, \otimes)$ is a field. Thus $(\mathbb{R} \times \{0\}, \oplus, \otimes)$ is a subfield of $(\mathbb{C}, \oplus, \otimes)$. Clearly the map $x \longmapsto (x, 0)$, from $\mathbb{R}$ to $\mathbb{R} \times \{0\}$, is a field isomorphism. Hence $\mathbb{R}$, the field of real numbers can be identified with the subfield $(\mathbb{R} \times \{0\}, \oplus, \otimes)$ of $\mathbb{C}$. In other words, $\mathbb{C}$ is a field extension of $\mathbb{R}$. ∎

7.2.6 Remark. In view of the above theorem, we shall identify $(x, 0) \in \mathbb{C}$ with $x \in \mathbb{R}$ and write $+$ for $\oplus$ and $Z_1 \otimes Z_2$ will be denoted by $Z_1 Z_2$. We shall write $Z^n = Z^{n-1} Z \ \forall \ n \geq 1$. We also denote $(0, 0) \in \mathbb{C}$ by 0 itself.

7.2 Construction of complex numbers

7.2.7 Theorem. *In the field $\mathbb{C}$, the equation $Z^2 + 1 = 0$ has a solution, namely $Z = i$.*

Proof. We note that

$$(i)^2 = (0,1)^2 = (0,1) \otimes (0,1) = (-1,0) = -1. \blacksquare$$

7.2.8 Remark. We shall prove later in section 7.6 that every polynomial over $\mathbb{C}$ has as many solutions in $\mathbb{C}$ as the degree of the polynomial (Fundamental Theorem of Algebra).

7.2.9 Theorem. *Every complex number $Z = (x,y) \in \mathbb{C}$ can be written uniquely as $x + iy$.*

Proof. For $(x,y) \in \mathbb{C}$,

$$\begin{aligned}(x,y) &= (x,0) \oplus (0,y) \\ &= (x,0) \oplus ((0,1) \otimes (y,0)) \\ &= x + iy.\end{aligned}$$

The uniqueness of the representation is obvious. $\blacksquare$

7.2.10 Definition. For $Z = x + iy \in \mathbb{C}$, x is called the **real part** of Z, denoted by $\text{Re}(Z)$, and y is called the **imaginary part** of Z, denoted by $\text{Im}(Z)$. A complex number whose imaginary part is zero is a real number. A complex number whose real part is zero is called a **purely imaginary** number. For the complex numbers Z_1 and Z_2, $Z_1 = Z_2$ iff $\text{Re}(Z_1) = \text{Re}(Z_2)$ and $\text{Im}(Z_1) = \text{Im}(Z_2)$.

7.2.11 Exercise. Write the following complex numbers in the form $x + iy$:

$$\left(\frac{3-i}{2+3i}\right)^2, \quad \frac{Z+1}{Z-1}, \quad \frac{1}{Z^2}.$$

7.2.12 Exercise. Let $Z_0 = \alpha + i\beta$. Show that the equation $Z^2 = Z_0$ has two solutions. Express these solutions in terms of α and β.

7.2.13 Exercise. Find four solutions of each of the equations $Z^4 + 1 = 0$ and $Z^4 - i = 0$.

In theorem 7.2.5 we showed that $\mathbb{C}$, the field of complex numbers as constructed in theorem 7.2.2, includes a copy of the field $\mathbb{R}$ of real numbers, and in $\mathbb{C}$ the equation $Z^2 + 1 = 0$ has a solution. The natural question arises: Is $\mathbb{C}$ the only field with these properties? The answer is given in the next theorem.

7.2.14 Theorem. *Let F be any field which includes $\mathbb{R}$ as a subfield and in which the equation $x^2 + 1 = 0$ can be solved. Then there is a subfield $\tilde{\mathbb{C}}$ of F isomorphic to $\mathbb{C}$.*

Proof. Let $j \in F$ be a solution of $x^2 + 1 = 0$. Since $(x^2 + 1) = (x + j)(x - j)$, the equation $x^2 + 1 = 0$ has at least two solutions j and $-j$. Let $\tilde{\mathbb{C}} := \{\alpha + j\beta \mid \alpha, \beta \in \mathbb{R}\}$. Note that for an element $x \in \tilde{\mathbb{C}}$, if $x = \alpha_1 + j\beta_1 = \alpha_2 + j\beta_2$, then $(\alpha_1 - \alpha_2) = -j(\beta_1 - \beta_2)$ and hence $(\alpha_1 - \alpha_2)^2 = -(\beta_1 - \beta_2)^2$. Since the square of a nonzero real number is always positive, it follows that $\alpha_1 = \alpha_2$ and $\beta_1 = \beta_2$. Thus every element in $\tilde{\mathbb{C}}$ has a unique representation as $\alpha + j\beta$ for some $\alpha, \beta \in \mathbb{R}$. Further $\tilde{\mathbb{C}}$ is a subfield of F and includes $\mathbb{R}$. The map $(\alpha + j\beta) \longmapsto (\alpha + i\beta)$, from $\tilde{\mathbb{C}}$ to $\mathbb{C}$ is one-one, as follows from the unique representation of elements of $\tilde{\mathbb{C}}$. It is easy to check that it is also onto and is in fact a field isomorphism. ■

7.2.15 Note. We have constructed complex numbers as ordered pairs of real numbers. Another way of constructing complex numbers is to consider the set $\mathcal{C}$ of all 2×2 real matrices of the form $\begin{bmatrix} a & -b \\ b & a \end{bmatrix}$, where $a, b \in \mathbb{R}$. On $\mathcal{C}$ the binary operation of addition and multiplication are given by the usual matrix addition and matrix multiplication. It is easy to check that $\mathcal{C}$ becomes a field under these operations and $a + ib \longmapsto \begin{bmatrix} a & -b \\ b & a \end{bmatrix}$ is a field isomorphism. For more details, refer to [Cop-2].

7.3 Impossibility of ordering complex numbers

In view of theorem 7.2.14, we can say that $\mathbb{C}$ is the smallest field including $\mathbb{R}$ as a subfield such that $x^2 + 1 = 0$ has a solution in $\mathbb{C}$. The field $\mathbb{R}$ has an order on it with respect to which $\mathbb{R}$ is a complete ordered field. Can we extend the order of $\mathbb{R}$ to $\mathbb{C}$ so as to make $\mathbb{C}$ a complete ordered field?

Suppose that the order of $\mathbb{R}$ can be extended to an order on $\mathbb{C}$ such that it becomes an ordered field. By theorem 3.6.14, $Z^2 > 0 \ \forall \ Z \in \mathbb{C} \setminus \{0\}$. In particular, $-1 = (i)^2 > 0$ which is not possible since $>$ is an extension of the order on $\mathbb{R}$. Thus we cannot extend the order of $\mathbb{R}$ to make $\mathbb{C}$ an ordered field too. In fact, we cannot make it an ordered field under any order. Suppose if possible, there exists an order $>$ on $\mathbb{C}$ such that it is an ordered field. Again it follows from theorem 3.6.14 that $Z^2 > 0 \ \forall \ Z \in \mathbb{C}$. In particular, $1^2 > 0$ and $i^2 > 0$. Thus $0 = i^2 + 1 > 0$, a contradiction. Thus $\mathbb{C}$ cannot be made into an ordered field under any order. Also in view of theorem 7.2.14, if F is any field which includes $\mathbb{R}$ as a subfield and in which $x^2 + 1 = 0$ has a solution, then F cannot be made into an ordered field including $\mathbb{R}$ as ordered subfield. Thus, we cannot extend $\mathbb{R}$ to a bigger field retaining order and demanding that the equation $x^2 + 1 = 0$ have a solution.

7.4 Geometric representation of complex numbers

The set of complex numbers can be represented geometrically as follows. Consider a plane in which a rectangular coordinate system has been chosen. The complex number $Z = (x, y) = x + iy$ is identified with the point P in the plane having coordinates (x, y) and conversely. The x-axis is called the **real axis** and the y-axis is called the **imaginary axis**. We can also identify Z with the vector $\overrightarrow{OP}$, the directed line segment with initial point O and final point P. The origin itself represents the degenerate line segment with initial and final points as O itself.

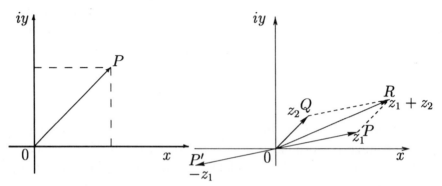

Figure 7.1 : Geometric representation.   Figure 7.2 : Addition.

For the point $Z = (x, y)$ given by the vector $\overrightarrow{OP}$, its additive inverse $-Z = (-x, -y)$ is given by the vector $\overrightarrow{OP'}$ where P' lies on the line POP' with length OP equal to the length OP'. The addition of complex numbers is represented by the familiar vector addition given by the parallelogram law as shown in figure 7.2. In order to understand the operation of multiplication of complex numbers geometrically, we make use of the polar coordinates of points in the plane. (Here we assume that the reader has some knowledge about trigonometric functions and their properties. Rigorous definitions of trigonometric functions and proofs of their properties require deeper knowledge of analysis. For details, the reader may refer to books on complex analysis as given in the list of references.) For every $Z = x + iy$, let

$$|Z| := \sqrt{x^2 + y^2},$$

called the **magnitude** or **the absolute value** of Z. Geometrically $|Z|$ is the distance of the point $Z = (x, y) \in \mathbb{R}^2$ from the origin. Thus the point $Z/|Z|$, say P, lies on the unit circle in the plane and

7.4 Geometric representation

has coordinates $(\cos\theta, \sin\theta)$, where $\theta \in [0, 2\pi)$ is the angle that the vector $\overrightarrow{OP}$ makes with the real axis in the counterclockwise direction. The number θ is called the **argument** of Z, denoted by $\arg(Z)$. For $Z = (0,0)$, its argument is not defined. Thus for $Z \in \mathbb{C}, Z \neq (0,0)$

$$\frac{Z}{|Z|} = (\cos\theta, \sin\theta) = \cos\theta + i\sin\theta.$$

Hence for $Z \neq (0,0)$,

$$Z = |Z|\frac{Z}{|Z|} = |Z|(\cos\theta + i\sin\theta).$$

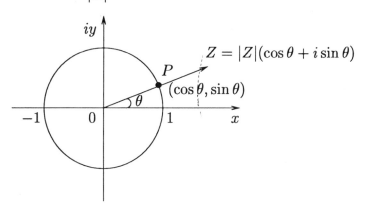

Figure 7.3 : Polar form.

For $Z \neq (0,0)$, $Z = |Z|(\cos\theta + \sin\theta)$ is called the **polar form** of the complex number Z. For complex numbers $Z, W \in \mathbb{C} \setminus \{(0,0)\}$, $Z = W$ iff they have the same magnitudes and arguments. Now using the addition formulas for the Sine and Cosine functions, if $Z = |Z|(\cos\theta + i\sin\theta)$ and $W = |W|(\cos\phi + i\sin\phi)$, then

$$\begin{aligned}
ZW &= |W||Z|[(\cos\theta + i\sin\theta) \times (\cos\phi + i\sin\phi)] \\
&= |Z||W|[(\cos\theta\cos\phi - \sin\theta\sin\phi) + \\
&\qquad\qquad i(\sin\theta\cos\phi + \cos\theta\sin\phi)] \\
&= |Z||W|[\cos(\theta + \phi) + i\sin(\theta + \phi)].
\end{aligned}$$

Since $0 \leq \theta + \phi \leq 4\pi$, we reduce $(\theta + \phi)$ modulo 2π. Thus for $Z, W \in \mathbb{C} \setminus (0,0)$, we have

$$|ZW| = |Z||W|$$

and

$$\arg(ZW) = (\arg(Z) + \arg(W))(mod\ 2\pi).$$

Thus for any $n \in \mathbb{N}$ and $Z \in \mathbb{C} \setminus \{(0,0)\}$ if $Z = |Z|(\cos\theta + i\sin\theta)$, we have $Z^n = |Z|^n[\cos n\theta + \sin n\theta]$, known as the **De Moivre's formula**.

For any complex number $Z = \alpha + i\beta$, its **conjugate** is defined as the complex number $\overline{Z} := \alpha - i\beta$. Geometrically, if $Z = |Z|(\cos\theta + i\sin\theta)$, then $\overline{Z} = |Z|(\cos\theta - i\sin\theta) = |Z|(\cos(-\theta) + i\sin(-\theta))$. Note that

$$Z\overline{Z} = |Z|^2. \tag{7.3}$$

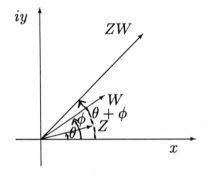
Figure 7.4 : Product ZW.

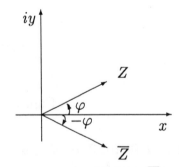
Figure 7.5 : $\overline{Z}$.

For $Z \in \mathbb{C} \setminus \{(0,0)\}$, Z^{-1} can also be expressed geometrically. If $Z = |Z|(\cos\theta + i\sin\theta)$, then from (7.3) we have

$$\begin{aligned}
Z^{-1} &= |Z|^{-2}\overline{Z} \\
&= |Z|^{-2}|Z|(\cos(-\theta) + i\sin(-\theta)) \\
&= \frac{1}{|Z|}(\cos(-\theta) + i\sin(-\theta)).
\end{aligned}$$

7.4 Geometric representation

Thus $|Z^{-1}| = |Z|^{-1}$ and $\arg(Z^{-1}) = -\arg(Z)$, as represented in figure 7.5.

Geometrically, it is also easy to locate the roots of a complex number using its polar form. For $n \in \mathbb{N}$ and $Z_0 \in \mathbb{C} \setminus \{(0,0)\}$, we want to find $Z \in \mathbb{C}$ such that $Z^n = Z_0$. Let $Z_0 = |Z_0|(\cos\theta_0 + i\sin\theta_0)$, then $Z = |Z|(\cos\theta + i\sin\theta)$ will satisfy $Z^n = Z_0$ iff

$$|Z|^n(\cos n\theta + i\sin n\theta) = |Z_0|(\cos\theta_0 + i\sin\theta_0).$$

Hence $|Z| = |Z_0|^{1/n}$ and $n\theta = \theta_0 + 2k\pi$ for some integer k. Thus $Z^n = Z_0$ has n distinct roots given by

$$|Z_0|^{1/n}\{\cos\frac{\arg(Z_0) + 2k\pi}{n} + i\sin\frac{\arg(Z_0) + 2k\pi}{n}\},$$

$k = 0, 1, \ldots, n-1$. These roots are equally spaced on the circle with center at the origin and radius $|Z_0|^{1/n}$ as shown in figure 7.6.

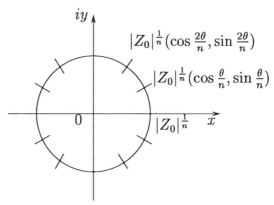

Figure 7.6 : Roots of Z.

7.4.1 Exercise. Show that the conjugation function $Z \longmapsto \overline{Z}$ from $\mathbb{C}$ to $\mathbb{C}$ has the following properties: for $Z_1, Z_2 \in \mathbb{C}$

(i) $\overline{0} = 0$.

(ii) $\overline{(Z_1 \pm Z_2)} = \overline{Z_1} \pm \overline{Z_2}$.

(iii) $\overline{Z_1 Z_2} = \overline{Z_1}\ \overline{Z_2}$.

(iv) $\overline{(\overline{Z_1})} = Z_1$.

(v) $\overline{(Z_1/Z_2)} = \overline{Z_1}/\overline{Z_2}$ if $Z_2 \neq 0$.

(vi) $Z + \overline{Z} = 2(\operatorname{Re}(Z))$ and $Z - \overline{Z} = 2i(\operatorname{Im}(Z))$.

7.4.2 Exercise. Let Z_0 be a solution of the polynomial equation
$$a_n Z^n + a_{n-1} Z^{n-1} + \cdots + a_1 Z + a_0 = 0,$$
where each $a_j \in \mathbb{C}$. Show that $\overline{Z_0}$ is a solution of the polynomial equation
$$\overline{a_n} Z^n + \overline{a_{n-1}} Z^{n-1} + \cdots + \overline{a_1} Z + \overline{a_0} = 0.$$
Hence deduce that if all the a_i's are real, then the solution of the above equation occur as pairs of complex numbers, conjugate to each other.

7.4.3 Exercise. Prove the following statements:

(i) Show analytically that the equation $Z^2 = \alpha + i\beta, \beta \neq 0$ has exactly two roots given by
$$\pm \left(\frac{\sqrt{\alpha + \sqrt{\alpha^2 + \beta^2}}}{2} + i\frac{\beta}{|\beta|} \frac{\sqrt{-\alpha + \sqrt{\alpha^2 + \beta^2}}}{2} \right).$$

(ii) Using (i) deduce that every quadratic $Z^2 + Z_1 Z + Z_2 = 0$, has exactly two solutions given by $Z = -\frac{1}{2}Z_1 \pm W$, where $W^2 = \frac{1}{4}Z_1^2 - Z_2$.

7.5 Cauchy completeness of $\mathbb{C}$

As we saw in section 7.3, the field of complex numbers includes $\mathbb{R}$ as a subfield and that $\mathbb{C}$ cannot be made an ordered field. Recall that the order in the field $\mathbb{R}$ was used to define the absolute value function on $\mathbb{R}$ with respect to which $\mathbb{R}$ is Cauchy complete. Even

7.5 Cauchy completeness of $\mathbb{C}$

though order cannot be defined in $\mathbb{C}$, we can extend the absolute value function of $\mathbb{R}$ to $\mathbb{C}$ and prove that $\mathbb{C}$ becomes Cauchy complete with respect to the metric given by this absolute value. We recall the absolute value of $Z \in \mathbb{C}$ as defined in the previous section.

7.5.1 Definition. For $Z = x + iy \in \mathbb{C}$, its **absolute value** is the nonnegative real number $|Z|$,

$$|Z| := \sqrt{x^2 + y^2}.$$

The properties of the absolute value function are given in the next theorem.

7.5.2 Theorem. *The absolute value function $Z \longmapsto |Z|$ has the following properties:*

(i) *If $Z \in \mathbb{R}$, then $|Z|$ is the same as the absolute value of the real number Z, as defined in chapter 4.*

(ii) $|Z| \geq 0 \; \forall \; Z \in \mathbb{C}$ *and* $|Z| = 0$ *iff* $Z = 0$.

(iii) $|Z| = |\overline{Z}|$ *and* $|Z|^2 = Z\overline{Z}$ *for every* $Z \in \mathbb{C}$.

(iv) $|Z_1 Z_2| = |Z_1||Z_2|$, *for* $Z_1, Z_2 \in \mathbb{C}$.

(v) $|Z_1/Z_2| = |Z_1|/|Z_2|$ *for* $Z_1, Z_2 \in \mathbb{C}, Z_2 \neq 0$.

(vi) $|Z_1 + Z_2|^2 + |Z_1 - Z_2|^2 = 2(|Z_1|^2 + |Z_2|^2)$ *if* $Z_1, Z_2 \in \mathbb{C}$. *This is known as the* **parallelogram identity**.

(vii) $-|Z| \leq \mathrm{Re}\,(Z) \leq |Z|$ *and* $-|Z| \leq \mathrm{Im}\,(Z) \leq |Z|, Z \in \mathbb{C}$.

(viii) $|Z_1 + Z_2| \leq |Z_1| + |Z_2|$ *for* $Z_1, Z_2 \in \mathbb{C}$. *This is known as* **triangle-inequality**.

(ix) *For* $Z_1, \ldots, Z_n \in \mathbb{C}$,

$$|Z_1 + \ldots + Z_n| \leq |Z_1| + |Z_2| + \ldots + |Z_n|.$$

(x) For $Z_1, \ldots, Z_n, W_1, \ldots, W_n \in \mathbb{C}$,

$$|\sum_{j=1}^n Z_j W_j|^2 \leq \left(\sum_{j=1}^n |Z_j|^2\right)\left(\sum_{j=1}^n |W_j|^2\right).$$

This is known as **Cauchy's inequality**.

Proof. (i) Follows from the fact that the real number x is identified with the complex number $(x, 0)$. The statements (ii) and (iii) are obvious. To prove (iv), using (iii) we have for $Z_1, Z_2 \in \mathbb{C}$,

$$|Z_1 Z_2|^2 = (Z_1 Z_2)\overline{(Z_1 Z_2)} = (Z_1 \overline{Z_1})(Z_2 \overline{Z_2}) = |Z_1|^2 |Z_2|^2.$$

Hence, either $|Z_1 Z_2| = |Z_1||Z_2|$ or $|Z_1 Z_2| = -|Z_1||Z_2|$. Since $|Z_1 Z_2| \geq 0$, it follows that $|Z_1 Z_2| = |Z_1||Z_2|$ if neither $Z_1 \neq 0$ nor $Z_2 \neq 0$. If either of them is zero, then also $|Z_1 Z_2| = 0 = |Z_1||Z_2|$. This proves (iv).

To prove (v), for $Z_1, Z_2 \in \mathbb{C}$ with $Z_2 \neq 0$, using (iv) we have

$$|Z_1| = |\frac{Z_1}{Z_2} Z_2| = |\frac{Z_1}{Z_2}| |Z_2|.$$

Hence $|Z_1/Z_2| = |Z_1|/|Z_2|$.

To prove (vi), for $Z_1, Z_2 \in \mathbb{C}$,

$$\begin{aligned}
|Z_1 + Z_2|^2 &= (Z_1 + Z_2)\overline{(Z_1 + Z_2)} \\
&= (Z_1 + Z_2)(\overline{Z_1} + \overline{Z_2}) \\
&= Z_1 \overline{Z_1} + Z_1 \overline{Z_2} + Z_2 \overline{Z_1} + Z_2 \overline{Z_2} \\
&= |Z_1|^2 + |Z_2|^2 + 2\mathrm{Re}\,(Z_1 \overline{Z_2}).
\end{aligned} \quad (7.4)$$

Similarly,

$$|Z_1 - Z_2|^2 = |Z_1|^2 + |Z_2|^2 - 2\mathrm{Re}\,(Z_1 \overline{Z_2}). \quad (7.5)$$

Adding (7.4) and (7.5) we have

$$|Z_1 + Z_2|^2 + |Z_1 - Z_2|^2 = 2(|Z_1|^2 + |Z_2|^2).$$

7.5 Cauchy completeness of $\mathbb{C}$

To prove (vii), for $Z = x + iy \in \mathbb{C}$,

$$|Z|^2 = x^2 + y^2 = |\text{Re}\,(Z)|^2 + |\text{Im}\,(Z)|^2.$$

From this it follows that

$$|\text{Re}\,(Z)| \leq |Z| \text{ and } |\text{Im}\,(Z)| \leq |Z|. \tag{7.6}$$

Hence (vii) follows. To prove (viii), from (7.4) and (7.6) we have for $Z_1, Z_2 \in \mathbb{C}$,

$$\begin{aligned} |Z_1 + Z_2|^2 &= |Z_1|^2 + |Z_2|^2 + 2\text{Re}\,(Z_1\overline{Z_2}) \\ &\leq |Z_1|^2 + |Z_2|^2 + 2|Z_1\overline{Z_2}| \\ &= |Z_1|^2 + |Z_2|^2 + 2|Z_1 Z_2| \\ &= (|Z_1| + |Z_2|)^2. \end{aligned}$$

Hence

$$|Z_1 + Z_2| \leq |Z_1| + |Z_2|.$$

Proof of (ix) follows from (vii) by induction.

(x) Note that the required inequality holds obviously if each W_i is zero. So suppose, at least one of them is not zero. Then for any complex number λ, using equation (2) we have

$$\begin{aligned} 0 &\leq \sum_{j=1}^{n} |Z_j - \lambda \overline{W_j}|^2 \\ &= \sum_{j=1}^{n} |Z_j|^2 + |\lambda|^2 \sum_{j=1}^{n} |W_j|^2 - 2\text{Re}\,(\overline{\lambda} \sum_{j=1}^{n} W_j Z_j). \end{aligned}$$

Choosing $\lambda = \left(\sum_{j=1}^{n} W_j Z_j\right) \Big/ \left(\sum_{j=1}^{n} |W_j|^2\right)$ we get

$$0 \leq \sum_{j=1}^{n} |Z_j|^2 + \frac{|\sum_{j=1}^{n} W_j Z_j|^2}{\sum_{j=1}^{n} |W_j|^2} - 2\frac{|\sum_{j=1}^{n} W_j Z_j|^2}{\sum_{j=1}^{n} |W_j|^2}.$$

Hence,
$$\left|\sum_{j=1}^{n} W_j Z_j\right|^2 \leq \left(\sum_{j=1}^{n} |Z_j|^2\right)\left(\sum_{j=1}^{n} |W_j|^2\right). \blacksquare$$

7.5.3 Exercise. For $Z, W \in \mathbb{C}$, show that
$$||Z| - |W|| \leq |Z - W|.$$

Next we show that with respect to the absolute value function, as defined above, every Cauchy sequence in $\mathbb{C}$ is convergent.

7.5.4 Theorem (Cauchy completeness). *For $Z, W \in \mathbb{C}$, let $d(Z, W) := |Z - W|$. Then $(\mathbb{C}, d)$ is a complete metric space.*

Proof. That $d(Z, W) := |Z - W|$ is a metric on $\mathbb{C}$ follows from (ii) and (viii) of theorem 7.5.2. To prove the completeness, let $\{Z_n\}_{n \geq 1}$ be a Cauchy sequence in $\mathbb{C}$. Let $Z_n = x_n + iy_n, n \geq 1$. Then using (vii) of theorem 7.5.2, we have $\forall\, n, m \in \mathbb{N}$,
$$|x_n - x_m| \leq |Z_n - Z_m| \text{ and } |y_n - y_m| \leq |Z_n - Z_m|.$$
Since $\{Z_n\}_{n \geq 1}$ is Cauchy, it follows that both $\{x_n\}_{n \geq 1}$ and $\{y_n\}_{n \geq 1}$ are Cauchy sequences in $\mathbb{R}$ and hence are convergent. Let
$$x_0 := \lim_{n \to \infty} x_n \text{ and } y_0 := \lim_{n \to \infty} y_n.$$
Let $Z_0 := x_0 + iy_0$. Then
$$\begin{aligned}
|Z_n - Z_0| &= |x_n + iy_n - x_0 - iy_0| \\
&\leq |x_n - x_0| + |i|\,|y_n - y_0| \\
&= |x_n - x_0| + |y_n - y_0|.
\end{aligned}$$
From this it follows that $\{Z_n\}_{n \geq 1}$ converges to Z_0. $\blacksquare$

7.5.5 Exercise. Let $\{Z_n\}_{n \geq 1}$ be a sequence of complex numbers such that both $\{\operatorname{Re}(Z_n)\}_{n \geq 1}$ and $\{\operatorname{Im}(Z_n)\}_{n \geq 1}$ are Cauchy sequences. Show that $\{Z_n\}_{n \geq 1}$ is also a Cauchy sequence in $(\mathbb{C}, d)$.

7.6 Algebraic completeness of $\mathbb{C}$

7.5.6 Note. The metric space $(\mathbb{C}, d)$ is the same as the metric space $\mathbb{R}^2$ with the Euclidean metric, as defined in example 6.2.12 (ii).

7.6 Algebraic completeness of $\mathbb{C}$

We saw in the previous sections that $\mathbb{C}$ is a field, includes $\mathbb{R}$ as a subfield and has the notion of absolute value which is an extension of the absolute value on $\mathbb{R}$ under which $\mathbb{C}$ is Cauchy complete. Also the equation $Z^2 + 1 = 0$ has solutions in $\mathbb{C}$. In fact we indicated that $Z^n = Z_0$ has exactly n solutions. We shall prove in this section that every polynomial of degree n in $\mathbb{C}$ has exactly n solutions. Let $f(Z)$ be a polynomial in $\mathbb{C}$ of degree n. Let

$$f(Z) := a_n Z^n + a_{n-1} Z^{n-1} + \ldots + a_1 Z + a_0, \qquad (7.7)$$

where $n \geq 1$, each $a_i \in \mathbb{C}$ and $a_n \neq 0$. Note that $f(Z)$ has a solution iff $\tilde{f}(Z)$ has a solution, where $\tilde{f}(Z) = Z^n + (a_{n-1}/a_n)Z^{n-1} + \ldots + (a_1/a_n)Z + (a_0/a_n)$. Thus we may assume that in (7.7) $a_n = 1$. We give the proof (due to R. Argand) of the fact that $|f(Z)| = 0$ for some Z by showing that for $f(Z)$, the function $|f(Z)|$ assumes its minimum and that this minimum cannot be nonzero.

7.6.1 Lemma. *Let* $f(Z) = Z^n + a_{n-1}Z^{n-1} + \cdots + a_1 Z_1 + a_0 = 0$. *Then there exists constant C and $r > 0$ such that*

$$|f(Z)| > C|Z|^n \quad for \ |Z| > r.$$

Proof. Using triangle inequality we have

$$\begin{aligned} |f(Z)| &= |Z^n + a_{n-1}Z^{n-1} + \ldots + a_1 Z + a_0| \\ &\geq |Z|^n - |a_{n-1}Z^{n-1} + \ldots + a_1 Z + a_0| \\ &\geq |Z|^n - \sum_{j=0}^{n-1} |a_j||Z|^j. \end{aligned}$$

Let $A := \max\{|a_j| \mid j = 0, 1, \ldots, n-1\}$. Then for $|Z| > 1$,

$$\begin{aligned} |f(Z)| &\geq |Z|^n - A \sum_{j=0}^{n-1} |Z|^j \\ &= |Z|^n - A\left(\frac{|Z|^n - 1}{|Z| - 1}\right) \\ &= |Z|^n\left(1 - \frac{A}{|Z|-1}\right) + \frac{A}{|Z|-1}. \end{aligned}$$

Thus, for $|Z| > 1$ we have

$$|f(Z)| \geq |Z|^n\left(1 - \frac{A}{|Z|-1}\right).$$

We choose $r > 1$ such that $A/(r-1) < 1$. Then for $|Z| > r$ we have

$$|f(Z)| \geq |Z|^n \left(1 - \frac{A}{r-1}\right).$$

The required inequality follows with $C := 1 - \dfrac{A}{r-1}$. ∎

7.6.2 Corollary. *For $f(Z) = Z^n + a_{n-1}Z^{n-1} + \ldots + a_n Z + a_0$, there exists a real number R such that $|f(Z)| > |a_0|$, for $|Z| > R$.*

Proof. By lemma 7.6.1, we can choose C and $r > 1$ such that $|f(Z)| > C|Z|^n$ for $|Z| > r$. Now choose $R > r$ such that $CR^n > |a_0|$, which is possible as $r > 1$ and the sequence $\{R^n\}_{n \geq 1}$ is unbounded (see lemma 5.2.3 (iii)). Then for $|Z| > R$, $|f(Z)| > |a_0|$. ∎

7.6.3 Theorem (Cauchy's minimum). *For every non-constant polynomial $f(Z) = a_0 + a_1 Z + \cdots + a_n Z^n$, there exists a point $Z_{min} \in \mathbb{C}$ such that $|f(Z_{min})| = \inf |f(\mathbb{C})|$.*

Proof. We can assume that $n \geq 1$ and $a_n \neq 0$. Further, we may assume that $a_n = 1$. To prove that $|f(Z)|$ attains its minimum

7.6 Algebraic completeness of $\mathbb{C}$

value, we note that by corollary 7.6.2, there exists a real number $R > 0$ such that

$$|f(Z)| > |a_0| = |f(0)|, \text{ for } |z| > R. \tag{7.8}$$

Since $\inf\{|f(Z)| \, |Z \in \mathbb{C}\} \leq |f(0)| = |a_0|, \inf\{|f(Z)| \mid Z \in \mathbb{C}\}$ can only be attained in the closed ball $\{Z \in \mathbb{C} \mid |Z| \leq R\}$. We first show that $|f|$ attains the value $\inf\{|f(Z)| \mid |Z| \leq R\}$. For this, we note that $|f(Z)|$ is a continuous function, by note 7.5.6, example 6.12.12 and exercise 6.12.13. Also it follows from note 7.5.6 and example 6.12.10 (iii) that $\{Z \mid |Z| \leq R\}$ is a compact subset of $\mathbb{C}$. Thus using theorem 6.12.14, $|f(Z)|$ attains its minimum value in $\{Z \mid |Z| \leq R\}$, say at Z_{min}. Then

$$|f(Z_{min})| = \inf\{|f(Z)| \mid |Z| \leq R\} \leq |f(0)|. \tag{7.9}$$

Also by (7.8)

$$|f(0)| \leq \inf\{|f(Z)| \mid |Z| > R\}. \tag{7.10}$$

From (7.9) and (7.10), we get $|f(Z_{min})| = \inf\{|f(z)| \, |Z \in \mathbb{C}\}$. ∎

To prove that for a non-constant polynomial $f(Z)$, the minimum of $|f(Z)|$ cannot be nonzero, we note that given $Z_0 \in \mathbb{C}$ with $f(Z_0) \neq 0$, we have to show that there exists a point $Z_1 \in \mathbb{C}$ such that $|f(Z_1)| < |f(Z_0)|$. For $Z \in \mathbb{C}$, let

$$g(Z) := \frac{f(Z + Z_0)}{f(Z_0)}, \quad Z \in \mathbb{C}.$$

Then $g(0) = 1$ and that there exists $Z_1 \in \mathbb{C}$ with $|f(Z_1)| < |f(Z_0)|$ will follow if we can show the existence of $\tilde{Z} \in \mathbb{C}$ such that $|g(\tilde{Z})| < 1$. For then $Z_1 := Z_0 + \tilde{Z}$ will be the required point for f. Note that $g(Z)$ can be written as $g(Z) = 1 + \sum_{\ell=n}^{m} a_\ell Z^\ell$. Thus we need to show that for $g(Z)$ as above, there exists a point $\tilde{Z} \in \mathbb{C}$ such that $|g(\tilde{Z})| < 1$. We need some elementary results to prove this.

7.6.4 Lemma. *For every natural number n, there is a number $Z \in \mathbb{C}$ such that*
$$\operatorname{Re}(Z^n) < 0 < \operatorname{Im}(Z^n).$$

Proof. (If we assume De Moivre's theorem, the proof is trivial: for $Z = \cos\theta + i\sin\theta$, $Z^n = \cos n\theta + i\sin n\theta$. Take $\theta = \dfrac{3\pi}{4n}$. Then $\dfrac{\pi}{2} < n\theta < \pi$ and hence $\operatorname{Re}(Z^n) < 0 < \operatorname{Im}(Z^n)$. However, since we are not assuming the polar representation of complex numbers - for the reasons given in section 7.4 - we give an alternate proof of the lemma.)

The case $n = 1$ is trivial. So let $n \geq 2$. We show that $Z := \left(1 + \frac{i}{n}\right)^2$ has the required property. Since $Z^n = \left(1 + \frac{i}{n}\right)^{2n} = \left[\left(1 + \frac{i}{n}\right)^n\right]^2$, to compute $\left(1 + \frac{i}{n}\right)^n$, let for any $m \in \mathbb{N}$,

$$(1 + \frac{i}{n})^m := u_m + iv_m,$$

where u_m and v_m are real. Then $u_1 = 1, v_1 = \frac{1}{n}$ and

$$\begin{aligned} u_{m+1} + iv_{m+1} &= (1 + \frac{i}{n})^m(1 + \frac{i}{n}) \\ &= (u_m + iv_m)(1 + \frac{i}{n}) \\ &= u_m - \frac{v_m}{n} + i(v_m + \frac{u_m}{n}). \end{aligned}$$

Thus
$$u_{m+1} = u_m - v_m/n \tag{7.11}$$

and
$$v_{m+1} = v_m + u_m/n. \tag{7.12}$$

From this using induction it is easy to show that $\forall\, 1 \leq m \leq n$,

$$1 - \frac{m}{n} < u_m \leq 1 \quad \text{and} \quad 0 < v_m \leq \frac{m}{n}. \tag{7.13}$$

7.6 Algebraic completeness of $\mathbb{C}$

From (7.11) and (7.13) we have $0 < u_{m+1} < u_m$ for $m = 1, 2, \ldots, n$. Thus using (7.12) and (7.13), we have

$$\frac{v_{m+1}}{u_{m+1}} > \frac{v_{m+1}}{u_m} = \frac{v_m}{u_m} + \frac{1}{n}, \quad m = 1, 2, \ldots, n-1.$$

Thus

$$u_n, v_m > 0 \quad \text{and} \quad \frac{v_n}{u_n} > \frac{v_1}{u_1} + \frac{(n-1)}{n} = 1. \tag{7.14}$$

Now

$$Z^n = (1 + \frac{i}{n})^{2n} = (u_n + iv_n)^2 = (u_n^2 - v_n^2) + 2iu_nv_n.$$

Thus using the inequality (7.14), we have $\operatorname{Re}(Z^n) = u_n^2 - v_n^2 < 0$ and $\operatorname{Im}(Z^n) = 2u_nv_n > 0$. ∎

7.6.5 Lemma. *For every $Z_0 \in \mathbb{C} \setminus \{0\}$ and every natural number n, there exists a number $Z_1 \in \mathbb{C}$ such that $\operatorname{Re}(Z_0 Z_1^n) < 0$.*

Proof. Using lemma 7.6.4, choose $Z \in \mathbb{C}$ such that $\operatorname{Re}(Z^n) < 0 < \operatorname{Im}(Z^n)$. Let $Z^n := u + iv$ and $Z_0 = a + ib$ where u, v, a and b are real numbers. Then $u < 0 < v$ and $a^2 + b^2 \neq 0$. If $a < 0$, choose $Z_1 = 1$. Then $\operatorname{Re}(Z_0 Z_1^n) = a < 0$. If $a \geq 0$ and $b \geq 0$, choose $Z_1 = Z$. Then $\operatorname{Re}(Z_0 Z_1^n) = au - bv < 0$, since at least one of a and b is not zero. Finally, if $a \geq 0$ and $b < 0$, choose $Z_1 = \overline{Z}$. Then $\operatorname{Re}(Z_0 Z_1^n) = au + bv < 0$. ∎

7.6.6 Lemma. *Let $g(Z) = 1 + a_n Z^n + \ldots + a_m Z^m$, where $1 \leq n \leq m$ and $a_n \neq 0$. Then there exists $\tilde{Z} \in \mathbb{C}$ such that $|g(\tilde{Z})| < 1$.*

Proof. Using lemma 7.6.5, choose $Z_1 \in \mathbb{C}$ such that $\operatorname{Re}(a_n Z_1^n) < 0$. Let $u := -\operatorname{Re}(a_n Z_1^n)$ and $v := \operatorname{Im}(a_n Z_1^n)$. Then for $0 < t \leq 1$,

$$\begin{aligned}|g(tZ_1)| &= |1 + a_n Z_1^n t^n + \ldots + a_m Z_1^m t^m| \\ &\leq |1 + a_n Z_1^n t^n| + |a_{n+1} Z_1^{n+1} t^{n+1}| + \cdots + |a_m Z_1^m t^m|\end{aligned}$$

$$= |1 - ut^n + ivt^n| + \sum_{j=n+1}^{m} |a_j Z_1^j t^j|$$

$$\leq |1 - ut^n + ivt^n| + t^{n+1} \left(\sum_{j=n+1}^{m} |a_j Z_1^j| \right).$$

Let $C := \sum_{j=n+1}^{m} |a_j Z_1^j|$ if $n < m$, and 0 otherwise. Then

$$|g(tZ_1)| \leq |1 - ut^n + ivt^n| + Ct^{n+1}. \qquad (7.15)$$

Also, using exercise 6.2.7, we have

$$\begin{aligned}
|1 - ut^n + ivt^n| &= \left\{ (1 - ut^n)^2 + v^2 t^{2n} \right\}^{1/2} \\
&= \left\{ 1 - 2ut^n + (u^2 + v^2) t^{2n} \right\}^{1/2} \\
&\leq 1 - ut^n + \frac{1}{2}(u^2 + v^2) t^{2n} \\
&\leq 1 - ut^n + \frac{1}{2}(u^2 + v^2) t^{n+1}. \qquad (7.16)
\end{aligned}$$

From (7.15) and (7.16) we have $\forall\, 0 < t \leq 1$,

$$|g(tZ_1)| \leq 1 - ut^n + \{\frac{1}{2}(u^2 + v^2) + C\} t^{n+1}.$$

In particular, if $t = \min\{1, \dfrac{u}{u^2 + v^2 + 2C}\}$, then

$$|g(tZ_1)| \leq 1 - \frac{1}{2} ut^n < 1.$$

Thus $\tilde{Z} := tZ_1$ satisfies the required property. ∎

7.6.7 Theorem (Argand's inequality). *Let $f(Z)$ be a non-constant polynomial. Then for every $Z_0 \in \mathbb{C}$ such that $f(Z_0) \neq 0$, there exists $Z_1 \in \mathbb{C}$ such that $|f(Z_1)| < |f(Z_0)|$.*

7.6 Algebraic completeness of $\mathbb{C}$

Proof. Since $f(Z)$ is non-constant and $f(Z_0) \neq 0$, consider $g(Z) := f(Z+Z_0)/f(Z_0)$. Then g is a non-constant polynomial and $g(0) = 1$. Thus we can write $g(Z)$ as

$$g(Z) = 1 + a_n Z^n + \ldots + a_m Z^m,$$

for some $1 \leq n \leq m$ and $a_n, a_{n+1}, \ldots a_m \in \mathbb{C}$ with $a_n \neq 0$. Now by lemma 7.6.6, there exists $\tilde{Z} \in \mathbb{C}$ such that $|g(\tilde{Z})| < 1$. Let $Z_1 := Z_0 + \tilde{Z}$. Then

$$|f(Z_1)| = |f(Z_0 + \tilde{Z})| = |f(Z_0)g(\tilde{Z})| < |f(Z_0)|. \blacksquare$$

7.6.8 Note. The proofs of theorem 7.6.7 and lemmas 7.6.4 to 7.6.6 are based on the arguments given in [Est]. In case one assumes the existence of solutions for $Z^k - Z_0 = 0$ (which we showed geometrically only), a simpler proof of theorem 7.6.7 can be given. For details refer to [Ebb].

7.6.9 Fundamental theorem of algebra. *Every non-constant polynomial with complex coefficients has a complex root.*

Proof. Follows from theorems 7.6.3 and 7.6.7. $\blacksquare$

7.6.10 Corollary. *Let $f(Z)$ be a polynomial of degree n with complex coefficients. Then there exists numbers $C_1, C_2, \ldots, C_n \in \mathbb{C}$ and $a \in \mathbb{C} \setminus \{0\}$ such that*

$$f(Z) = a(Z - C_1)(Z - C_2)\ldots(Z - C_n).$$

In particular, f has exactly n roots in $\mathbb{C}$.

Proof. We use induction on n. The case $n = 1$ is obvious. Assume it holds for polynomials of degree $\leq n$. Consider $f(Z)$ a polynomial of degree n given by

$$f(Z) = a_0 + a_1 Z + \ldots + a_n Z^n,$$

where $a_0, a_1, \ldots, a_n \in \mathbb{C}$ and $a_n \neq 0$. By theorem 7.6.8, there exists $C_1 \in \mathbb{C}$ such that $f(C_1) = 0$. Since for every $k \in \mathbb{N}$,

$$Z^k - C_1^k = (Z - C_1)g_k(Z),$$

where $g_k(Z) = Z^{k-1} + Z^{k-2}C_1 + \ldots + C_1^{k-1}$, we have

$$\begin{aligned} f(Z) &= f(Z) - f(C_1) = \sum_{k=1}^{n} a_k(Z^k - C_1^k) \\ &= (Z - C_1)g(Z), \end{aligned}$$

where $g(Z) := a_1 g_1(Z) + \ldots + a_k g_k(Z)$. Note that the polynomial g is of degree $(n-1)$. By induction hypothesis, $g(Z) = a(Z-C_2)(Z-C_3)\ldots(Z-C_n)$ for some complex numbers $C_2, \ldots, C_n$ and $a \neq 0$. Hence $f(Z) = a(Z - C_1)(Z - C_2)\ldots(Z - C_n)$. ∎

By exercise 7.4.2, if we treat a polynomial $f(x)$ with real coefficients as a polynomial over $\mathbb{C}$, then the non-real roots of $f(x)$ occur in pairs. In fact, we can say something stronger.

7.6.11 Definition. Let $f(z)$ be a polynomial with coefficients from $\mathbb{C}$. We say $\lambda \in \mathbb{C}$ is a **root of order** k of $f(z)$ if there exists a polynomial $g(z)$ such that $g(\lambda) \neq 0$ and

$$f(z) = (Z - \lambda)^k g(z).$$

7.6.12 Lemma. *Let $f(z)$ be a non-constant polynomial with real coefficients. If $\lambda \in \mathbb{C}$ is a root of $f(z)$ of order k then $\overline{\lambda}$ is also a root of $f(z)$ of order k.*

Proof. We prove the required claim by induction on the degree of f. The result obviously holds if $\text{degree}(f) = 1$. Then f has only real root. Assume the required claim holds for polynomials with degree $< n$. Let $f(z)$ be a non-constant polynomial of degree n and $\lambda \in \mathbb{C}$ be a root of z of order k. Without loss of generality, let λ be non-real. Then by exercise 7.4.2, $\overline{\lambda}$ is also a root of $f(z)$ and

7.6 Algebraic completeness of $\mathbb{C}$

by corollary 7.6.10, $(Z - \lambda)(Z - \bar{\lambda})$ is a factor of $f(z)$. Hence there exists a polynomial $g(z)$ such that

$$f(z) = (Z - \lambda)(Z - \bar{\lambda})g(z), \quad Z \in \mathbb{R}. \tag{7.17}$$

Note that degree$(g) = n - 2$, λ is a root of order $(k-1)$ of g, and

$$f(x) = (x - \lambda)(x - \bar{\lambda})g(x), \quad x \in \mathbb{R}.$$

Since λ is non-real, $(x - \lambda)(x - \bar{\lambda}) = x^2 - 2\text{Re}\,(\lambda)x + \lambda^2$ is never zero and we have

$$g(x) = \frac{f(x)}{x^2 - 2\text{Re}\,(\lambda)x + |\lambda|^2}, \quad x \in \mathbb{R}. \tag{7.18}$$

Let $g(z) = \alpha_0 + \alpha_1 Z + \ldots + \alpha_{n-2} Z^{n-2}$, for some $\alpha_0, \ldots, \alpha_{n-2} \in \mathbb{C}$. Then by (7.18), $g(x) \in \mathbb{R}$ for $x \in \mathbb{R}$. Hence $\forall\, x \in \mathbb{R}$

$$0 = \text{Im}\,(g(x)) = \text{Im}\,(\alpha_0) + \text{Im}\,(\alpha_1)x + \ldots \text{Im}\,(\alpha_{n-2})x^{n-2}.$$

Thus $\text{Im}\,(\alpha_j) = 0$, $\forall\, 0 \le j \le n-2$, and hence $g(z)$ is a polynomial with real coefficients. By induction hypothesis, since λ is a root of order $(k-1)$ of $g(x)$, $\bar{\lambda}$ is a root of order $(k-1)$ of $g(x)$. Hence it follows from (7.17) that $\bar{\lambda}$ is a root of order k for f. ∎

7.6.13 Theorem (Factorization of real polynomials). *Let $f(x) = \alpha_0 + \alpha_1 x + \ldots + \alpha_n x^n$, where $\alpha_0, \ldots, \alpha_n \in \mathbb{R}$ and $\alpha_n \ne 0$. Then there exists nonnegative integers j, k and real numbers $C, \lambda_1, \ldots, \lambda_j, a_1, \ldots, a_k, c_1, \ldots, c_k$ such that the following hold:*

(i) $2k + j = n$.

(ii) $a_r^2 < c_r^2 \quad \forall\, 1 \le r \le k$.

(iii) $f(x) = C(x - \lambda_1) \ldots (x - \lambda_j)(x^2 + 2a_1 x + c_1^2) \ldots (x^2 + 2a_k x + c_k^2)$.

Proof. Consider f as a polynomial in $\mathbb{C}$. Then by corollary 7.6.10, there exist $C, \lambda_1, \ldots, \lambda_n \in \mathbb{C}$ such that

$$f(z) = C(z - \lambda_1) \ldots (z - \lambda_n).$$

Let us rearrange $\lambda_1, \ldots, \lambda_n$ such that $\lambda_1, \ldots, \lambda_j$ are real and $\lambda_{j+1}, \ldots, \lambda_n$ are non-real. Then, by lemma 7.6.12, we can pair off $\lambda_{j+1}, \lambda_{j+2}, \ldots, \lambda_n$ into k pairs of the type $\lambda, \bar{\lambda}$. Let these be $b_1 + id_1, b_2 + id_2, \ldots, b_k + id_k$ where $2k = n - j$. Thus

$$\begin{aligned} f(z) &= C(Z - \lambda_1) \ldots (Z - \lambda_j)(Z - (b_1 + id_1))(Z - (b_1 - id_1)) \ldots \\ &\qquad \ldots (Z - (b_k + id_k))(Z - (b_k - id_k)) \\ &= C(Z - \lambda_1) \ldots (Z - \lambda_j)(Z^2 + 2a_1 Z + c_1^2) \ldots \\ &\qquad \ldots (Z^2 + 2a_k Z + c_k^2), \end{aligned}$$

where $c_i^2 := b_i^2 + d_i^2$ and $a_i := -b_i$ for $1 \leq i \leq k$. In particular, $\forall\, x \in \mathbb{R}$

$$f(x) = C(x - \lambda_1) \cdots (x - \lambda_j)(x^2 + 2a_1 x + c_1^2) \ldots (x^2 + 2a_k x + c_k^2).$$

Since this holds for every $x \in \mathbb{R}$ and $f(x)$ is real, we get $C \in \mathbb{R}$. Further each quadratic factor $x^2 + 2a_r x + c_r^2$ has only non-real roots. Hence $a_r^2 < c_r^2$. ∎

7.7 Beyond complex numbers

We saw in section 7.4 that complex numbers can be represented geometrically as points/directed line segments in the plane $\mathbb{R}^2$. The binary operation of addition on $\mathbb{C}$ is the same as translation in the plane: $(\alpha, \beta) + (x, y) = (\alpha + x, \beta + y)$; and the binary operation of multiplication by an element of absolute value one in $\mathbb{C}$ is the same as a rotation in $\mathbb{R}^2$. Thus the field of complex numbers provides an algebraic formulation for geometry in the plane. Can the same be done in $\mathbb{R}^3$ to represent vectors in space? Purely in algebraic terms, the problem can be stated as follows: Can one define binary operations of addition and multiplication on the ordered triples of real numbers such that it extends the corresponding binary operations of addition and multiplication on $\mathbb{C} = \mathbb{R}^2$? Here we understand that $\mathbb{R}^2 \subseteq \mathbb{R}^3$ via the identification: $(a, b) \in \mathbb{R}^2$ is identified with

7.7 Beyond complex numbers

$(a, b, 0) \in \mathbb{R}^3$. Hamilton, after having defined complex numbers, thought that this could be done and tried to do so. However he was not successful in achieving this. In fact one can prove that this cannot be done. Hamilton considered the 'numbers' of the form $z := a + bi + cj$ where a, b, c are real numbers and i, j are certain symbols. We can identify these numbers with triples $(a, b, c) \in \mathbb{R}^3$. Also numbers $z = a + 0i + 0j$ can be identified with $\mathbb{R}$ and numbers $z = a + bi + 0j$ can be identified with $\mathbb{R}^2$ or $\mathbb{C}$. The problem is to analyze the possibility of defining addition and multiplication on $\mathbb{R}^3$ such that the following hold:

(1) $a_1 + b_1 i + c_1 j = a_2 + b_2 i + c_2 j$ iff $a_1 = a_2, b_1 = b_2, z_1 = z_2$.

(2) Addition and multiplication restricted to $\mathbb{R}$ and $\mathbb{R}^2 (= \mathbb{C})$ coincides with the usual addition and multiplication.

(3) $(az_1)(bz_2) = (ab)(z_1 z_2) \ \forall \ a, b \in \mathbb{R}$ and $z_1, z_2 \in \mathbb{R}^3$.

(4) Multiplication distributes over addition, i.e.

$$z_1(z_2 + z_3) = z_1 z_2 + z_1 z_3,$$
$$(z_2 + z_3) z_1 = z_2 z_1 + z_3 z_1.$$

Suppose that it is possible to define binary operations of addition and multiplication with the above properties. Let $e_1 := 1 + 0i + 0j, e_2 := 0 + i + 0j$ and $e_3 := 0 + 0i + j$. Then by (1) every element $z \in \mathbb{R}^3$ has a unique representation

$$z = ae_1 + be_2 + ce_3, \quad \text{for some } a, b, c \in \mathbb{R}.$$

Also, assuming that the multiplication is associative, $e_2 e_2 = -e_1$ (corresponding to the property of complex numbers $i^2 = -1$), and $e_2 e_1 = e_2$ (corresponding to the fact that $i \times 1 = i$), we have

$$e_2(e_2 e_3) = (e_2 e_2) e_3 = -e_3. \tag{7.19}$$

Suppose

$$e_2 e_3 = a + bi + cj, \quad \text{for some } a, b, c \in \mathbb{R}.$$

Then $e_2e_3 = ae_1 + be_2 + ce_3$. Using the relations $e_2e_2 = -e_1$ and $e_2e_1 = e_2$, we get

$$\begin{aligned} e_2(e_2e_3) &= e_2(ae_1 + be_2 + ce_3) \\ &= a(e_2e_1) + b(e_2e_2) + c(e_2e_3) \\ &= ae_2 - be_1 + c(ae_1 + be_2 + ce_3) \\ &= (ac - b)e_1 + (a + bc)e_2 + c^2 e_3. \end{aligned} \quad (7.20)$$

From (7.19) and (7.20) we have for a real number c, $c^2 = -1$, which is not possible.

Hence it is not possible to extend the binary operations of addition and multiplication from $\mathbb{R}^2(=\mathbb{C})$ to $\mathbb{R}^3$ with the usual properties. It was proved by Hamilton in 1843 that it is possible to almost achieve this on $\mathbb{R}^4$. To be precise, on $\mathbb{R}^4$ one can define binary operations of addition, $\oplus$, and multiplication, $\odot$, such that these are extensions of the corresponding binary operations on $\mathbb{R}$ and $\mathbb{R}^2(=\mathbb{C})$, and $(\mathbb{R}^4, \oplus, \odot)$ satisfies axioms of a field except for commutativity of $\odot$, as described in example 7.1.2. We describe this in detail.

7.7.1 Definition. A triple $(X, \oplus, \odot)$ is called a **skew-field** if X is a set and $\oplus, \odot$ are two binary operations on X with the following properties:

(i) $(X, \oplus)$ is a commutative group,

(ii) $(X \setminus \{0\}, \odot)$ is a group, where 0 is the identity for $\oplus$.

(iii) $x \odot (y \oplus z) = (x \odot y) \oplus (x \odot z)$ and $(y \oplus z) \odot x = (y \odot x) \oplus (z \odot x)$.

A skew-field is also called a **division ring**. $\mathbb{R}$ and $\mathbb{C}$ with their usual binary operations of addition and multiplication are division rings.

7.7 Beyond complex numbers

7.7.2 Example. (Hamilton's Quaternions). Consider three symbols i, j, k which are multiplied according to the rules

$$i^2 = j^2 = k^2 = -1,$$
$$ij = k, jk = i, ki = j,$$
$$ji = -k, kj = -i, ik = -j.$$

Figure 7.7 helps us to remember these rules:

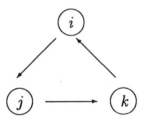

Figure 7.7 : Multiplication of i, j, k.

The product of any two of the symbols in the clockwise order is the third and the product of any two symbols in the anti-clockwise order is negative of the third.

We conceive elements of $\mathbb{R}^4$ as $a = \alpha + xi + yj + zk$, where $\alpha, x, y, z \in \mathbb{R}$. Now addition and multiplication can be defined with the usual rules. To be precise, for $a_r = \alpha_r + x_r i + y_r j + z_r k, r = 1, 2,$ define

$$\begin{aligned}
a_1 + a_2 &= (\alpha_1 + x_1 i + y_1 j + z_1 k) + (\alpha_2 + x_2 i + y_2 j + z_2 k) \\
&:= (\alpha_1 + \alpha_2) + (x_1 + x_2)i + (y_1 + y_2)j + (z_1 + z_2)k. \\
a_1 a_2 &= (\alpha_1 + x_1 i + y_1 j + z_1 k)(\alpha_2 + x_2 i + y_2 j + z_2 k) \\
&= \alpha_1 \alpha_2 + \alpha_1(x_2 i) + \alpha_1(y_2 j) + \alpha_1(z_2 k) \\
&\quad + (x_1 i)\alpha_2 + (x_1 i)(x_2 i) + (x_1 i)(y_2 j) + (x_1 i)(z_2 k) \\
&\quad + (y_1 j)\alpha_2 + (y_1 j)(x_2 i) + (y_1 j)(y_2 j) + (y_1 j)(z_2 k) \\
&\quad + (z_1 k)\alpha_2 + (z_1 k)(x_2 i) + (z_1 k)(y_2 j) + (z_1 k)(z_2 k)
\end{aligned}$$

$$:= (\alpha_1\alpha_2 - x_1x_2 - y_1y_2 - z_1z_2)$$
$$+ (\alpha_1 x_2 + x_1\alpha_2 + y_1 z_2 - z_1 y_2)i$$
$$+ (\alpha_1 y_2 - x_1 z_2 + y_1\alpha_2 + z_1 x_2)j$$
$$+ (\alpha_1 z_2 + x_1 y_2 - y_1 x_2 + z_1\alpha_2)k.$$

Clearly, if we identify $\mathbb{R}^2 (= \mathbb{C})$ with elements of $\mathbb{R}^4$ of the form $\alpha + ix + 0j + 0k$, then addition and multiplication defined above is the same as that of complex numbers. It is not difficult, only tedious, to check that the set

$$\mathbb{H} := \{\alpha + x_i + y_j z_k |\ x, y, z \in \mathbb{R}\}$$

under the above binary operations of addition and multiplication forms a division ring. For example to check that $a_1(a_2 a_3) = (a_1 a_2)a_3$, i.e., multiplication is associative, one will get on each side 64 terms. Anyway, the identity for addition is $0 + 0i + 0j + 0k$, or $(0, 0, 0, 0)$ if elements of $\mathbb{H}$ are written as quadruples. The additive inverse for (α, x, y, z) is $(-\alpha, -x, -y, -z)$. The multiplicative identity is $(1, 0, 0, 0)$ and for $(\alpha, x, y, z) \neq (0, 0, 0, 0)$ the multiplicative inverse is $(\alpha/\beta, x/\beta, y/\beta, z/\beta)$ where $\beta := \alpha^2 + x^2 + y^2 + z^2$. It is also easy to see that multiplication on $\mathbb{H}$ is not commutative, for example $(0, 0, 1, 0)(0, 0, 0, 1) = (0, 1, 0, 0)$ but $(0, 0, 0, 1)(0, 0, 1, 0) = -(0, 1, 0, 0) = (0, -1, 0, 0)$. The division ring $\mathbb{H}$ obtained above is called the division ring of **quaternions**. Since $\mathbb{H}$ includes the field $\mathbb{C}$, under the identification $(a, b) \longmapsto (a, b, 0, 0)$, elements of $\mathbb{H}$ are also known as **hyper-complex numbers**. One can also define the conjugate of elements of $\mathbb{H}$ as follows: for $a = (\alpha, x, y, z)$, its conjugate is defined as $\bar{a} := (\alpha, -x, -y, -z)$. It is easy to show that for $a, b \in \mathbb{H}$

$$\overline{a+b} = \bar{a} + \bar{b} \quad \text{and} \quad \overline{ab} = \bar{b}\,\bar{a}.$$

Further $\forall\, a \in \mathbb{H}$ with $a = (\alpha, x, y, z)$, $a\bar{a} = \alpha^2 + x^2 + y^2 + z^2$ and one can define **absolute value** of $a \in \mathbb{H}$ to be

$$|a| := \sqrt{a\bar{a}} = \sqrt{\alpha^2 + x^2 + y^2 + z^2}.$$

7.7 Beyond complex numbers

7.7.3 Note. The skew field of quaternions find applications in geometric representation of vectors in space. Let $a \in \mathbb{H}, a = \alpha + xi + yj + zk, \alpha, x, b, z \in \mathbb{R}$. We call α the **scalar part** of a and $xi + yj + zk$ the **vector part** of a. Given two vector-quaternions $a_r = x_r i + y_r j + z_r k$, $r = 1, 2$, their product is

$$a_1 a_2 = -(x_1 x_2 + y_1 y_2 + z_1 z_2) + (y_1 z_2 - z_2 y_1)i \\ + (z_1 x_2 - z_2 x_1)j + (x_1 y_2 - x_2 y_1)k.$$

Thus

$$-(\text{scalar part } (a_1 a_2)) = a_1 \cdot a_2$$
$$\text{vector part } (a_1 a_2) = a_1 \times a_2,$$

when $a_1 \cdot a_2$ is the **scalar product** or the **dot product** of the vectors a_1, a_2 and $a_1 \times a_2$ is the **vector product** or **cross product** of the vector a_1 with a_2. It is more popular in vector algebra and physics to represent vectors with respect to the three unit vectors i, j, k and treat their dot product and cross product separately. Quaternions can also be used to describe rotations in space.

From the algebraic point of view, the discovery of quaternions is of great significance. As a consequence of this discovery, mathematicians were free to construct algebraic structures with binary operations that need not be commutative. In 1845 Arthur Cayley (1821–95) was able to define 'compatible' addition and multiplication on $\mathbb{R}^8$ such that multiplication is neither commutative nor associative. These are known as **Cayley numbers**. If a division ring X includes $\mathbb{R}$ as a sub-ring, then X can be considered as a 'vector space' over $\mathbb{R}$. It was proved by Georg Frobenius (1849–1917) in 1878 that if a division ring has finite dimension as a vector space over $\mathbb{R}$, then it must be (up to isomorphism) either $\mathbb{R}, \mathbb{C}$ or $\mathbb{H}$. For more details refer to [Ebb], [Kan].

Suggestions for further reading

History of Mathematics and general reading

[Cou], [Dav], [Kli], [Kri], [Ste]

Set Theory

[End], [Hal], [Sto], [Sup]

Real Analysis

[Apo], [Bar], [Ber], [Boa], [Gau]

Metric spaces

[Cop-1], [Gof]

Complex Analysis

[Ahl], [Con], [Pon], [Rem], [Rud]

Algebra

[Art], [Fra], [Lang]

Symbol Index

Chapter 1

$/$	:	not, 9
$\Rightarrow$	:	implies, 9
$\Leftrightarrow$	:	if and only if, iff, 9
$\wedge$	:	and, 9
$\vee$	:	or, 9
$\forall$	:	for all, 9
$\exists$	:	there exists, 9
$\exists!$	:	there exists exactly one, 9
$\in$	:	member of, 9
$=$	:	equal, 9
$\emptyset$	:	empty set, 11
$\subseteq$	:	subset, 11
$\cup$	:	union, 12
$\mathcal{P}(X)$	:	power set of X, 14
(a, b)	:	ordered pair, 14
$A \times B$	:	Cartesian product, 16
$\cap$	:	intersection, 17
$A \setminus B$	:	relative complement, 18
$\mathcal{D}(F)$	:	domain of F, 19
$\mathcal{R}(F)$	:	range of F, 19
Id_A	:	identity function, 19
F^{-1}	:	inverse function, 20
$G \circ F$	:	composite function, 20

aRb	:	relation, 21
$\subsetneq$	:	proper subset, 23
S^+	:	successor of S, 24

Chapter 2

$(\mathbb{N}, 1, \phi)$	:	natural number system, Peano system, 35
$\mathbb{N}$	:	set of natural numbers, 41
$<$	:	order on natural numbers, 49
$[1, n]$	:	the set $\{1, 2, \ldots, n\}$, 53
$lub(S), \sup(S)$	:	least upper bound of $S \subseteq \mathbb{N}$, 55
$glb(S), \inf(S)$	:	greatest lower bound of $S \subseteq \mathbb{N}$, 55
$\mathbb{N}_c$	:	even natural numbers, 57
$\mathbb{N}_o$	:	odd natural numbers, 57
A^n	:	n-fold Cartesian product of A, 70
(X, ϕ)	:	semigroup, group, 76
S_n	:	symmetric group of permutations, 77

Chapter 3

$\mathbb{Z}$	:	the set $\mathbb{N} \times \mathbb{N}/\sim$, 85
$\mathbb{N}_\mathbb{Z}$	:	natural numbers in $\mathbb{Z}$, 86
$\mathbb{Z}$	:	set of integers, 86
$gcd(n, m)$	:	greatest common divisor of n and m, 97
$(X, \oplus, \odot)$	:	ring, 106
$C(\mathbb{Z})$	:	set of sequences in $\mathbb{Z}$, 106
$\mathbb{Z}_n$	:	set of integers modulo n, 107
$\mathcal{P}[X]$	:	the ring of polynomials over X, 108
$\mathbb{Z}_e$	:	ring of even integers, 110
$\mathcal{P}[\mathbb{Z}]$	:	ring of polynomials over $\mathbb{Z}$, 114
$\mathcal{L}(X)$	:	formal Laurent series over X, 114
$(X, +, \cdot, <)$	:	ordered integral domain, 116
1_X	:	identity for multiplication in X, 117
$\mathbb{N}_X$	:	set of natural numbers of X, 118
$\mathbb{Z}_X$	:	set of integers of X, 119

Symbol index

Chapter 4

$\mathbb{Q}$	:	the set of rationals, 133
$\mathbb{Z}_\mathbb{Q}$	:	integers of $\mathbb{Q}$, 136
$\mathbb{Q}$	:	field of rational numbers, 136
(A, B)	:	gap in $\mathbb{Q}$, 148
$|r|$	:	absolute value of rationals, 152
$\{r_n\}_{n \geq 1}$	:	sequence of rationals, 153
$\lim_{n \to \infty} x_n$	:	limit of the sequence $\{x_n\}$, 157
$C(\mathbb{Q})$	:	set of Cauchy sequences in $\mathbb{Q}$, 171
$\mathbb{Q}_F$	:	fractions of a field F, 174

Chapter 5

(α, β)	:	cut of rationals, 182
$\mathbb{R}$	:	set of all cuts, 186
α_r	:	cut induces by r, 186
$\{\alpha_n\}_{n \geq 1}$	:	sequence of reals, 200
$\{r_n\}_{n \geq 1} \sim \{s_n\}_{n \geq 1}$	:	equivalence of rational sequences, 206
$[\{r_n\}]$	:	equivalence class containing $\{r_n\}_{n \geq 1}$, 208
$\mathbb{R}_D$	:	real numbers by Dedekind's method, 224
$\mathbb{R}_C$	:	real numbers by Cantor's method, 225

Chapter 6

$[x]$	:	integral part of x, 239
$-S$	:	$\{-s | s \in S\}$, 240
$\sqrt[r]{\alpha}$,	:	positive n^{th} root, 244
$|\alpha|$,	:	absolute value of a real number, 246
$C_\mathbb{R}$,	:	set of all real convergent sequence, 249
e	:	Euler's number, 250
π	:	pi, 253

$(a, \infty), [a, +\infty),$
$(a, b) [a, b], [a, b),$
$(a, b], (-\infty, b],$
$(-\infty, b),$
$(-\infty, +\infty)$
$\Big\}$: intervals, 261

$\overline{A}$: closure of a set A, 264
A° : interior of a set A, 266
$\lambda(I)$: length of an interval, 273
$A \sim B$: equivalence of sets, 280
$\aleph_0$: aleph-nought, 284
c : cardinality of the continuum, 284
$2^{\aleph_0}$: 287
$\lim_{x \to u} f(x)$: limit of a function, 298
$f_1 \wedge f_2$: maximum of functions, 299
$f \vee f_2$: minimum of functions, 299
$d_\alpha(x, y)$: 308
$d_1(x, y)$: 309
$d_\infty(x, y)$: uniform metric, 309
$d_2(x, y)$: Euclidean metric, 309
$\langle \mathbf{x}, \mathbf{y} \rangle$: inner product, 309
$\rho_1(f, g)$: L_1-metric, 311
$\rho_\infty(f, g)$: 311
$B(x, r)$: open ball, 313
$\overline{B}(x, r)$: closed ball, 313

Chapter 7

$\mathbb{C}$: complex numbers, 322
i : imaginary unit, 323
$\mathrm{Re}(Z),$: real part of Z, 325
$\mathrm{Im}(z),$: imaginary part of Z, 325
$|Z|$: absolute value of Z, 328
$\overline{Z}$: conjugate of Z, 330
$\mathbb{H}$: quaternions, 350

References

[Art]   Artin, M.: "Algebra", Prentice-Hall Int., Inc., Englewood Cliffs, N.J., 1991.

[Bar]   Bartle, R.G. and Sherbert D.R.: "Introduction to Real Analysis" (Second Edition), John Wiley and Sons Inc., New York, 1994.

[Bel]   Bell, E.T.: "Men of Mathematics", Simon & Schuster, Inc. New York, 1965.

[Ber]   Berberian, S.K.: "A First Course in Real Analysis", Springer-Verlag, New York, 1994.

[Boa]   Boas, Jr., Ralph P.: "A Primer of Real Functions", Mathematical Association of America, 1981.

[Boy]   Boyer, Carl B. and Merzback Uta C.: "A History of Mathematics", (2nd Edition) John Wiley and Sons, New York, 1989.

[Bur]   Burril, C.W.: "Foundations of Real Numbers", McGraw Hill, Inc., New York, 1951.

[Coh-1]   Cohen, L.W. and Eherlich, G.: "Structure of the Real Number System", D. Van Nostrand Company, Inc. Princeton, N.J., 1963.

[Coh-2]   Cohen, P.J.: "Set Theory and the Continuum Hypothesis", W.A. Benjamin, Inc., New York, 1966.

[Con]   Conway, John B.: "Functions of One Complex Variable" (Second Edition), Springer-Verlag, New York, 1973.

[Cop-1]   Copson, E.T., "Metric Spaces", Cambridge University Press, New York, 1968.

[Cop-2]   Copson, E.T.: "Introduction to the Theory of Functions of a Complex Variable", Oxford University Press, New York, 1992.

[Cou]   Courrant, R. and Robbins, H.: "What Is Mathematics", Oxford University Press, New York, 1941.

[Dav]   Davis, Philip J. and Hersh Reuben: "The Mathematical Experience", Penguin Books Ltd., 1984.

[Die]   Dieudonne, J.: "Mathematics – The Music of Reasons", (English translation by H.G. and S.C. Dales), Springer-Verlag, New York, 1987.

[Dra]   Drake, F.R. and Singh D.: "Intermediate Set Theory", John Wiley & Sons, New York, 1996.

[Ebb]   Ebbinghaus, H.-D. et.al.: "Numbers", Springer-Verlag, New York, 1990.

[End]   Enderton, H.B.: "Elements of Set Theory", Academic Press, New York, 1977.

[Est]   Estermann, T.: "On the Fundamental Theorem of Algebra", J. Lond. Math. Soc. (31) 1956, pp. 238-240.

[Eve]   Eves, Howard: "An Introduction to the History of Mathematics", (Fourth Edition), holt, rinehart and winston, New York, 1976.

[Fra]	Fraleigh, John B.: "A First Course in Abstract Algebra", Addison-Wesley Publishing Co., Inc., New York, 1987.
[Gau]	Gaughan Edward D.: "Introduction to Analysis" (3rd Ed.), Brooks/Cole Publishing Co., California, 1987.
[Gof]	Goffman, C. and Pedrick, G.: "First Course in Functional Analysis", Prentice-Hall Int., 1965.
[Hal]	Halmos, P.R.: "Naive Set Theory", Springer-Verlag, New York, 1974.
[Hen]	Henle, James M.: "An Outline of Set Theory". Springer-Verlag, New York, 1986.
[Kan]	Kantor, I.L. and Solodovmikov, A.S.: "Hypercomplex Numbers, An Elementary Introduction to Algebras", Springer-Verlag, New York, 1989.
[Kli]	Kline, M.: "Mathematics Thoughts from Ancient to Modern Times", Oxford University Press, New York, 1972.
[Kri]	Krishnamurthy, V.: "Cultural Excitement and Relevance of Mathematics", Wiley Eastern Ltd., India, 1990.
[Lan]	Landau, E.: "Foundations of Analysis", Chelsea Publishing Co., New York, 1951.
[Lang]	Lang, S.: "Algebra" (3rd edition), Addison-Wesley Pub. Co. Inc., New York, 1993.
[Mao]	Maor, E.: "To Infinity and Beyond – A Cultural History of the Infinite", Birkhäuser, Boston, 1987.
[Men]	Mendelson, E.: "Number Systems and the Foundations of Analysis", Academic Press, New York, 1973.

[Niv]   Niven, I.M.: "Numbers : Rational and Irrational", Random House, New York, 1961.

[Pon]   Ponnusamy, S.: "Foundations of Complex Analysis", Narosa Publishers, India, 1995.

[Rem]   Remmert, R.: "Theory of Complex Functions". Readings in Mathematics, Springer-Verlag, New York, 1991.

[Rud]   Rudin, W.: "Principles of Mathematical Analysis" (3rd Edition), McGraw Hill Inc., 1976.

[Ste]   Stewart, Ian and David Tall: "The Foundations of Mathematics". Oxford University Press, 1977.

[Sto]   Stoll, R.: "Set Theory and Logic", Dover Publications, New York, 1963.

[Sup]   Suppes, P.: "Axiomatic Set Theory", Van Nostrand Reinhold, 1961.

[Tre]   Trewin, A.H.: "Mathematics with a Difference", Macmillan of Australia, 1968.

Index

A
Abel, Niels Hernik 180
abelian group 76
absolute value, 199, 219
    of a complex number 328
    of quaternions 350
addition,
    of complex numbers 322
    of cuts 188
    of integers 86
    of natural number 42
    of polynomials 108
additive inverse 86
additive inverse of a cut 189
amicable 33
Archimedean 138
Archimedean ordered integral
    domain 176
Archimedes 255
Archimedean property of
    the reals 239
Argand's inequality 342
Argand, Jean Robert 320
argument 329
associative 43, 45, 75
atomic formulae 9
axiom, 7
    of choice 26, 74
    of comprehension 15
    of empty set 10
    of extensionality 10
    of pairing 11
    of power set 14
    of regularity 27
    of replacement 25
    of separation 15
    of union 12

B
base 2, 104
Bernays, Paul 28
Bernoulli, Daniell 233
Bhaskara 255
bijective 19
binary,
    expansion 259
    operation 42, 75
    representation 104
Binomial,
    coefficients 141
    inequality 141
    theorem 141
Bolzano, Bernard 2, 23, 57, 130, 234
Bolzano-Weierstrass property 202, 245
Bombelli, Rafael 319
Borel, Emile 235

bounded,
    above 54, 72
    below 55, 72
    sequence 200
    set 264, 298
Brahmagupta 81

C

cancellation law,
    for addition 44
    for multiplication 47, 51
    in integral domains 109
Cantor-Bernstein-Schröder
    theorem 282
Cantor's ternary set 260
Cantor, Georg 2, 234
Cardano, Geronimo 319
cardinal numbers 284
cardinality,
    aleph-nought 284
    of a set 283
    of power set 286
    of the continuum 284
Cartesian product 17
Cartesian product, n-fold 70
Cauchy,
    complete ordered field 178
    completeness 202
    sequence 171, 219
Cauchy's,
    inequality 309, 334
    minimum theorem 348
Cauchy, Augustin Louis 180, 234, 236
Cayley numbers 351
Cayley, Arthur 351
center 314

characteristic, 119
    zero 119
choice function 26
Chung-chi, Tsu 255
closed,
    ball 314
    interval 261
    set 262
closure 264
codomain 19
coefficient leading 108
coefficients 108
Cohen, Paul 3
common divisor 97
commutative, 45
    binary operation 75
    group 76
    law 43
    semigroup 76
compact set 269, 316
complete,
    ordered field 231
    metric space 312
completeness, axiom of 238
complex numbers 320, 324
composite 20, 96
congruent modulo 107
conjugate 330
connected 291
constant, 8
    function 297, 317
    sequence 162, 208
    term 108
continuous, 296, 317
    at a point 296
    everywhere 317
continuum 284

Continuum Hypothesis 285
convergent sequence 158, 220
countable 57
countably finite 57
Cousin, Pierre 235
cross product 351
cut 182, 185
    induced 186
    bigger than 195
    rational 186
    smaller than 195

D
d'Alembert, Jean le 233
De Moivre's formula 330
decimal, 151, 256
    rational 152
    representation 104, 149, 256
    periodic 256
Dedekind, Richard 33 180
degree 108
denseness of irrationals 199
denseness of rationals 199, 218
Descartes, René 81
Diophantos 319
Dirichlet, Peter Gustav Lejeune 180, 234
disconnected 291
discontinuity 296
distributive, 45
    property 45
division,
    algorithm 93
    ring 348
domain 19, 21
dot product 351

E
empty,
    function 19
    set 11
equivalence,
    class 22
    relation 21
equivalent 84, 206, 281
Eudoxus 130
Euler's number 250
Euler, Leonhard 233, 320
even number 57, 101
Existence of $\mathbb{N}$ 36
exponentiation, 47, 48
    laws of 47

F
factor 96
factorial 250
field, 111
    of fractions 174
    of rationals 137
    ordered 173
finite,
    cardinals 284
    set 23
formula 9
Fourier, Joseph 234
fractional part 140
Fraenkel, Abraham A. 3
Frege 33
Frobenius, Georg 351
function, 19
    constant 297, 317
    successor 34
fundamental theorem of algebra 320, 343

G

Gödel, Kurt 3, 29
gap 148
Gauss, Carl Friedrich 320, 321
General Principle of Induction 67
Girard, Albert 319
greater than, 49
    or equal 50
greatest,
    common divisor 97
    lower bound 55, 72
    lower bound property 74
group, 76
    abelian 76
    commutative 76

H

Hamilton, Sir William Rowan 320, 321, 347, 348
Hankel, Hermann 130
Harriot, Thomas 319
Heine, Eduard 234
Heine-Borel theorem 271
Hermite, Charles 255
Hippasus 124
hyper-complex numbers 350

I

identity, 86
    for multiplication 44
    function 19
imaginary,
    axis 327
    part 325
    unit 324
index 278
indicator function 298

Induction,
    General Principle 67
    Modified Principle 67
    Principle of Mathematical 35
inductively 40
infinite,
    cardinals 285
    set 23
injective 19
integer system 87
integers 88, 119, 173
integral,
    domain 92, 109
    part 140, 239
interior point 266, 314
Intermediate value theorem 302
intersection 17
interval, 260
    closed 261
    open 261
inverse 75
irrational number 198, 244
irrational points 147
irreflexive 21
isomorphic, 79
    fields 120
    integral domains 120
    ordered field 120
    ordered integral domain 120
isomorphism of,
    field 120
    groups 79
    integral domain 120
    ordered field 120
    ordered integral domain 120
    semigroups 79

Index

J
Jordan, Camilla 235

K
Kronecker, Leopold 82, 181
Kummer, Ernst Eduard 181

L
L_1-metric 311
Lagrange 236
Lambert, Johann Heinrich 255
Laplace 236
larger than 93
Laurent series 114
leading coefficient 108, 114
least upper bound, 55, 72
    property 74
Lebesgue, Henri 237
left-distributive 78
left-end point 260
Leibniz, Gottfried Wilhelm 320
length 310
less than 49, 138
less than or equal 50
limit,
    of a function 298
    of a sequence 158
    point 266
Lindermann, Ferdinand 255
linear,
    order 23, 73
    polynomials 108
linearly ordered set 73
lower bound 55, 72
lub-property 238

M
magnitude 328
Mahaviracharya 81
maximum, 72, 294
    value 294
member 7
metric, 308
    space 308
    L_1 311
    uniform 311
    complete 312
minimum, 72, 294
    value 294
Modified Principle of Induction 67
monotonically,
    decreasing 201
    increasing 201
multiplication of,
    complex numbers 322
    integers 89
    of natural numbers 47
    of polynomials 108

N
natural number system 35
natural numbers, 34, 41
    of an integral domain 118
negative, 138
    cut 189
    integers 88
    sequence 214
neighborhood 305
Nested interval property 274
nested intervals 273
Neumann, von 28
Newton, Isaac 320

nonempty set 16
norm 310
null sequence 207
number,
    even 101
    real algebraic 245
    composite 96
    irrational 199
    odd 97
    rational 198
    prime 96
    purely imaginary 325
    rational 133
    transcendental 255
numbers,
    perfect 33
    real 207, 225
    complex 320
    natural 34, 41
    perfect 33

O

odd number 57, 97
Ohm, Martin 130
one 34
one-one 19
onto 19
open,
    ball 314
    cover 270
    interval 261
    set 267
order, 23, 73
    complete 75, 178, 230
    linear 23, 73
    partial 23
    total 23, 73

ordered,
    n-tuple 71
    field 173
    integral domain 113
    pair 14
    set 73
    subfield 173
    triple 71

P

parallelogram identity 333
partition 291
partial order 23
partially ordered 23
Peano system 35
Peano, Giuseppe 33
perfect numbers 33
periodic decimal 151, 256
permutation 77
pi 253
Pigeonhole principle 66
Poincaré, Henri 235
point of discontinuity 296
polar form 329
polynomial 108
positive, 115, 137
    n^{th} root 244
    cut 189
    integers 88
    sequence 214
power set 14
prime,
    decomposition 100
    number 96
primitive 7
Principle of Mathematical
    Induction 35

Index

product of,
    cuts 189
    natural numbers 44
proper subset 23
propositional connectives 8
purely imaginary number 325
Pythagoras 32
Pythagorean triplets 33

Q
quadratic polynomials 108
quantifiers 8
quaternions 350
quotient 95

R
radius 314
range 19, 21
rational,
    cut 186
    number 133, 198
real,
    algebraic number 245
    axis 327
    line 260
    number 225
    numbers 207
    part 325
recursively 40
reflexive 21
relation 21
relative complement 18
relatively prime 99
remainder 95
Riemann, Bernhard 234
right,
    distributive 78

    end point 260
ring, 92, 106
    division 348
    of integers modulo n 107
    of polynomials 109
    with identity 106
Roots of real polynomials 304
Russell, Bertrand 3

S
scalar,
    part 351
    product 351
semigroup 76
separated 290
separation 291
sequence, 66, 107, 153
    Cauchy 171, 219
    constant 162
    convergent 158, 220
    negative 214
    null 207
    positive 214
set, 7
    compact 316
    bounded 264
    closed 262
    infinite 23
    linearly ordered 73
    open 267
    ordered 73
    singleton 11
    totally ordered 73
    compact 269
    well ordered 73
    countable 57
    countably finite 57

finite 23
  linearly ordered 23
  partially ordered 23
  successor 24
singleton set 11
skew field 348
Skolem, T. 3
smaller than 93
Steinitz, Ernst 130
Stifel, M. 180
Stolz, Otto 180
subdomain 120
subfield, 173
  ordered 173
subsequence 201
subset, 11
  proper 23
successor, 24
  function 34
sum of cuts 188
surjective 19
symmetric, 21
  difference 107
  group of permutations 77
symmetry 308
system,
  natural number 35
  Peano 35

T
tens place 104
total order 23, 73
totally ordered set 73
transcendental number 255
transitive 21
triangle-inequality 152, 199, 308, 333

trichotomy law 50

U
uncountable set 57
uniform metric 311
union 12
units place 104
upper bound 54, 72

V
variable 8, 108
vector part 351
vector product 351

W
Wallis, John 130
Weber, Heinrich 181
Weierstrass, Karl Theodore 130, 234, 235
well-ordered set 73
well-ordering, 55
  of IN 56
  principle 74
Wessel, Caspar 320
Whitehead, Alfred North 3

Z
Zermelo, Ernst 3
zero, 88
  divisors 92